Ius Gentium: Comparative Perspectives on Law and Justice

Volume 58

More information about this series at http://www.springer.com/series/7888

William van Caenegem · Jen Cleary
Editors

The Importance of Place: Geographical Indications as a Tool for Local and Regional Development

Springer

Editors
William van Caenegem
Faculty of Law
Bond University
Robina, QLD
Australia

Jen Cleary
Centre for Global Food and Resources
University of Adelaide
Adelaide, SA
Australia

ISSN 1534-6781 ISSN 2214-9902 (electronic)
Ius Gentium: Comparative Perspectives on Law and Justice
ISBN 978-3-319-53072-7 ISBN 978-3-319-53073-4 (eBook)
DOI 10.1007/978-3-319-53073-4

Library of Congress Control Number: 2016963738

Printed on acid-free paper

This Springer imprint is published by Springer Nature
The registered company is Springer International Publishing AG
The registered company address is: Gewerbestrasse 11, 6330 Cham, Switzerland

Contents

About the Editors

William van Caenegem is a professor of Law at the Faculty of Law, Bond University. He specializes in intellectual property law and comparative law. William wrote some of the first articles in Australia concerning the legal protection of geographical indications of origin, in the context of the WTO/TRIPS negotiations on the subject. He is a co-author with Jen Cleary and Peter Drahos of the RIRDC Report 'Provenance of Australian food products: is there a place for Geographical Indications?' (2015) and with Jen Cleary, Madeline Taylor and Brenda Marshall of the RIRDC Report 'Collective Bargaining in the Agricultural Sector' (2015). He has published on many areas of intellectual property law, including IP and innovation and also IP and the protection of commercial reputation.

Jen Cleary is a human geographer with research interests in regional development and agricultural value chain analysis. Jen is based within the Centre for Global Food Studies at the University of Adelaide and holds an honorary position as adjunct associate professor at the Institute for Governance and Policy Analysis at the University of Canberra. Jen has served as chair of Regional Development Australia, Far North (RDAFN) from 2009–14 and was appointed in 2015 as chair of the RDA National Reference Group, established by former Deputy Prime Minister Warren Truss. In 2013, Jen joined the National Steering Committee of SEGRA (Sustainable Economic Growth for Regional Australia), a national, independent body that fosters collaboration and formulates policy advice across rural, regional and remote Australia.

Abbreviations

ACCC	Australian Competition and Consumer Commission
AFPQCA	Agricultural and Fishery Products Quality Control Act
AOC	Controlled denomination of origin
ASEAN	Association of Southeast Asian Nations
CAP	European Union Common Agricultural Policy
CECA	Comprehensive Economic Cooperation Agreement
ChAFTA	China–Australia Free Trade Agreement
CIRAD	French Agricultural Research and International Cooperation Organization
CJEU	Court of Justice of the European Union
COFRADEP	French Committee for the defence and qualitative promotion of AOC wines and spirits
EC	European Union Council
ECJ	European Court of Justice
EU	European Union
FITFIR	First in time, first in right
FTA	Free Trade Agreement
GATT	General Agreement on Trade and Tariffs
GI	Geographical Indications
INAO	National Institute for Signs of Origin and Quality
INAO (*fr.*)	National Institute of Quality and Origin (*fr.*)
INTA	International Trademark Association
IP	Intellectual Property
JICEPA	Japan-Republic of India Comprehensive Economic Partnership Agreement
JPEPA	Japan–Peru Free Trade Agreement
KOREU	European Union–Korea Free Trade Agreement
KORUS	USA–Korea Free Trade Agreement
NGO	Non-governmental Organization
OAPI/AIPO	African Intellectual Property Organization

OHIM	Office for Harmonization in the Internal Market
OP	Origin Product
PCT	Patent Cooperation Treaty
PDO	Protected Designation of Origin
PGI	Protected Geographical Indication
TLT	Trademark Law Treaty
TM	Trade Mark
TPP	Trans Pacific Partnership
TRIPS	Agreement on Trade-Related Aspects of Intellectual Property Rights
TSG	Traditional Specialty Guaranteed
TTIP	Transatlantic Trade and Investment Partnership
UNESCO	United Nations Educational, Scientific and Cultural Organization
US/USA	United States of America
VCLT	Vienna Convention on the Law of Treaties
VDQS	Delimited wine of superior quality
WIPO	World Intellectual Property Organization
WTO	World Trade Organization

Introduction

Geographical indications of origin may be the poor relation of intellectual property law, but it is a subject that continues to grow in interest and importance. In international trade terms, it remains high on the agenda, in particular in agreements and negotiations to which the European Union is a party. National registration schemes for GIs have proliferated around the world, including Asia.

The motivation for this new book arose out of the research conducted by the editors on the role, potential and place of GIs in the Australian agricultural context. In undertaking this research, we concluded that the domestic debate on GIs has been framed almost exclusively in terms of gains and losses in international trade. It ignored the question of the potential for GIs to serve as rural and regional development tools in Australia, something we highlighted in our research. This led us to speculate about the perceived role of GIs as rural policy instruments globally. What has been the expectation and experience with GIs as agricultural development tools in both the Old World and the New? Are GIs seen to contribute to the protection of high-value products for smaller countries whose volume production is increasingly under threat by larger producing nations? How significant is rural economic impact compared to other GI policy goals? We found that in the Old World, for instance, cultural and historical dimensions are more central to the consideration of GIs than in Australia. However, rural economic impact has always been a significant theme in debates around the world.

Although our Australian work focused on the rural development potential of GIs, we also understood that our conclusions would have implications for Australia's international trade negotiation stance. The rural development question cannot be considered in isolation because Australian agricultural policy is inescapably linked to global trade and treaty systems. Therefore, in Part I of this volume we consider recent developments in international agreements and their implications on our rural development question. The chapters by Geuze and Gervais et al. in this part provide contrasting perspectives on the Geneva revision of the Lisbon Agreement, which aimed to increase the number of participating countries. O'Connor focuses on the positionality of the EU and the USA in trade negotiations with South Korea and the impact that entrenched positions on GIs continue to have.

This is important, because what policy analysis might suggest with regard to the adoption of GI registration schemes is readily discounted at the negotiating table. Taylor and Taylor show, however, that with smart negotiation, the rural development promise of GIs could be delivered domestically.

Part II of this book then considers our rural development conclusions in specific national and local settings, both in the Old World and the New. Cleary and van Caenegem firstly present what they learned from the Australian research and frame this in the context of both agricultural policy and broader rural social policy. The chapter by Teil then examines 150 years of winemaking in place, the contentionality of constructed *versus* natural links to *terroir*, and their impact on the development in France. Donner et al. use France and Morocco to contrast GIs with collective place brands as legal tools and their different effects on rural development. Overton addresses the long-standing debate in New Zealand about GIs for wine and concludes that a GI registration scheme offers tangible advantages over existing legal mechanisms in rural development terms. Marie-Vivien takes us further by asking whether GIs have a role beyond food in supporting the development of local handicrafts.

Finally, this book concludes with two succinct case studies—one from the Old World and the New. The first considers the actual impact of the GI registration of Sorana beans in Tuscany, Italy, and identifies the benefits for local farmers. The second case study considers the rural development on King Island, Australia. In the first instance, Thorn identifies the existing challenges for rural development on the island. She then discusses the potential of GIs for maximizing returns to local farmers from already well-known foods produced there.

We hereby thank contributors for their broader consideration of the rural development question that we raised in the Australian context. We would also like to acknowledge Leyla-Denisa Obreja for her work in organizing the correspondence with both the contributors and the publisher, and developing and editing the manuscript. She worked tirelessly, enthusiastically and efficiently to draw together the work presented here.

Part I
International Trade: Perspectives on Politics and Place

Chapter 1
Protecting Geographical-Origin-Brands Abroad: The Geneva Act of the Lisbon Agreement

Matthijs Geuze

Abstract The present chapter outlines the advantages of the Geneva Act of the Lisbon Agreement for the protection of geographical-origin-brands abroad. Thanks to the flexibilities the Act contains, its international registration and protection system should be attractive for any country that has an interest in allowing right holders in geographical-origin-brands for products from its territory to defend their rights in the jurisdictions of multiple actual and potential export markets.

1.1 Introduction

Elsewhere in this publication, the question is addressed whether geographical-origin-branding of products would be suitable for Australian producers and, if so, how such geographical-origin-brands might be protected in Australia against misappropriation. The present Chapter addresses another aspect that needs to be taken into account in this connection, namely how geographical-origin-branding is dealt with in the jurisdictions of actual and potential export markets for Australian products. This is important due to the fact that the acquisition of intellectual property rights in respect of geographical-origin-brands in other countries may be subject to quite different requirements or procedures compared to those Australia may establish.

However, it is also important because certain geographical names that were originally used as the indicator of a specific product from a particular geographical area may no longer be available as geographical-origin-brands, as they have become the generic indicator of a particular type of product not necessarily originating from a particular geographical area.

Examples of place names that have become generic in this sense include names that are even generic in the country where the place is situated, such as Brie or

M. Geuze (✉)
Formerly Protection of Appellations of Origin and Their International Registration, Geneva, Switzerland
e-mail: matthijs.geuze77@gmail.com

W. van Caenegem and J. Cleary (eds.), *The Importance of Place: Geographical Indications as a Tool for Local and Regional Development*, Ius Gentium: Comparative Perspectives on Law and Justice 58, DOI 10.1007/978-3-319-53073-4_1

Edam, but also names that, although still acting as an indicator of a specific product from a particular geographical area in the country of origin, are considered generic in (some) other countries, such as Cognac or Feta. Moreover, issues may also arise if the name of a geographical area used and protected as the indicator of a specific product from the area is the same as the name of another geographical area used and protected as the indicator of a specific product from that other area; or if the name conflicts with a sign for which a trademark has been acquired.

In principle, a geographical name used as an indicator of a specific product from a particular geographical area cannot be protected as a trademark, as the name will be considered descriptive of the product and should not be monopolized for the benefit of one of its producers only. Protection, therefore, takes place through a right that belongs to a collectivity of producers, i.e., a brand based on the geographical origin of the product. Such brands exist in different forms and under different terms. Depending on the law of the country in question, the right acquired is called a geographical indication, an appellation of origin, a collective mark or a certification mark. Over the years, many bilateral agreements have been concluded, under which countries have agreed to protect (a number of) each other's geographical-origin-brands. However, there are also a number of multilateral treaties dealing with the protection of such brands. One of those treaties is the Lisbon Agreement for the Protection of Appellations of Origin and their International Registration, which is administered by the World Intellectual Property Organization (WIPO). In view of the need to adapt the system established by this treaty ("the Lisbon System") to the different approaches that exist around the world for the protection of geographical-origin-brands, a comprehensive review of its provisions recently resulted in the conclusion, on May 20, 2015, of the Geneva Act of the Lisbon Agreement on Appellations of Origin and Geographical Indications ("the Geneva Act"). Entry into force of the Geneva Act can be expected in a few years. Consequently, it may be useful, if Australia were to opt for geographical-origin-branding, to also take into consideration its possible accession to the Geneva Act.

1.2 Procedures Facilitating the Acquisition of Rights Abroad

Under the Geneva Act, the Lisbon System will be a one stop solution for registering geographical-origin-brands in multiple countries, whether they are protected in their country of origin as geographical indications, as appellations of origin, as collective marks, as certification marks or under a specific term the country may choose to designate and protect them. Once protection is obtained in the country of origin under national law, or under the law of an intergovernmental organization of which the country is a member (e.g., the European Union or the African Intellectual Property Organization), a single application procedure can be used to extend protection to the territories of the other members of the System. It suffices to file one application, in one language (English, French or Spanish), and pay one set of fees in Swiss francs, to obtain an international registration that will be effective in multiple territories within a set time-limit of 12 months.

Protection of the internationally registered geographical-origin-brand will be governed by the laws that apply in the territories of the members of the System, which will have to be in conformity with the standards stipulated in the Geneva Act. Thus, the international application is the equivalent of a bundle of national applications, but saves time and money, as there is no need to pay for translations into multiple languages or spend the time required under the national or regional procedures of the various countries or intergovernmental organizations for deciding whether protection can be granted or not—a period of time that is usually considerably longer than the 12 month time-limit that applies under the international registration procedures and which will in any case be different *per* national or regional system.

1.3 Who Can File the Application, How, and What Does It Need to Indicate?

International applications are to be presented to the International Bureau of WIPO by the national or regional authority that the country of origin of the geographical-origin-brand has identified for the exchange of communications with WIPO under the procedures of the Lisbon System. Usually, this will be the agency responsible for the grant of protection in the country or intergovernmental organization concerned. Consequently, a procedure needs to be available in each member of the System for right holder(s) to submit a request to that agency for filing an international application with WIPO. The Geneva Act is flexible as to who are the right holders that can submit such a request and specifies that an application can be filed in the name of: (i) the beneficiaries, i.e. those having the right to use the geographical-origin-brand; or (ii) a natural person or legal entity having legal standing under the law of the country of origin to assert the rights of the beneficiaries or other rights in the geographical-origin-brand, i.e. whoever is entitled to exercise control over the use of the geographical-origin-brand or to enforce rights in it.

The Geneva Act also provides for the possibility that such beneficiaries or such a natural person or legal entity can file the application with WIPO directly. However, such possibility only applies if the country of origin has made a declaration that its legislation allows for direct filings by right holders in respect of geographical-origin-brands. In regards to the content of an application, the Regulations under the Geneva Act distinguish between mandatory indications, declaration-based indications and optional indications.[1] As for mandatory indications, it is obvious that the

---

[1]Rule 5 of the Regulations under the Geneva Act of the Lisbon Agreement on Appellations of Origin and Geographical Indications (as adopted on May 20, 2015).

geographical-origin-brand itself needs to be mentioned in the application. However, also the good(s) to which the brand applies have to be indicated, as well as the applicant(s), the country of origin (or regional organization) and the beneficiaries or the natural person or legal entity referred to above. In addition, an indication needs to be given of the boundaries of the geographical area where the good(s) are made[2] and the identifying details of the national or regional registration (or other decision) by virtue of which protection was granted to the geographical-origin-brand in the country of origin (or regional organization). If the national or regional registration (or other decision) explicitly mentions that one or more elements of the brand are not covered by the protection granted, these elements have to be indicated in the application as well.[3]

The application may also need to contain other indications. That is because, under the Geneva Act, a country or regional organization that a member of the System may notify the Director General of WIPO that, for protection of the geographical-origin-brand in its territory, the application must be signed by a natural person or legal entity having legal standing under the law of the country of origin to assert the rights of the beneficiaries or other rights in the geographical-origin-brand. A member of the System may also notify the Director General that the application must be accompanied by a declaration of intention to use; or that the application must indicate particulars concerning the quality, reputation or characteristics of the good and the connection thereof with the good's geographical origin. If such a declaration-based requirement is not met, the International Bureau of WIPO will not reject the application, but the geographical-origin-brand that is the subject of the resulting international registration will not be protected in a country or regional organization that has notified the requirement in question. Protection can, however, be extended to such a country or regional organization as from a later point in time, once the requirement is met.[4] The same will apply in respect of a country or regional organization that has notified the Director General that, for protection of the geographical-origin-brand in

[2]Such a geographical area of origin may consist of the entire territory of the country of origin or a region, locality or place in it, or even a trans-border geographical area (Article 2(2) of the Geneva Act). Descriptions of geographical areas of origin vary widely. For example, in respect of Lisbon registration No. 892 for the Serbian appellation of origin Homoljski med (honey), the geographical area of origin is described as "geographical area in Eastern Serbia". In respect of Lisbon registration No. 80 for the French appellation of origin Bordeaux (wine), the geographical area of origin is described as "delimited territory within the department of Gironde". However, in respect of, for example, Lisbon registrations Nos. 893 and 931, the description of the geographical area of origin is very detailed covering about a page of text.

[3]By way of example, reference could be made to the registration of the cheese "Edam Holland" as a protected geographical indication (PGI) in the European Union. The decision by virtue of which protection was granted to this PGI in Commission Regulation (EU) No. 1121/2010 of 2 December 2010 explicitly mentions that the PGI protects "Edam Holland", but that "Edam" remains free for use as a generic term for a particular kind of cheese.

[4]Rule 16(2) of the Regulations under the Geneva Act of the Lisbon Agreement on Appellations of Origin and Geographical Indications (as adopted on May 20, 2015).

its territory, a fee has to be paid to cover the cost of substantive examination of the international registration by its competent authority. Such a country or regional organization may also require an administrative fee relating to the use by the beneficiaries of the geographical-origin-brand in its territory.[5]

As per the costs for the acquisition and maintenance of an international registration, it should be noted that no fee amounts have been established in the Geneva Act or its Regulations. It is for the Assembly of the members of the Lisbon System to make a decision on this prior to the entry into force of the Geneva Act. However, for comparison, under the current Lisbon Agreement, following a decision by the Assembly in October 2015, the fee to be paid for an international registration is 1000 Swiss francs and for any subsequent modification of the international registration 500 Swiss francs. International registration is not subject to renewal. Furthermore, as mentioned in the previous paragraph, the Geneva Act will introduce the possibility for its members to require an individual fee to cover the cost of substantive examination of the international registration by its competent authority as well as an administrative fee relating to the use by the beneficiaries of the geographical-origin-brand in its territory.

1.4 Definition(s)

Under the Geneva Act, geographical-origin-brands can be registered when they meet either of the following definitions:[6]

(i) "any denomination protected in the Contracting Party of Origin consisting of or containing the name of a geographical area, or another denomination known as referring to such area, which serves to designate a good as originating in that geographical area, where the quality or characteristics of the good are due exclusively or essentially to the geographical environment, including natural and human factors, and which has given the good its reputation"; or
(ii) "any indication protected in the Contracting Party of Origin consisting of or containing the name of a geographical area, or another indication known as referring to such area, which identifies a good as originating in that geographical area, where a given quality, reputation or other characteristic of the good is essentially attributable to its geographical origin".

In the Geneva Act, these geographical-origin-brands are referred to as "appellations of origin" and "geographical indications", respectively. However, the Geneva Act does not require member countries and intergovernmental organizations to use

[5]Article 7(4) of the Geneva Act of the Lisbon Agreement on Appellations of Origin and Geographical Indications (as adopted on May 20, 2015).

[6]Article 2(1) of the Geneva Act of the Lisbon Agreement on Appellations of Origin and Geographical Indications (as adopted on May 20, 2015).

the same terminology. As mentioned before, the Geneva Act provides for the registration and protection of geographical-origin-brands whether such protection is provided under national or regional law in the form of an appellation of origin, a geographical indication, a certification mark, a collective mark or whatever other title as may exist under national or regional law for geographical-origin-brands.[7]

Moreover, there will be no obligation for member countries and intergovernmental organizations to make a distinction in their legislation on the basis of these two definitions.[8] They are free to provide protection on the basis of one definition only—a definition that corresponds to the broader definition mentioned under (ii)—in which case they have to deal with denominations, as referred to under (i), as if they are indications, as referred to under (ii).

1.5 Protection

The content of the protection to be granted in respect to geographical-origin-brands in member countries and intergovernmental organizations has to be in accordance with the level of protection required by the provisions of the Geneva Act in that respect. As with every rule, there are also exceptions. However, before addressing the possible exceptions, the following sections will first explain what protection member countries and intergovernmental organizations have to provide under their national or regional law.

The Geneva Act makes a distinction between any use of the geographical-origin-brand in respect of goods of the same kind as those to which the brand applies and any use in respect of goods that are not of the same kind or services.[9] In respect to goods of the same kind, protection has to be available for right holders against any use of the brand in respect to goods that do not originate in the geographical area of origin specified in the international registration. In addition, right holders must have the opportunity to take legal action against any use of goods that do originate in that geographical area but do not comply with other requirements that may apply for use of the brand—for example, if production of the goods has not taken place in conformity with the requirements specified in respect of the geographical-origin-brand.[10] Regarding goods that are not of the same kind, or services, protection has to be available against use that would indicate or suggest a connection between those goods or services and the beneficiaries of the internationally registered geographical-origin-brand and would be likely to damage their interests. In addition, protection may be available in a given member country or

[7]Ibid, Article 10(1).

[8]Ibid, Article 9.

[9]Ibid, Article 11(1)(a).

[10]For example, the product specification (*cahier de charge*) of an appellation of origin or the regulations governing use of a certification mark.

intergovernmental organization against any use that would be likely to impair or dilute in an unfair manner, or take unfair advantage of, the reputation that the brand has in that member.

The Geneva Act further specifies that "use of the geographical-origin-brand" is not limited to its use in exactly the same form, but also covers use of the brand that amounts to imitation of the brand, in particular use of the brand with slight differences, use in translation, use in combination with a delocalizing term, or in combination with terms such as "style", "kind", "type", "make", "imitation", "method", "as produced in", "like", "similar" or the like.[11] Thus, any use of the geographical-origin-brand as a generic term for a particular kind of good can, in principle, be stopped by the right holders. However, as we will see below, exceptions may apply in this regard.

Unfair competition can also exist in the form of other practices in the presentation of goods than those addressed in the previous paragraphs. In that sense, the Geneva Act specifies that right holders must have the opportunity to take legal action against such practices, if these are liable to mislead consumers as to the true origin, provenance or nature of the goods.[12]

Issues may also arise when trademark rights are claimed in respect of a sign consisting of or containing the geographical-origin-brand. As we will see below, exceptions may apply in this regard in case the trademark predates the geographical-origin-brand. However, if the geographical-origin-brand is the earlier right, the Geneva Act specifies that registration of the later trademark must be refused or invalidated in a member country or intergovernmental organization where the geographical-origin-brand was first, if use of the trademark would result in one of the situations addressed in the previous paragraphs.[13] The purpose of this provision is to prevent the registration of trademarks that consist of or contain an internationally registered geographical-origin-brand for goods that do not originate in the geographical area of origin or, if they do originate there, are not in conformity with the requirements for use of the geographical-origin-brand. The use of trademarks containing a geographical-origin-brand is allowed, provided they are used in respect of goods originating in the geographical area of origin of the brand and complying with any other applicable requirements for using the brand.[14]

Member countries or intergovernmental organizations may also provide more extensive protection than stipulated in the Geneva Act and the Geneva Act does not affect any other protection that members may accord under a bilateral, pluri-lateral

[11]Article 11(2) of the Geneva Act of the Lisbon Agreement on Appellations of Origin and Geographical Indications (as adopted on May 20, 2015).

[12]Ibid, Article 11(1)(b).

[13]Ibid, Article 11(3).

[14]It is recalled that the term "geographical-origin-brand" is used in this chapter to refer to any indicator of a specific product from a particular geographical area. Depending on the applicable national or regional law, these are called appellations of origin, geographical indications, collective marks or certification marks.

or multilateral agreement. Thus, nothing in the Geneva Act shall derogate from any obligations that members have in respect to each other, and nothing shall prejudice any rights they have, under, for example, the WTO TRIPS Agreement.[15]

1.6 Obstacles May Exist for Obtaining Protection in (Some) Countries

By their accession to the Geneva Act, member countries and intergovernmental organizations commit themselves to protect geographical-origin-brands registered under the Geneva Act on their territory in accordance with the terms of the Geneva Act.[16] They can do so within their own legal system and practice, which means that they are not obliged to use the same wording and terminology as used in the Geneva Act, as long as they meet the substantive requirements of the provisions of the Geneva Act.[17] Moreover, as already mentioned in the previous section, obstacles may exist for obtaining protection in (some) foreign countries. The commitment for members to protect geographical-origin-brands registered under the Geneva Act of the Lisbon Agreement is subject to any refusal, renunciation, invalidation or cancellation that may become effective with respect to the territory of a member country or intergovernmental organization.[18] As the Geneva Act is not yet in force, let me illustrate this on the basis of the first ever geographical-origin-brand registered under the Lisbon Agreement itself, i.e., the appellation of origin Plzeň (or Pilsner, Pilsener, Pilsen Pils) for beer originating from that city in the Czech Republic.[19] In respect of this appellation of origin, eight members of the Lisbon Agreement have sent a notification to the International Bureau of WIPO stating that they could not ensure its protection. Following Mexico's withdrawal of its refusal,[20] the appellation of origin is under protection in 20 Lisbon members (other than the country of origin) and not under protection in the seven countries whose refusals remain effective.

Refusals have to be submitted by the country's competent authority within one year from the receipt of the notification of a new international registration, or from the date on which a country's accession becomes effective,[21] and they must indicate the ground on which the refusal is based. In regards to the appellation of origin

[15]Article 10(2) and (3) of the Geneva Act of the Lisbon Agreement on Appellations of Origin and Geographical Indications (as adopted on May 20, 2015).

[16]Ibid, Article 9.

[17]Ibid, Articles 9 and 10(1).

[18]Ibid, Article 9.

[19]International registration No. 1, dated November 11, 1967.

[20]May 25, 1980.

[21]Under the Geneva Act, a member may declare that it will apply two years for the examination of the existing stock of international registrations.

Plzeň (or Pilsner, Pilsener, Pilsen Pils), the refusals issued show four different grounds of refusal. Bosnia and Herzegovina, Montenegro, Serbia and The former Yugoslav Republic of Macedonia based their refusal on the fact that, in their territories, the denomination Plzeň (or Pilsner, Pilsener, Pilsen Pils) is a generic indication for a certain type of beer. Peru and Mexico based their refusal on the existence of prior trademark rights. France refused the term "Pils" on the ground that it does not constitute an appellation of origin. The Islamic Republic of Iran based its refusal on the fact that, under Iranian law, the production, distribution and consumption of alcoholic beverages is prohibited and appellations that are against religious values, ethics or public order are not protected.

Under the Geneva Act, refusals can, just as under the Lisbon Agreement, be based on any ground. Nevertheless, the practice under the Lisbon Agreement shows that, in respect of the majority of international registrations, the International Bureau of WIPO has not received any refusal. Even in the case of those international registrations that were refused by several countries, still the majority of countries did not submit a refusal. In this regard, reference could be made to international registration No. 837 for the Czech appellation of origin Budweiser Burgerbräu, for which 11 refusals are effective—which means that the appellation of origin is under protection in 16 Lisbon members (other than the country of origin).[22]

1.7 Safeguards

The Geneva Act specifies that member countries and intergovernmental organizations have to provide safeguards in respect of prior trademark rights, personal names in business and rights based on a plant variety or animal breed denomination. Other safeguards are contained in two Agreed Statements by the Diplomatic Conference that adopted the Geneva Act. Regarding prior trademark rights, the Geneva Act recognizes that, among the existing national and regional systems, some allow, in certain circumstances, for coexistence of the geographical-origin-brand and the prior trademark, while other systems do not allow for such exceptions. In that sense, the Geneva Act specifies that, in principle, the provisions of the Geneva Act shall not prejudice prior trademark rights acquired in accordance with the law of the member country or intergovernmental organization concerned. However, where a member country or intergovernmental organization, which allows for coexistence, provides a limited exception to the rights conferred by a trademark to the effect that such a prior trademark may, in certain circumstances, not entitle its owner to prevent the grant of protection in respect of the

[22]The content of the International Register of the Lisbon System can be consulted on-line on the WIPO website at http://www.wipo.int/ipdl/en/lisbon/.

geographical-origin-brand or its use in the territory of that member, protection of the geographical-origin-brand shall not limit the rights conferred by that trademark in any other way. This is meant to ensure that it is up to each member to provide rights on the basis of the principle of *first-in-time first-in-right* or on the basis of coexistence of the geographical-origin-brand as a limited exception to the trademark right.[23]

Under the Geneva Act, geographical-origin-brands that are the subject of an international registration are protected against becoming generic.[24] However, it is understood that a geographical-origin-brand may not be protectable, in whole or in part, in other members than the country of origin, as the brand may contain, or consist of, a term that, at the time of international registration of the brand, is generic in such member State, for example because the brand, or part of it, is identical with a term customary in common language as the common name of a good or service, or is identical with the customary name of a grape variety.[25] In addition, it is understood that, where certain elements of the brand have a generic character in the member country of origin, their protection shall not be required in the other countries. Reference could be made to the Edam Holland example mentioned above. Refusal or invalidation of a trademark or a finding of infringement in a member country or intergovernmental organization cannot be based on a component that has a generic character.[26] Whether a term, element or component can be considered to have a generic character in accordance with the provisions of the Geneva Act is to be determined in legal proceedings in the member country or intergovernmental organization concerned,[27] within that member's own legal system and practice.[28]

1.8 Conclusion

The present chapter outlines the advantages of the Geneva Act of the Lisbon Agreement for the protection of geographical-origin-brands abroad. Thanks to the flexibilities the Act contains, its international registration and protection system should be attractive for any country that has an interest in allowing right holders in geographical-origin-brands for products from its territory to defend their rights in the jurisdictions of multiple actual and potential export markets.

---

[23] Article 13(1) and (4) of the Geneva Act of the Lisbon Agreement on Appellations of Origin and Geographical Indications (as adopted on May 20, 2015).

[24] Ibid, Article 12.

[25] Agreed Statement, adopted by the Diplomatic Conference, to Article 12 of the Geneva Act.

[26] Agreed Statement, adopted by the Diplomatic Conference, to Article 11(2) of the Geneva Act.

[27] Article 14 of the Geneva Act of the Lisbon Agreement on Appellations of Origin and Geographical Indications (as adopted on May 20, 2015).

[28] Ibid, Article 9.

Bibliography

WIPO Publication No. 239E (ISBN 978-92-805-2626-4), entitled "Geneva Act of the Lisbon Agreement on Appellations of Origin and Geographical Indications and Regulations Under the Geneva Act of the Lisbon Agreement", WIPO 2015.

Author Biography

Geuze was in charge of the administration of the International Register under the WIPO-administered Lisbon Agreement for the Protection of Appellations of Origin and their International Registration from 2007 until his retirement in February 2016. In this capacity, he has been the Secretary of the Working Group on the Development of the Lisbon System that prepared the Geneva Act of the Lisbon Agreement on Appellations of Origin and Geographical Indications, which was negotiated between 2008 and 2015 and adopted by a Diplomatic Conference at WIPO headquarters in May 2015.

Chapter 2
The Geneva Act of the Lisbon Agreement: Controversial Negotiations and Controversial Results

Daniel J. Gervais and Matthew Slider

Abstract This chapter examines the historical process that led to the adoption of the Geneva Act of the Lisbon Agreement for the Protection of Appellations of Origin, and some of the debates that may arise as ratification and implementation of the new Act progress. The chapter also considers whether common law jurisdictions are likely to eventually join the updated Lisbon system and the potential motivations for doing so, as at the time of this writing, none were part of this system yet. The chapter observes that the bridge between Lisbon and the common law system has not yet been built because of the different approaches of the Old World and New World representatives to appellations of origin as governed by the Geneva Act. Old World representatives are normatively rigid and concerned with preserving traditional elements such as *terroir* while New World representatives are concerned with elements of genericness and economic value. Thus, the adoption of the Geneva Act of the Lisbon Agreement represents a clear chance to reconcile the Lisbon system doctrinally and render it more appealing to a larger array of countries.

2.1 Introduction

Geographical indications (GIs) matter on several levels. They have roots in the *terroir,* a French word designed to encapsulate a blend of land, tradition, and human know-how (Hughes 2006). *Terroir* matters to many producers and many countries,

---

A US-focused version of part of this chapter is published in the Houston Law Review as Daniel Gervais, 'Irreconcilable Differences? The Geneva Act of the Lisbon Agreement and the Common Law' (2015) 53 Houston Law Review 339. Another part is based in part on the following book chapter: Daniel Gervais, "A Cognac after Spanish Champagne? Geographical Indications as Certification Marks", in *Intellectual Property at the Edge* (Jane C. Ginsburg and Rochelle Dreyfuss, eds) (Cambridge Univ. Press, 2014) 130–155.

D.J. Gervais (✉) · M. Slider
Vanderbilt University Law School, Nashville, USA
e-mail: daniel.gervais@Law.Vanderbilt.Edu

W. van Caenegem and J. Cleary (eds.), *The Importance of Place: Geographical Indications as a Tool for Local and Regional Development*, Ius Gentium: Comparative Perspectives on Law and Justice 58, DOI 10.1007/978-3-319-53073-4_2

and not just in Europe. But in parts of the 'Old World', it is not an exaggeration to say that some countries link *terroir* to national identity (Guy 2009; Creditt 2009, 427). So-called 'New World' producers see things differently. While they also recognize the economic value of geographic origin for certain products (e.g., in the United States, Napa Valley wines, Vidalia onions, Wisconsin cheese and Idaho potatoes), they are concerned about possible restrictions on the use of terms considered generic (meaning terms that describe a type of product not its geographic origin) (McCarthy 1999, §19:92.50). Developing country producers have concerns of their own, notably about the protection of foreign terms that have become or may become generic in their country or region. Several developing nations also see GIs as a way of protecting and globally marketing rural and traditional products at a higher price, which they assert should lead to 'development from within', that is, 'an alternative development strategy that prioritizes local autonomy and broad, community-wide development goals' (Bowen 2010, 232; Gervais 2012a, 121–47). GIs tend to focus production on a nation's comparative advantage in making a product whose origin infuses it with a higher market value (Bramley et al. 2009, 121). GIs may have environmental significance and thus form an increasingly relevant part of agricultural and food policy (van Caenegem 2004, 172–73). It is not surprising, therefore, that the debate surrounding the protection of GIs has captured the attention of a number of consumer groups, many of which insist on proper labeling of products, notably to indicate their origin.

At an international level, a sui generis regime of protection (that is, outside of trademark law) for 'appellations of origins' has been in place since the adoption of the 1958 Lisbon Agreement.[1] In May 2015, a number of World Intellectual Property Organization (WIPO) member States met in Geneva and adopted a new version or 'Act' of the Lisbon Agreement.[2] In this chapter, we consider the process that led to the adoption of this Geneva Act and some of the controversies that may arise as the ratification and implementation of the new Act is underway, and whether common law jurisdictions are likely to join the updated Lisbon system.

As of this writing (January 2017), no common law jurisdiction is party to the Lisbon Agreement (WIPO 2013).[3] There are a number of reasons that explain this lack of enthusiasm. Several common law jurisdictions use trademarks, collective

---

[1]Lisbon Agreement for the Protection of Appellations of Origin and their International Registration 1958 (Lisbon Agreement). The notions of appellations of origin and geographical indications are closely related. One of the authors discussed the differences in detail in Daniel Gervais, 'Reinventing Lisbon: The Case for a Protocol to the Lisbon Agreement (Geographical Indications)' (2010) 11 Chicago Journal of International Law 67, 83–87.

[2]Geneva Act of the Lisbon Agreement on Appellations of Origin and Geographical Indications, 2015 (Geneva Act). A previous update of the original 1958 Lisbon Agreement was adopted on July 14, 1967.

[3]As of February 2016, the parties to the Lisbon Agreement are: Algeria, Bosnia and Herzegovina, Bulgaria, Burkina Faso, Congo, Costa Rica, Cuba, Czech Republic, Democratic People's Republic of Korea, France, Gabon, Georgia, Haiti, Hungary, Iran, Israel, Italy, Mexico, Montenegro, Nicaragua, Peru, Portugal, Republic of Moldova, Serbia, Slovakia, the former Yugoslav Republic of Macedonia, Togo, and Tunisia.

marks, and certification marks to protect geographic symbols and names, instead of a sui generis regime. This has a number of both normative and administrative implications, including use requirements, possible loss or diminution of right due to acquiescence and abandonment or genericness, and the payment of maintenance fees, to mention just the main ones. The question of the compatibility of *terroir* recognition in law with modern international trade rules has also been raised in the United States and elsewhere. The chapter proceeds as follows. In Part I, we review the adoption process, which was somewhat unusual. In Part II, we review the outcome, namely the main features of the Geneva Act. In Part III, we review future options, including whether the Geneva Act is likely to gain global traction as a vehicle to protect GIs. In Part IV, we discuss briefly what implications this may have for other multilateral initiatives.

2.2 Procedural Controversies

2.2.1 Stated Objectives at Odds with the Willingness to Compromise

According to WIPO Director General Francis Gurry, a primary objective of the adoption of the Geneva Act was to 'produce a Lisbon system that is attractive to the full membership of the Organization' by providing a 'basis for the expression of the Lisbon system beyond the *historically rather low level of participation* amongst the members states of the WIPO'.[4] Such a statement recognizes that, despite over 50 years of existence, the Lisbon Union was perceived as unattractive to the full membership of WIPO. As of early 2017, it only had 28 members. By comparison, the Madrid Union (international registration of trademarks) had 98 members, the Hague Agreement (Industrial Designs) had 66 contracting parties, and the Patent Cooperation Treaty (PCT) had 151 contracting parties.[5] Placed in this context, the Lisbon system appears to be an IP 'niche'. Although the stated objective of the Geneva Act was to widen the scope of the Lisbon Agreement, that objective was not reflected in the procedure chosen to negotiate and adopt the New Act. Instead of

[4]*Diplomatic Conference for the Adoption of a New Act of the Lisbon Agreement*, World Intell. Prop. Org., available at http://www.wipo.int/webcasting/en/index.jsp (as consulted 1 Feb 2015) (to find the Director General's comments, select on 'Videos on Demand,' choose LI/DC: Diplomatic Conference for the Adoption of a New Act of the Lisbon Agreement for the Protection of Appellations of,' select LI/DC—Mon 11—Opening Ceremony and Plenary, remarks begin at 4: 21 min in.).

[5]*Status on January 15, 2017*, http://www.wipo.int/export/sites/www/treaties/en/documents/pdf/hague.pdf(as consulted Jan. 17, 2017) (document listing Hague members); Contracting Parties > Madrid Protocol, WIPO, http://www.wipo.int/treaties/en/ShowResults.jsp?lang=en&treaty_id=8 (as consulted Jan. 17, 2017) (webpage listing Madrid members); Contracting Parties > Patent Cooperation Treaty, http://www.wipo.int/treaties/en/ShowResults.jsp?lang=en&treaty_id=6 (as consulted Jan. 17, 2017) (webpage listing PCT members).

granting all WIPO members full participation rights, the Lisbon Union kept voting rights for all ten sessions of the working group and then the Diplomatic Conference restricted to its current 28 member states. This relegated the majority of WIPO members who were not also Lisbon member states to observer status, able only to participate by making statements (only after the list of member States requesting the floor had been exhausted) and submitting proposals but unable to vote. This move infuriated a number of non-member countries, especially the United States. Critics of the Geneva Act's procedure alleged that it was a divisive procedure that called into question the very legitimacy of the process, especially in view of the fact that rules of procedure adopted by the Lisbon Assembly bucked some precedents at WIPO that would have supported opening up a treaty revision fully to all members (*General Report* 2016, p. 38). Lisbon members held firm that the rules and procedures for both the working group meetings and the Diplomatic Conference were not only valid and fair, but indeed were legally required. In a joint letter distributed to all non-members, Lisbon members stated that they were 'bound by international law' to restrict full participation. Invoking the Vienna Convention on the Law of Treaties (VCLT), they argued that to allow only contracting parties to a treaty to 'decide on the conditions of its amendment' was not only possible but mandated by the VCLT (Saez 2015a). The letter also argued that Article 13 of the 1958 Lisbon Agreement stipulated that the text 'may be revised by conferences held between the delegates of the countries of the Special Union'.

These assertions are questionable. First, the Director of WIPO's Office of Legal Counsel, Edward Kwakwa, stated that neither the VCLT nor Article 13 of the 1958 Lisbon Agreement explicitly *required* Lisbon members to restrict voting rights to member states (Saez 2015c). In short, Lisbon members had the choice. The WIPO Secretariat, including Director General Francis Gurry and Deputy Director General Binying Wang,[6] apparently concur on this point. At the sixth working group meeting, they noted that:

> '[T]he Working Group would recommend the Lisbon Union Assembly to call a diplomatic conference for the adoption of a Revised Act of the Lisbon Agreement, the Rules of Procedure adopted at the Conference would in turn determine who would have a right to vote' (*Report* 2013, 7).

Furthermore, as already noted a number of precedents favored full participation. For example, a U.S. official cited 'a longstanding tradition' at WIPO to include all members in the process of revising WIPO-based treaties, citing three recent cases, namely the Marrakesh Treaty to Facilitate Access to Published Works for Persons Who Are Blind, Visually Impaired or Otherwise Print Disabled adopted in 2013; the Beijing Treaty on Audiovisual Performances adopted in 2012; and the 1999 Geneva Act of the Hague Agreement Concerning the International Registration of Industrial Designs (Saez 2015b). In all three examples all WIPO members had full

[6]The WIPO Secretariat also included Matthijs Geuze, Head, Lisbon Registry, Brands and Designs Sector; Florence Rojal, Legal Officer, Lisbon Registry, Brands and Designs Sector.

participation rights.[7] It can be said that the first two were new instruments and not a new act of an existing treaty, though they certainly had to fit within the framework of extant instruments including the Berne Convention for the Protection of Literary and Artistic Works and other copyright-related instruments to which not all WIPO Members are members. The Hague revision is directly on point, however, it was a revision of an exiting agreement. The argument that Lisbon itself *required* a closed vote (open only to existing members) is thus dubious. The relevant language of the Lisbon Agreement did not *foreclose* opening the process to all WIPO Members. Article 13(2) of the Lisbon Agreement is clear: the 'Agreement *may* be revised by conferences held between the delegates of the countries of the Special Union.' Indeed, the 1999 Geneva Act to the Hague Agreement contains similar language in Article 25(1) and was the product of a revision process open to all members. On its face, language such as '*may* be revised' allows for flexibility to keep participation limited in cases of minor revisions. The phrase also leaves the door open to include all members in the case of complete revision or the adoption of a new treaty. The fact that the 1958 Lisbon Agreement shares similar language with other WIPO-administered instruments strengthens the argument that the 1958 Lisbon Agreement is not unique in this regard.

The 2006 Diplomatic Conference for the Adoption of a Revised Trademark Law Treaty (TLT) opened participation to all WIPO members.[8] That conference resulted in the Singapore Treaty on the Law of Trademarks and contrasts sharply with the Geneva Act to the Lisbon Agreement. Perhaps one could argue that the Singapore revision was different because the Geneva Act extended Lisbon Agreement protection beyond appellations of origin to the broader notion of geographical indications (*Geneva Act* 2015).[9] By contrast the Singapore Treaty only widened the scope of protected marks (*Regulations Under Singapore Treaty* 2011).[10] As one of the Authors demonstrated in a previous publication, however, there was little difference between AOs and GIs to begin with (Gervais 2010a, 83–86). Moreover, in discussing the draft rules and procedures for the Geneva Act, the delegation of

[7]Rules of Procedure for the Diplomatic Conference to Conclude a Treaty to Facilitate Access to Published Works by Visually Impaired Persons and Persons with Print Disabilities *available at* http://www.wipo.int/edocs/mdocs/diplconf/en/vip_dc/vip_dc_2.pdf; Preparatory Committee of the Diplomatic Conference on the Protection of Audiovisual Performances, *available at* http://www.wipo.int/edocs/mdocs/copyright/en/avp_pm/avp_pm_3.pdf; Diplomatic Conference For The Adoption Of A New Act Of The Hague Agreement Concerning The International Deposit Of Industrial Designs - Draft Rules Of Procedure, *available at* http://www.wipo.int/edocs/mdocs/diplconf/en/h_dc/h_dc_2.pdf *(as all consulted on March 21st 2016).*

[8]Rule 33 of the rules of procedure stated, '[e]ach Ordinary Member Delegation shall have the right to vote. An Ordinary Member Delegation shall have one vote, may represent itself only and may vote in its name only'.

[9]Article 11 covers protections of both appellations of origin and GIs and therefore constitutes an expansion over the 1958 Lisbon Agreement, which only covered appellations of origin.

[10]Rule 3 of the regulations allows for the protection of, among others, a 'Hologram Mark', 'Motion Mark', 'Color Mark', and 'Position Mark', expanding the scope of the Trademark Law Treaty.

Switzerland reasoned that appellations of origin were a sub-category of GIs and therefore the Lisbon Agreement already de facto essentially covered GIs (Saez 2015a). Besides, if this logic held, then the contracting parties for the Trademark Law Treaty could have put forth a very similar argument in regards to a revisionary Diplomatic Conference. By stating that the previous Trademark Law Treaty covered specific types of trademarks, each a sub-category of 'trademarks as a whole,' contracting parties could have asserted that the Trademark Law Treaty merely needed to be updated to cover 'trademarks as a whole' just as the Lisbon Agreement would be updated to cover GIs 'as a whole'. Nevertheless, the contracting parties to the TLT chose to open the voting on the Singapore Treaty to all WIPO members. This reinforces the notion that the Lisbon Union made a conscious decision to deny all non-Lisbon member states the right to vote.

Bolstering the impression that the Geneva Act to the Lisbon Agreement represented a 'new' treaty—and thus necessitated extending the right to vote to all member states at the Diplomatic Conference—was a Secretariat's statement during the fifth session of the working group. This statement concerned the substance of the proposed revisions as they began to take place. Following a discussion in which several delegations questioned whether a Diplomatic Conference would be required to implement the revisions and the EU delegation suggested that the term 'Revision Conference' could serve as an alternative, the WIPO Secretariat[11] issued a clarifying statement (*Report* 2012b, pp. 44–45). It was recorded as follows:

> 'In response to the question raised by the Delegation of Iran (Islamic Republic of) as to whether a Protocol would be a new treaty or where it would fit in the text which only referred to a revision of the Lisbon Agreement, the Secretariat said that the answer to that question would depend on the nature of the Protocol itself. Hence, if in substance the Protocol went beyond the ambit of the original Lisbon Agreement then the Protocol would be considered to be a new treaty, whereas, if the Protocol stayed within the ambit of the original Agreement, it would be seen as a Protocol amending the Lisbon Agreement. *The Secretariat pointed out that the working documents focused more on a Protocol as a new treaty* rather than an amending Protocol to an original treaty.'

This statement from the WIPO Secretariat suggests that the substance of the revisions at the fifth working group necessitated a new treaty and full participation of all WIPO members. As the overall substance of the revisions were not drastically changed or scaled back to stay within the ambit of the original Lisbon Agreement, a diplomatic conference open to all WIPO members would have been required in order to enact a new treaty. Yet, while a diplomatic conference was indeed called, the rules and procedures withheld voting rights from non-Lisbon member states.

---

[11]The WIPO Secretariat at the fifth session of the working group consisted of the following: Francis Gurry, Director General; Binying Wang, Deputy Director General; Matthijs Geuze, Head, Lisbon Registry, Brands and Designs Sector; Florence Rojal, Legal Officer, Lisbon Registry, Brands and Designs Sector; Jessica Van Weelde, Consultant, Lisbon Registry, Brands and Designs Sector.

2.3 Actions and Results Reveal Motivations Behind Decisions

The decision by the Lisbon Union to keep voting closed to non-members reveals the Lisbon member states' motivation. The representative of CEIPI,[12] Francois Curchod, a former Deputy Director-General of WIPO, offered a prescient statement in the fourth session of the working group, recorded as follows:

> 'The Representative of CEIPI recalled that he had indeed raised the question of the apparent contradiction between the desire of a majority of the member States of the Lisbon Union to maintain or even strengthen the current level of protection and to apply it in a uniform fashion to both appellations of origin and geographical indications on the one hand, and on the other hand *the fact that such a position might compromise the purpose of the review, which was to make the system more attractive* for countries which were not in the system at present' (*Report* 2012a, 35–36).

The decision to opt for a high level of protections at the expense of making the Lisbon Agreement more attractive becomes all the more apparent when one looks at the statements and positions from the working groups on the development of the Lisbon system. In June of 2014, the participating parties at the ninth session of the working group discussed draft Article 11 concerning the protection contracting parties would be required to extend to a registered GI or AO (*Report* 2014a, 29–35). The minimal level of protection afforded to rights holders was a crucial topic and thus ardently debated. The proposed Article 11(3) was particularly controversial, as it provided that a state could declare that protection guaranteed by 11(1) were 'not compatible with its legal system' and lower level of protection otherwise available under 11(1).

The U.S. delegation spoke early in the proceedings and argued in favor of 11(3), explaining how 11(1) was incompatible with its common law trademark system. The Australian delegation supported the U.S. by proposing that 11(3) should actually replace the level of protections afforded in 11(1) as the former was 'more inclusive and consistent with existing international standards'. The delegation of the EU initially responded by stating that it 'had some difficulties in understanding the idea that the Revised Lisbon Agreement would establish two standards for the required level of protection.' The EU then stated concisely that it 'failed to understand' the argument that 11(1) might not be compatible with the laws of potential contracting parties. These statements illustrate how large the intellectual gap and level of disagreement was between the two delegations most representative of different trademark (and GI) systems. While the U.S. argued vigorously that the draft was incompatible with its current system, the EU neglected to offer any counterargument detailing how a trademark system could coexist with higher levels

---

[12]CEIPI is an French acronym for the Strasbourg-based 'Centre d'Etudes Internationales de la Propriété Intellectuelle' otherwise written in English as 'Centre for International Intellectual Property Studies.' The president of the Board was, until recently, WIPO Director General Francis Gurry.

of GI protections. Rather, the EU effectively threw its hands up and pretended that it simply could not understand what the U.S. was talking about, a statement that the authors find somewhat hard to take at face value. While it may be argued that non-Lisbon members could have done more, they obviously do not bear all the blame for this failure to compromise. They had no voting right and held no legal leverage in the negotiations.

While the withholding of voting power eventually became a controversial issue, the subject was not discussed in the early sessions of the working group. Indeed, no delegation mentioned voting until the fifth session. The United States did not formally participate on the record until the sixth session. Did the United States and other common law countries simply expect a full vote as a 'right' afforded all WIPO members, or can they be blamed for failing to ask early enough? The U.S. delegation addressed this matter in the eighth working group session, claiming that they had not participated much in the early stages because it was the delegation's understanding that the working group's mandate was limited to making procedural changes, as opposed to creating a new treaty entirely (*Report* 2014a, 4–5).[13] This may be correct. Yet it can also be argued that by the *fifth* working group session it was clear that a new substantive treaty was the preferred outcome of many Lisbon members (*Report* 2012b).

This brings up the convoluted nature of the issue. While Lisbon member States wanted to keep the revision 'in-house' and exclude common law members from full participation, the U.S. for its part might not have truly been interested in participating substantively when it saw that the desired result was a new treaty increasing the attractiveness of the Lisbon system and its sui generis approach to GI protection. To be fair, one can intimate from the early working groups' reports that Lisbon members made very few efforts to craft a new agreement attractive to a common law nations. It is only when the picture of a thoroughly revised Lisbon system started to emerge that a number of non Lisbon members started to give this process their full attention. This late interest was perhaps due to a sincere desire to contribute and ultimately participate in the new system, but some saw it as an effort to limit the long-term attractiveness of the new system.

Despite its lack of regard for the concerns of several non-Lisbon members, the Geneva Act might result in a more attractive Lisbon Union for a limited though not insignificant number of nations. For example, Germany and Switzerland stated that

[13]As quoted in the Report of the fifth working group: '[t]he Delegation of the United States of America went on to say that, since the subject-matter of the Lisbon Agreement, namely appellations of origin, featured in only a handful of protection systems around the world, the Delegation had not seen any reason to intervene in the work of the Lisbon Working Group by engaging with the text as such, all the more so as its initial understanding had been that the mandate of the Lisbon Working Group was limited to making procedural changes to the Lisbon Agreement, in order to accommodate the needs of certain Lisbon member States. However, the situation was very different now, because of the inclusion of geographical indications as subject matter. This meant that the Working Group was not merely dealing with a revision of the Lisbon Agreement but, instead, with an entirely new treaty. From the Delegation's perspective, this represented a radical expansion of scope and subject matter.'

they were interested in joining the amended Lisbon system (Saez 2015a) and many French-speaking African countries are reportedly set to join the system (Gervais 2015, 368). Whether that added attractiveness builds a far-reaching stable of member states remains to be seen. Any such optimism must be tempered by the reality that whatever added appeal exists is limited at best. At worst, a mildly appealing amended Lisbon system may be ironically counterproductive to the goal of achieving a comparatively more attractive registration system in the mold of the Hague, Madrid, and PCT registration systems (*Report* 2012b, 38–39).[14]

What if non-Lisbon member states had been fully at the table? Perhaps bold proposals necessitating substantive compromises would have been more heavily debated and possibly included in the final draft, including provisions on laches, abandonment and acquiescence. However, this presupposes a diplomatic conference in which a compromise on the whole was feasible. In this hypothetical scenario, one must imagine that a coalition of trademark countries would partake in the stated goal of creating a broadly attractive system. This, in retrospect, was probably a fantasy.

2.4 Retribution at the General Assembly

The procedure leading to the adoption of the Geneva Act has inspired consequences beyond the Diplomatic Conference. Many common law jurisdictions, including the United States, committed substantial effort, time and energy at the 2015 WIPO General Assembly to find a solution to the Lisbon system's long-term financial sustainability. Leaving no doubt for confusion, the U.S. delegation acknowledged in the opening plenary that it was bringing up financial stability as a result of being excluded from participation in the 'illegitimate' adoption of the Geneva Act in addition to opposing the outcome of the Geneva Act itself (*General Report* 2016, 38).[15] Thus, the U.S.' efforts to force the Lisbon system into financial sustainability

[14]Indeed, a statement from Director General Gurry at the fifth working group advocated for broadly increasing the attractiveness of the Lisbon system rather than achieving limited progress for the sake of preserving the strength of the system. The Director General noted that the most recent Madrid Protocol revision fell short of broadly increasing the attractiveness of the system after the Madrid member states chose to adopt terms in the best interest of member states, consequently setting back any chance of real expansion by ten years. In comparison, the Lisbon System was in even more dire need of boldly increasing attractiveness of the system.

[15]The U.S. delegation is recorded as stating, '[f]aced with serious questions as to why and how United States fees and contributions were being required to subsidize a system the adoption of which the State had opposed and which was so harmful to national trade, the Delegation could not support a framework under which the Lisbon System was permitted to carry a deficit covered by other Unions.' … 'It supported the use of PCT revenue to support the overall activities of the Organization because those activities had had the continuous support of the full membership. It objected only to the use of such revenue to support the Lisbon System, which, by its own Agreement, was required to be self-funding and which had recently been expanded without the broad consultations for which WIPO had long been respected.'

was at least in part a direct result of the procedure chosen for the 2015 Diplomatic Conference for the Revision of the Lisbon Agreement. More surprising, however, were the lengths to which the U.S. went to force the Lisbon Union to alter its method of financing. The U.S. first reiterated its position taken at previous meetings of the Program and Budget Committee that it would not support a budget for 2016–2017 unless a number of conditions relating to the financial independence of the Lisbon system were met (*General Report* 2016, p. 117). This hardline stance was further advocated for and bolstered within various sessions of the General Assembly as the U.S. added proposals regarding Lisbon financial sustainability to the agendas of the Lisbon Union, Madrid Union, PCT Union, and the WIPO Coordination Committee (WIPO 2016a).

In the Madrid and PCT sessions, the U.S. stated that no registration system should allow surpluses to be used by the Lisbon system to cover its deficit. Somewhat boldly, the U.S. proposed that, rather than letting it be used by Lisbon system, the Madrid Union should distribute its surplus among Madrid member states. Although supported by Australia, a coalition led by Switzerland, Hungary, and France rejected the proposal. Yet the U.S. succeeded by increasing awareness of the issue and opening up many nations to the idea of cutting the Lisbon Union off from access to surpluses generated by other registration and application systems (*Report* 2016a, b).

U.S.-applied pressure paid off. At the end of the 2015 General Assembly a compromise was reached on the Lisbon Union's portion of the 2016-2017 biennial WIPO budget. Just before midnight on the final day of the Assembly the Lisbon Union agreed 'to adopt measures by the 2016 Assemblies to eliminate the Lisbon Union's projected biennial deficit' (WIPO International Bureau 2015a, 8). Moreover, the General Assembly approved a *loan* from the other contribution-financed unions to the Lisbon Union to cover its impending deficit. While Lisbon members may be satisfied with the deal in the short-term, the U.S. will likely continue to put pressure on the Lisbon Union in the future. If the Lisbon Union is to make good on the promise to eliminate its deficit, then it must look at the prospect of raising fees yet again, likely much more substantially At the 2015 General Assembly, the Lisbon Union doubled registration fees to 1000 Swiss francs for an international application. Unfortunately, this improved fee schedule is not close to the 3350 Swiss franc fee recommended for registration under Director General Francis Gurry's proposal (WIPO International Bureau 2015, 5).[16] Furthermore, the fee increase pales in comparison to the Secretariat's summary of options in order to reach financial sustainability, which states that if single fees alone are to fund the operations of the Lisbon Union, the international registration fee required would

[16]The Director General's complete fee recommendation stated, '[i]n view of the preceding considerations, it is proposed that the Fee Schedule under Rule 23 of the Lisbon Agreement be updated, so as to reflect the following amounts: (i) an international registration fee of 3350 Swiss francs; (ii) a fee of 1500 Swiss francs for the modification of an international registration; (iii) a fee of 150 Swiss francs for providing an extract from the International Register; (iv) a fee of 100 Swiss francs for providing an attestation or any other information given in writing concerning the contents of the International Register'.

need to be anywhere from 54,750 to 80,357 Swiss francs (WIPO Secretariat 2015b). At that fee rate the system would be at risk of receiving no applications, thus endangering its very survival. Taken together, the Lisbon Union's commitment to erasing the deficit and taking loans are appropriately concerning. It may be necessary for Lisbon members to collect contributions from Lisbon member states in order to bridge the gap, but that would go against the current WIPO practice of the unitary contribution system. It also would be highly unpopular in certain circles. Regardless, the Lisbon Union will be under increasing pressure to balance its books, possibly taking measures that will make the system as a whole less attractive to WIPO member states and producers within those states. If the Lisbon system becomes less attractive over the next few years due to increased fees or obligatory member state contributions, then the Lisbon Union's aforementioned move to limit voting rights at the Diplomatic Conference would likely represent a mistake and a long-term setback for the Lisbon Union.

2.5 A Look at the Geneva Act

2.5.1 Registration Issues

A fuller comment on the substantive aspects of Geneva Act appears in Matthijs Geuze's chapter of the present volume. Nevertheless, a few features of the new Act deserve mention here in light of the kerfuffle over the process that led to the adoption of the new Act. The Geneva Act made several improvements to the administrative operation of the Lisbon register and generally modernized the system of registration of both geographical indications and appellations of origin, thus solving the terminological issue due to the reluctance that a number of countries had to using the notion of 'appellation of origin'. Appellations of origin are often considered to be tied to sui generis European systems, such as the well-known French and Italian systems of *appellation d'origine contrôlée* (AOC) et *denominazione di origine controllata*, respectively (Gervais 2010a, 108–09). Without offering an exhaustive list of such improvements, the Geneva Act (a) made the role of national offices much clearer; (b) allows members to charge a per country fee (for applications that seek protection in their territory); and (c) allows per-country fees to use the protected indication, though not the maintenance fees common to most trademark systems.[17] The Geneva Act also introduced a clearer system to refuse

[17]The issue of maintenance fees was the topic of heated debate during the conference. Maintenance fees had been mentioned in square brackets in early versions of the draft Geneva Act but did not make it into the final version. *Compare* Geneva Act 2015, arts 7–8 (explaining the fee structure, which does not include a maintenance fee), with WIPO Document LI/DC/3 of 14 November 2014 on the Basic Proposal for the New Act of the Lisbon Agreement on Appellations of Origin and Geographical Indications, art 7–8. WIPO doc. LI/DC/3 available at: http://www.wipo.int/edocs/mdocs/mdocs/en/li_dc/li_dc_3.pdf (as consulted March 21st 2016).

new GIs (and then withdraw refusals in whole or in part) at the time of application and a system allowing interested parties to 'request the Competent Authority to notify a refusal in respect of the international registration' (*Geneva Act* 2015, art 15–16, rule 19). By contrast, the 1958 Act did not mention grounds for refusal of a GI; did not explicitly allow partial refusals (though they had happened); and contained no detailed provision on withdrawals.[18] This reflects the nature of an administrative 'clean-up' rather than a true substantive change, however. Under Article 5 of the 1958 Agreement, a court or other competent authority in the country where protection is claimed could invalidate an appellation. The Geneva Act maintains this system more or less but includes both much clearer rules and specific processes.[19]

The issue of fees was a major discussion during the Diplomatic Conference and, as mentioned at the end of Part I, at the WIPO General Assembly that followed its adoption. Under the 1958 Lisbon system, applicants pay a *single fee once* to WIPO, none to the members' individual Offices, and their application is deemed valid (subject to refusal by individual members) essentially forever in all Lisbon member states where it was not initially refused or later invalidated (WIPO 2010, 2015).[20] This fee (which was 500 Swiss francs, but doubled as of 1 January 2016) is paid to, and retained by WIPO. The absence of country designation and 'per-country' fees stand in sharp contrast to the Patent Cooperation Treaty (PCT), the Madrid system for trademarks, and the Hague system for designs (WIPO 2010, 2015a; WIPO 2015b, 2016b). In addition, no renewal fee is ever required under the Lisbon Agreement.[21] This is administratively different from the situation with several other IP rights, and also arguably incompatible with the current practice of common law jurisdictions that protect GIs as trademarks. Trademarks can be considered abandoned and canceled for non-use, and a registration renewal or maintenance fee is typically payable.[22] Under the Geneva Act, part of the gap between Lisbon and those other systems was filled, but not all. Country designation was introduced and per-country fees are now possible for the initial registration and for use of the indication, but not maintenance fees. This is likely to continue to be an issue going forward.

---

[18]*Compare* rules 9-10 of Geneva Act to the Lisbon Agreement rules 9–10 with Article 5 of 1958 Lisbon Agreement.

[19]*See* Article 15–16, rules 9–11 of the Geneva Act.

[20]*See* Articles 5 and 7 of the Lisbon Agreement.

[21]The Lisbon Agreement specifically provides for the absence of renewal fees under Article 7.

[22]In the United States, since 1989, the renewal term has been 10 years. McCarthy, § 19:142, 19-422. Also, under US law, cancelation is possible after three years of nonuse. *See* 15 USC § 1127 (2012).

2.6 Genericness Issues

The 1958 Lisbon Agreement was flexible in several respects but not on genericide. Genericide is the loss of an appellation when it becomes generic, and is thus unable to function as an indication of geographic origin in a given market (Kemp et al. 2006, 266; Gervais 2010b, 96–97). The 1958 Agreement prevents invalidation for genericness in the country where protection is claimed, *unless the appellation has become generic in its country of origin.*[23] This is a rare mandatory application of *lex originis* in international intellectual property (Dinwoodie 2009, 731). A country under the Lisbon system has one year to determine whether it will refuse to protect any previously registered appellations *at the time of joining.*[24] For appellations registered *after* a country has joined, there is a 12-month period to make a similar decision. After that period has lapsed, it becomes much more complicated to refuse protection. Invalidation (e.g., by a court) is possible. An issue might surface if a common law jurisdictions were to join Lisbon, if the owner of a Lisbon GI does nothing while the GI was used with his knowledge but without his consent by a third party. This might trigger a finding of abandonment, laches, or lead to a finding of genericness of the GI is used descriptively.

It is a well-established principle of international intellectual property that the law of the country of protection (*lex loci protectionis*) typically governs. Thus, a Court or other competent authority determines the validity of a copyright, trademark, or patent in its own jurisdiction. The Paris Convention makes it clear that this does not directly affect the same trademark or a patent on the same invention in other jurisdictions—a principle known as the independence of patents and trademarks.[25] In the case of infringement, the law of the country where protection is claimed (*lex loci delicti*) typically applies.[26] It would be strange indeed if a Court could not find a patent or mark invalid unless it had been found invalid in the paten holder or trademark owner's *country of origin.* Yet that is the system under the 1958 Lisbon Agreement. Implementing this provision functionally requires a sui generis system because countries that protect GIs using trademarks assess genericness *in their territory* not in the mark's country of origin. Unfortunately, the Geneva Act, while it uses language that differs from the 1958 text, essentially maintains a version of the *lex originis* regime, especially in its Article 12: 'Subject to the provisions of this Act, registered appellations of origin and registered geographical indications cannot be considered to have become generic in a Contracting Party'.[27]The Geneva Act, like the 1958 Lisbon Agreement, allows a Lisbon member to reject a GI at the time

---

[23]*See* Article 6 of the Lisbon Agreement.

[24]*See* Lisbon Agreement art 14.

[25]*See* Paris Convention for the Protection of Industrial Property 1979, arts 4*bis*, 6.

[26]*Ibid*, arts. 6*quinquies*(B), 10*ter*(2); Berne Convention for the Protection of Literary and Artistic Works 1979, art. 6*bis*.

[27]Further language maintaining *lex originis* is also in art 8(1).

(within 12 months) of registration if it is generic in their territory.[28] It will also allow its members to maintain coexistence of a GI and a trademark and to protect prior trademark rights. However, if the GI gets protection under the Lisbon system and later becomes generic in a Lisbon member, for example because of inaction or even acquiescing by the holder of the GI, then it is not clear under either the 1958 Agreement or the Geneva Act how that Lisbon member could find this GI generic if the GI is not generic in its country of origin without violating Article 12.[29] Perhaps a remedy could be refused to the trademark owner in such a case. Very unfortunately, an Agreed Statement to that effect which had been initially accepted by the Chair of the group drafting the substantive part of the Geneva Act was then rejected after 'consultations' the details of which are not known (Gervais 2015, 356).

2.7 Conflicts with Prior Trademarks and Generic Terms

The 1958 Lisbon Agreement allows its Members to adopt or continue to use one of three approaches in managing conflicts between trademarks and GIs (however those are protected under national law): (1) a 'first in time, first in right' approach; (2) a 'coexistence approach' (that is, a GI and trademark with similar legal effect); or (3) a 'GI superiority approach', under which the GI wins the conflict, except perhaps where the previous trademark is considered well-known (Montén 2006, 328–30).[30]

'First in time, first in right' (FITFIR) usually is understood to mean first *in the territory of the jurisdiction concerned* (Mostert 2004, 2–34; Gangjee 2007, 1262; Rhee 2012, 48).[31] One could see a degree of irony in the case of GIs because GIs very often have been in use for decades or more before the registration of a trademark that would defeat the use of such GIs in a country other than the country of origin (Patent and Trademark Office n.d.). Yet, as a matter of trademark law, the analysis focuses on each domestic market and is largely a question of the domestic consumer's perception—although this perception may include knowledge gained through advertising of a famous mark not in use or not yet widely used in the

---

[28] *See* Lisbon Agreement art 5(3), (4); Geneva Act art 15(1).

[29] *See* Lisbon Agreement art 6; Geneva Act art 12.

[30] For example, while the EU has a GI superiority approach in most cases, its regulations provide protection for prior, well-known marks. Council Regulation No 510/2006 of 20 March 2006 on the Protection of Geographical Indications and Designations of Origin for Agricultural Products and Foodstuffs [2006] OJ L93, art 3(4): (providing that a 'designation of origin or geographical indication shall not be registered where, in the light of a trademark's reputation and renown and the length of time it has been used, registration is liable to mislead the consumer as to the true identity of the product.'). In sum, EU law provides that prior well-known trademarks prevail over GIs and allows coexistence of prior trademarks not considered to be well-known. ibid.

[31] A number of trade agreements have enshrined this principle. Eg, United States-Australia Free Trade Agreement, Austl-US, art 17.2(4), of May 18, 2004.

territory concerned.[32] The Geneva Act allows, but does not mandate, a member State to apply the 'first in time, first in right' principle in resolving GI–trademark disputes.[33]

What is the situation under the TRIPS Agreement? First, a WTO Member must refuse or invalidate the registration of a trademark that contains or consists of a geographical indication if: (a) the goods do not originate in the territory indicated; and (b) use of the indication in the trademark for such goods in the territory of the 'Member' concerned is *of such a nature as to mislead* the public as to the true place of origin.[34] For GIs used in connection with wines and spirits, however, deception (misleading the public as to the true place of origin) does not need to be present.[35] Second, TRIPS Article 24.5 then allows a WTO Member to protect those who were using or applied for registration of a trademark (or were 'grandfathered') in the WTO Member concerned either before the TRIPS Agreement became applicable in the Member concerned, or before the indication in question was protected in its country of origin.[36] It is also worth noting that TRIPS Article 24.6 provides that WTO Members may decide not to protect a geographical indication used in connection with foreign goods or services. This choice to withhold protection is available when the relevant indication 'is identical with the term customary in common language as the common name for such goods or services in the territory of that Member'.[37]

In *EC–Trademarks and Geographical Indications I*, a WTO dispute-settlement panel explained that the coexistence of a protected indication and a trademark can be considered, under certain circumstances a limited exception justifiable under TRIPS Article 17.[38] In other words, the coexistence of a protected indication and a

[32]This situation is reflected in TRIPS Article 16, which provides that, in determining whether a trademark is well-known for the purposes of applying Article 6*bis* of the Paris Convention (which only applies to identical or similar goods but which TRIPS extends to both dissimilar goods and to services), WTO 'Members shall take account of the knowledge of the trademark in the relevant sector of the public, including knowledge in the Member concerned which has been obtained as a result of the promotion of the trademark'. *See* TRIPS Agreement art 16; Paris Convention (n 25) art 6*bis*.

[33]*See* Geneva Act art 13(1).

[34]*See* TRIPS Agreement art 22.3.

[35]Ibid, art 23.2.

[36]Ibid art 24.5; see also WIPO Document SC/5/3 of 8 June 2000 on the Possible Solutions for Conflicts Between Trademarks and Geographical Indications and for Conflicts Between Homonymous Geographical Indications, 11–12.

[37]See TRIPS Agreement art 24.6.

[38]WTO Report of the Panel WT/DS174/R of 15 March 2005 on European Communities—Protection of Trademarks and Geographical Indications for Agricultural Products and Foodstuffs: Complaint by the United States, 7.688 (Complaint by the US); WTO Report of the Panel WT/DS290/R of 15 March 2005 on European Communities—Protection of Trademarks and Geographical Indications for Agricultural Products and Foodstuffs: Complaint by Australia, 7.644–7.686 (Complaint by Australia).

non-famous trademark is a permitted exception to *trademark rights*.[39] In sum, TRIPS allows—but does not mandate—continued application of 'first in time, first in right' by each WTO Member.[40]

The Geneva Act is clearer on this point than the 1958 Lisbon Agreement in that it provides specific conflict rules, which mostly allow for prior trademark rights to be protected and, like the 1958 text, allows GIs to be rejected for genericness in a member *at the time that member joins the system*. The genericide prohibition in the Geneva Act would make it difficult, however, to reconcile with 'first in time, first in right' in the case of a GI that was used but later becomes generic as a geographic term in a country, or a similar term was used by a third party as a trademark. A country or intergovernmental organization (such as the European Union) joining the Geneva Act may thus limit its option to use to its full extent the TRIPS exception regime as interpreted in the WTO dispute-settlement process.[41]

2.8 The Future of the Geneva Act

2.8.1 Geographical Indications and a Free Market Economy

Attaching an intangible, yet measurable (that is, a higher price due to the IP rent) value to the identification of the geographic origin of a product seems to postulate the existence of a correlative, measurable difference in *actual quality*; that is, an objectively quantifiable difference between products of different origin but similar composition—say, a wine produced from Pinot Noir grapes in Napa Valley, New Zealand or in Bourgogne (Burgundy) (Bird 2005, 1–16). One could posit that this measurable difference—assuming that one *can* measure it—lies in natural factors such as soil and climate. Which brings us close to the *terroir*. Indeed, a system of protection for denominations of geographic origin emphasizes the cluster of factors traditionally amalgamated under the term *terroir:* tradition, know-how and a link between product and land (Calboli 2015, 434). *Terroir* can be used as a marketing

[39]And thus allowed under Article 17 of TRIPS, stating '[m]embers may provide limited exceptions to the rights conferred by a trademark…'. The report of the dispute-settlement panel mentioned in the previous note concluded, 'with respect to the coexistence of GIs with prior trademarks, the Regulation is inconsistent with Article 16.1 of the TRIPS Agreement but, on the basis of the evidence presented to the Panel, this is justified by Article 17 of the TRIPS Agreement.' Complaint by Australia 7.686; Complaint by the US 7.688. In countries that use certification marks and/or the torts of 'unfair competition' and 'passing off' in this context, such a conflict would be between two trademarks, not between a trademark and 'something else', such as a GI.

[40]Essentially, under TRIPS Article 24.5, a prior trademark continues to be registrable and its owner may continue to use the mark. This must be interpreted together with Article 16.1.

[41]State practice subsequent to the adoption of a treaty is one of the relevant factors in that treaty's interpretation. Vienna Convention on the Law of Treaties (1969) 1155 UNTS 331, reprinted in 8 ILM 679, *see* art 31(3)(b) (entered into force 27 January 1980).

tool to extract additional rents. A consumer who does not share the history and culture of the French, Italian, or Spanish *terroir* may not easily identify with the conventions and practices that were used to define the quality of a product (say, a red wine) at its point of origin. That same consumer *can* nonetheless attribute a higher value or quality to a product for a different set of reasons. Even if the higher price resulting from the linkage between a product and its origin is arguably or even demonstrably irrational—say, because no measurable objective quality differentiates the product and its non-GI equivalent—that is not a fatal argument against GI protection: trademarks also sometimes perform an 'irrational' yet well-accepted function in guiding consumer behavior.[42] Put differently, and more concretely, using a consumer's belief that *brie* cheese will be not just different but *better* because it was produced in Meaux[43] (France), as opposed to in an unknown location in some industrial plant or even in a known cheese producing location (such as the US state of Wisconsin), is not incompatible with the principles of a 'free' market economy.[44]

2.9 Can GIs Be Protected Only as Marks?

Geographical indications, like trademarks, depend on reputation and associated goodwill.[45] To paraphrase Professor Gangjee (2007, 1262), they share the same epistemological backcloth. Reputation is an element mentioned in both the Lisbon Agreement and the TRIPS Agreement.[46] 'Normal' trademarks, collective marks and certification marks can be used as symbols of or ways to suggest geographic origin.

A common feature of common law based trademark law is the so-called 'first-in-time, first-in-right' approach according to which the first user in a given territory will have senior rights. Using the US system as an example, a geographic designation of origin must not clash with a pre-existing trademark if it is to be protected under a trademark system.[47] Similarly, a geographic term that is generic

---

[42]McCarthy (1999), McCarthy on Trademarks and Unfair Competition. 2nd edn. New York: West Group, §2:38.

[43]'Meaux' is the name of a town near Paris, France. 'Brie de Meaux' is a protected indication, but 'Brie' is not. BRIE DE MEAUX, No 686 www.wipo.int/ipdl/en/lisbon/key.jsp?KEY=686 (as consulted 13 January 2016). Wisconsin also produces 'brie' cheese. Wisconsin Milk Marketing Board, 'Brie' (2016) www.eatwisconsincheese.com/wisconsin-cheese/article?cid=9 (as consulted 12 January 2016).

[44]Unless the perception is based on deceptive marketing practices, but that is typically handled under a different area of law.

[45]*Supra*, note 43, McCarthy § 2:38, 2–76.

[46]*See* Lisbon Agreement art 2(2); TRIPS Agreement art 22.1.

[47]*Supra*, note 43, McCarthy § 16:1.

in the United States cannot acquire trademark status.[48] A term that is merely descriptive can be protected as a mark once it acquires secondary meaning.[49] In the case of a geographic *certification mark*, there is a *collective goodwill* linked to a region or locality. This notion of collective goodwill emerged in UK in the so-called 'Drinks' cases.[50] Collective goodwill is goodwill shared by a group of producer. Any producer part of the group may file a claim in civil courts to protect the goodwill—without having to show control over or agreement of the collective.

United States courts have recognized that goodwill could be shared but they have generally taken a less liberal approach on the right of action by individual users. They have given the holder of a certification mark more control over its use and enforcement. For example, in *State of Idaho Potato Comm'n v. G & T Terminal Packaging, Inc.*, the Ninth Circuit found that selling genuine Idaho potatoes using the name IDAHO constituted counterfeiting of the certification mark IDAHO for potatoes because the mark owner had not had the opportunity to exercise quality control.[51] The owner of the mark 'acts as the representative of the mark users'.[52] Hence, as a matter of trademark law, the owner of a certification mark—not the State—controls the certification standards. Examples of US geographic certification marks include: FAMOUS IDAHO POTATOES, FAMOUS POTATOES GROWN IN IDAHO, GROWN IN IDAHO IDAHO POTATOES, and PREMIUM PACKED IDAHO POTATOES for (unsurprisingly) Idaho potatoes; PARMIGIANO-REGGIANO, ROQUEFORT, STILTON, and REAL CALIFORNIA for cheese; PARMA for ham; DARJEELING for tea; WASHINGTON for apples; and THE FLORIDA SUNSHINE TREE for citrus.[53] Certification marks have also been used in a more typical 'Lisbon' context: booze. The three US 'Cognac' cases serve

---

[48]Ibid § 14:18.

[49]Ibid § 11:25.

[50]*Bollinger v Costa Brava Wine Co. Ltd*, [1961] RPC 116 (Ch); *Vine Prods Ltd v Mackenzie & Co Ltd*, [1969] RPC 1 (Ch), 4 (citing *Bollinger v. Costa Brava Wine Co Ltd*, [1960] RPC 16. These cases have a progeny in the UK. *John Walker & Sons Ltd v Henry Ost & Co Ltd*, [1970] RPC 151 (Ch); *HP Bulmer Ltd v Bollinger SA*, [1978] RPC 79 (CA); *Erven Warnink Besloten Vennootschap v J Townend & Sons (Hull) Ltd*, [1979] AC 731 (HL). The same happened in other common law jurisdictions such as Canada and New Zealand. *Institut National des Appellations d'Origine des Vins et Eaux-de-Vie* v *Andre Wines Ltd*, (1990) 30 CPR (3d) 279; *Dairy Bureau of Canada* v *Annable Foods Ltd*, (1993) 46 CPR (3d) 289 at 309; *See* also *Comite Interprofessionel du Vin de Champagne v Wineworths Group, Ltd* [1991] 2 NZLR 432.

[51]State of Idaho Potato Comm'n v G & T Terminal Packaging, Inc, 425 F3d 708, 721–22 (9th Cir 2005).

[52]*Supra*, note 43, McCarthy § 19:92.50.

[53]FAMOUS IDAHO POTATOES FAMOUS POTATOES GROWN IN IDAHO, Registration No 2,403,069; PARMIGIANO-REGGIANO, Registration No 1,896,683; ROQUEFORT, Registration No 571,798; STILTON, Registration No 1,959,589; REAL CALIFORNIA CHEESE, Registration No 1,285,675; PARMA HAM, Registration No 2,014,628; DARJEELING, Registration No 1,632,726; WASHINGTON, Registration No 1,528,514; THE FLORIDA SUNSHINE TREE, Registration No 932,033. The vast majority of certification marks in the United States are not for geographic origin. This majority includes marks registered by unions and various standardization bodies, such as Underwriters' Laboratories (Belson 2002, 26–28).

as milestones on the road to GI protection via trademark law. The first case dates back to 1944.[54] The case involved applications for the trademarks COLOGNAC and CALOGNAC for brandy. Oppositions to those applications were filed by a 'genuine' (French) Cognac maker.[55] The oppositions were not made on the basis of pre-existing registrations for other marks (such as COGNAC). Instead they were made under the terms of the statute in force at the time, which allowed certain parties (this would emerge in the case, as we shall see) to oppose the use of descriptors as marks if those descriptors would confuse or mislead the consumer. Reversing both the Examiner of Trade-Mark Interferences—who had not recognized the opposer's right to battle the applications—and the Commissioner of Trademarks, the Court of Customs and Patent Appeals confirmed the opposer's standing and agreed with the opposer.[56] The Court stated the following in a passage that has the unmistakable aroma of an 'Old World' GI perspective:

> Cognac is a name applied to a type of brandy distilled from wines made from grapes grown in a limited territorial region of France, often referred to as the Cognac district, the boundaries of which are defined by French law … there is a certain quality in the soil of the region which gives to the grapes there grown a particular character or flavor, which enters into the brandy made from them, and that that quality of soil is not found elsewhere in France, nor, it is claimed, in any other part of the world. It is recognized as a superior brandy.[57]

The second case, four decades later, dealt with applications for two similar derivatives of Cognac, namely COLOGNAC and CALOGNAC.[58] The case evidently bears strong resemblance to the previous one, but here the opposer was not an individual producer of Cognac, but rather an entity responsible under French law for defending the word Cognac.[59] Between the two cases, Congress had adopted the 1946 Trademark (Lanham) Act, which contained a specific provision for certification marks.[60] Indeed, one of the Board's first steps in this second case was to note that the statute defined certification marks and included marks used to denote a geographic origin.[61] The 1946 statute also contained the now familiar prohibition of the registration of a mark that, 'when applied to the goods of the applicant,

[54]The date is significant because, although the case is silent on that point, sympathy for France may have been a factor at that particular juncture in US history.

[55]Otard, Inc, v Italian Swiss Colony, 141 F2d 706 (CCPA 1944).

[56]Ibid, 709–11.

[57]Ibid, 708.

[58]Bureau Nat'l Interprofessionnel du Cognac v Int'l Better Drinks Corp, 6 USPQ2d 1610, 1615 (TTAB 1988).The application was filed in 1984, Ibid, 1610.

[59]Ibid, 1610–11.

[60]15 USC §§ 1052(d)–(e), 1054 (2012) (Trademark Act).

[61]Bureau Nat'l, *supra* note 59, 1614. The terminology was new, but not the notion of a collectively owned mark. Section 62 of the Trade-Marks Act of 1905 referred to standardization marks. Trade Marks Act 1905 (9 Edw 7 c 15), § 62. The statute was amended in 1936 and again in 1938 to provide for registration of collective marks owned by foreign associations.

is primarily geographically descriptive or deceptively misdescriptive of them'.[62] The Board's approach was consonant with trademark law principles, putting the focus squarely on consumer deception:

> [I]f a mark is the name of a place known generally to the public, purchasers who encounter goods bearing the mark would think that the goods originate in that place [i.e., purchasers would make a 'goods-place association' …], [if] the goods do not come from the named place, and the deception is material to the purchasing decision, the mark is deceptive under Sect. 2(a); if the deception is not material to the purchasing decision, the mark is primarily geographically deceptively misdescriptive under Sect. 2(e)(2) of the Act.[63]

In the third and final case, decided a decade later, the opposer, the French *Institut National des Appellations d'Origine*, was also a 'GI defending entity' responsible for defending several protected wine and spirit appellations of origin in France.[64] The Board applied Sect. 2(d) of the Lanham Act, which deals with likelihood of confusion with a preexisting mark, and decided that the test was the same as for other marks:

> There is nothing in the language of Sect. 2(d) which mandates or warrants application of one level of likelihood of confusion analysis … in cases where the plaintiff's mark is a trademark or service mark, but a different and more limited likelihood of confusion analysis in cases where the plaintiff's mark is a certification mark.[65]

In other words, a showing of deception was unnecessary. Certification marks should be treated for purposes of an infringement analysis in the same way as ordinary trademarks. While the opinion parallels the previous case, it is worth noting that the Board also recognized protection of the COGNAC mark under *labeling regulations*.[66]

Overall, the picture that emerges from the Cognac cases shows a willingness to recognize collective goodwill in the form of (common law) certification marks.[67] The application of trademark law by US courts and other authorities suggests that common law can protect symbols of geographic origin used in trade beyond the sui generis mechanism used for alcohol labels. Unfortunately for the future of the Lisbon system, using trademark law to protect GIs implies the ability to remove a

---

[62]Trademark Act, *supra* note 64, § 1052(e)(2).

[63]Bureau Nat'l, *supra* note 64, 1615.

[64]Institut Nat'l des Appellations d'Origine v Brown-Forman Corp, 47 USPQ2d 1875, 1877 (TTAB 1998). Apparently, the applicant's concoction actually contained some French brandy. Ibid, 1880.

[65]Ibid, 1890.

[66]Ibid, 1878, 1885–96. The regulation is 27 CFR § 5.22(d)(2) (2014), (recognizing French governmental authority to designate and regulate Cognac grape brandy distilled in the Cognac region of France).

[67]This actually started with Pillsbury-Washburn Flour Mills Co v Eagle, 86 F 608 (7th Cir 1898). For a fuller discussion of this topic (from which this part of the Article is derived), see Gervais, Daniel. 2014. Cognac after Spanish Champagne ? Geographical Indications as Certification Marks, *see* n.1 above.

mark not used by its owner. More generally, trademark law implies (1) a requirement of use to register, (2) the payment of renewal or maintenance fees, and (3) a level of protection tied to the existence of a likelihood of confusion except for famous or well-known marks.[68] None of these are allowed, at least explicitly, under the Geneva Act even though they were raised on several occasions by delegations at the Diplomatic Conference. Fortunately, the Geneva Act does allow member countries to impose examination and use fees, however. Yet a number of potential areas of incompatibility between GIs—or a certain conception of GIs at least—and trademarks remains. It seems that it is that particular conception that informed many of the members that negotiated the Geneva Act and, as a result, it may not be easy to get countries that look at GIs differently to join.

2.10 Will the Geneva Act Succeed?

As of the date of the Diplomatic Conference that adopted the Geneva Act of the Lisbon Agreement in May 2015, the Special Union (of Lisbon members) had 28 member states (WIPO 2013).[69] Many of those members will likely join the Geneva Act, which is also open to intergovernmental organizations such as the European Union and *Organisation Africaine de la Propriété Intellectuelle* (OAPI).[70] Even if several OAPI members join Lisbon, how much they will use it (that is, how many GIs they will actually register) is unclear. Indeed some OAPI members are already members of the 'old' Agreement and have not used it much.[71] North African countries such as Morocco and Tunisia, together with Algeria, which is already party to the 'old' (1958/1967) Act, may also join the Geneva Act membership. A few Latin American countries might do so as well, adding to the existing five members of the old Agreement (Costa Rica, Cuba, Mexico, Nicaragua and Peru).[72] More importantly, European membership will likely broaden to include most or all EU members (including Germany, who is not currently a Lisbon member) and probably Moldova and Russia, among others. This means a potential of approximately 55 member States. If the EU were to join backed by all its member states,

[68]On the protection of well-known marks, TRIPS Agreement Article 16.2 requires WTO Members to take account 'of the knowledge of the trademark in the relevant sector of the public, including knowledge in the Member concerned which has been obtained as a result of the promotion of the trademark.'

[69]The term 'Special Union' (that is, the group of parties to the Lisbon Agreement) is used in Article 1 of the Lisbon Agreement, and in Article 21 of the Geneva Act.

[70]Geneva Act, art 22(2). The OAPI has 17 members, most of which are French-speaking sub-Saharan African countries (OAPI 2010, *see*: www.oapi.int/index.php/en/aipo/etats-membres. (as consulted January 12, 2016).

[71]The following countries are members of both agreements: Burkina Faso, Congo, Gabon, and Togo.

[72]Haiti is also a member (WIPO 2013), *Supra* note 72.

it would potentially have a majority vote in the Special Union (which it could exercise as a single vote on behalf of all 28 EU members-subject to Brexit of course).[73] Yet that would leave a sore spot: the complete absence of common law jurisdictions.

The denouement of the Geneva Act story may be more complicated, however. True, despite the strength of the opposition manifested at the closing plenary by a number of common law jurisdictions, the art of trade negotiations is such that anything is possible. Smaller common law jurisdictions might find it harder to resist EU trade pressure to join Lisbon. The Trans-Pacific Partnership (TPP)[74] is intesresting.in this context. Clearly, it will affect the trademark laws of the various countries[75] and is currently the subject of widespread commentary and controversy (Dreyfuss et al. 2015, 569–71). Relevant here is the TPP's requirement that contracting parties ratify or accede to either the Madrid Protocol on international trademark registration or the Singapore Treaty on the Law of Trademarks.[76] If member states follow the treaty's obligations, the TPP would represent a breakthrough in the Latin American and Southeast Asian countries that had previously been difficult areas for the trademark registration systems to crack (Lince 2015). As more countries reflect the strong trademark system of the U.S., the ratcheting up of trademark protection in more countries gradually makes a universally attractive multilateral GI registration system all the more difficult to conceive.

More importantly, some see the TPP as containing normative countermeasures to Lisbon, such as definitions of prior rights or genericness standards. The TPP text is not easy to reconcile with Article 12 of the Geneva Act (providing for genericide only in the country of origin of the GI): it provides that a GI may become generic *in each country* that protects it, leading to its potential invalidation in such country.[77] Though not expressly permitted by Lisbon, it seems that court invalidation is, however, possible under Lisbon so that this may not be as formidable an obstacle at may seem at first glance. More generally, a renewed emphasis on the role of certification marks to protect GIs has surfaced in various fora. Is the TPP an anti-Lisbon tool? That is far from obvious. Some of the nations that negotiated the Trans-Pacific Partnership (TPP) recently introduced sui generis GI legislation that may make it easier for them to agree to join Lisbon, for example, as part of a trade

---

[73]Lisbon Agreement, art 22(b)(4)(ii) allows any Contracting Party that is an intergovernmental organization to vote in place of its members.

[74]The TPP is an extensive trade agreement between the U.S., Canada, Japan, Australia, Malaysia, Mexico, Peru, Vietnam, Chile, Brunei, Singapore, and New Zealand concerning a wide range of IP and trade issues. *See* Trans-Pacific Partnership, Oct. 5, 2015.

[75]TPP art. 18.

[76]TPP, art. 18.7(2).

[77]TPP, art 1832.3(3) states, '[n]o Party shall preclude the possibility that the protection or recognition of a geographical indication may be cancelled, or otherwise cease, on the basis that the protected or recognized term has ceased meeting the conditions upon which the protection or recognition was originally granted in that Party.'

deal with the EU.[78] The list of bilateral EU trade agreements with a GI focus might soon include India. In 2014, Canada and the EU signed a trade agreement (reportedly in trouble as of this writing due to resistance to its investor-state dispute provisions) that contains a detailed set of provisions on GIs.[79] CETA provides that, subject to exceptions, interested parties can prevent the use of certain GIs for products that fall within a class linked to that GI.[80] This protection is provided even when the true geographic origin of is indicated, the GI is used in translation, and even when the GI is accompanied 'by expressions such as "kind", "type", "style", "imitation" or the like.' In effect, the EU negotiated an extension of TRIPS Article 23 protection for wines and spirits.[81] Furthermore, registration of a trademark that contains or consists of a GI listed in CETA 'shall be refused or invalidated.' Although not without flaws, CETA delivers an adequate and improved level of protection for GIs exported from the EU.

The EU achieved protection for names that could be considered generic in Canada and therefore not capable of protection under trademark law. Naturally, however, the EU made concessions to reach a compromise. CETA provides that Canada is not obligated to prevent the use of certain GIs: Asiago, Feta, Fontina, Gorgonzola and Munster when they are accompanied by the terms '"kind", "type", "style", "imitation" or the like' in combination with a visible indication of the true origin.[82] Moreover, CETA Article 7.6.5 is a coexistence clause, stating that where trademarks have been applied for or registered in good faith prior to the date of the signing of the Agreement, those trademarks are valid and owners shall have the right to use them even if they are identical to a GI in Annex I. While this type of agreement, including those enumerated cheeses, amount to TRIPS protection extended beyond wine and spirits, the protection of previously registered trademarks-, could represent a rough blueprint for compromising those difficult GIs regarded as specific in the EU but generic in other countries. Though the agreement has many flaws, the end product delivers a system of protection that leaves both parties better off than they were before. Whether Canada and other common law jurisdictions that may sign similar deals will consider that it is compatible with Lisbon has not been publicly discussed.

The matter is more complex still. With or without the common law world in Lisbon, companies based in non-Lisbon members—including all those common law countries—using terms they consider generic or descriptive on their products will find Lisbon members in the path of their trade transactions with other nations.

---

[78] Geographical Indications of Goods (Registration and Protection) Act, 1999, Act of Parliament No 48 (India); Geographical Indications Act, 2000, Act 602 (Malaysia); Geographical Indicators Regulation, 2007, Government Regulation No 51 (Indonesia).

[79] See Comprehensive Trade and Economic Agreement, Can.-EU, Consolidated CETA Text, ch. 22, Intellectual Property, Sept. 26, 2014, available at http://trade.ec.europa.eu/doclib/docs/2014/september/tradoc_152806.pdf. (as consulted March 21st 2016).

[80] Ibid, 7.4.

[81] Ibid.

[82] Ibid art. 7.6.

Members of the Association of Southeast Asian Nations (ASEAN), where the cost of protecting foreign GIs (and implementation of the Geneva Act) could be low, may be substantial beneficiaries in trade dealings with the EU and be tempted to join Lisbon, though it may come at a cost in their relations with the United States and others.[83]

The Geneva Act may thus be considered not just as opposing common law and civil law but also plurilateral and multilateral approaches to the international protection of intellectual property. Countries have shifted their preferred forum for intellectual property negotiations from open-door multilateral negotiations to, initially, closed-door bilateral negotiations and, eventually, closed-door plurilateral negotiations—often referred to as 'country clubs' (Gervais 2012b)—seen in the TPP and Transatlantic Trade and Investment Partnership (TTIP) in addition to the Transpacific Partnership Agreement (TPP). The risk from this shift in forums is the emergence of a tapestry of different and possibly incompatible norms (Sell 2011). So far, bilateral trade negotiations have been rather successful, showing progress is possible by finding specific compromises on international protections of domestic GIs without overly infringing on trademark owners' expected rights (Yu 2004). Since 1994, the EU has concluded a series of agreements on the mutual protection of specifically handpicked GIs for wines and spirits, including those with the Australia, Mexico, South Africa, Canada, and even the USA (O'Connor et al. 2012, 39). More recently, a variety of EU bilateral agreements have protected a broader range of goods.[84] Yet they too add to this risk of incompatible norms.

Agreements like CETA show that some degree of compromise is possible. Meanwhile, the languishing multilateral picture especially at the WTO make broad progress in trade law unlikely. This is either good for the Geneva Act (as an alternative to the WTO) or bad because it shows reluctance to engage at the multilateral level on matters already covered in regional or plurilateral agreements. Our best guess is that the former argument is not going to be very convincing because the countries that were slow to engage at the WTO are precisely those that now feel left out of the Geneva Act. Compounded with an understanding of the differences between sui generis GI and trademark systems, this sets the stage for grasping why the rules of procedure for the Diplomatic Conference for the Geneva Act excluded non-Lisbon members from full participation. At the end of the day, it may all come down to money. As alluded to earlier, it appears that the progress on Lisbon fees and financing at the 2015 General Assembly is unlikely to be adequate: the recent doubling of fees will not be enough to balance Lisbon's budget, leading to further deficits

[83]According to an April 2015 press release announcing the restart of EU-ASEAN trade talks, '[t]he EU is ASEAN's second-largest trading partner with total trade amounting to $248 billion in 2014 and was Southeast Asia's largest source of foreign direct investment in 2014 with $29.1 billion, or 21.3% of foreign inflows to the region.' EurActiv.com. 2015. EU and ASEAN to Jumpstart Trade Agreement Talks. www.euractiv.com/sections/trade-society/eu-and-asean-kick-start-free-trade-agreement-talks-314100 (as consulted12 Jan 2016).

[84]The EU Korea Free Trade Agreement, EU Colombia Peru Agreement, and the EU Central America Agreement and the EU Singapore Agreement to name a few.

(WIPO Secretariat 2015b). Moreover, the Lisbon Union's loan agreement, due in large part to pressure from the U.S., will likely be a source of disputes if or when the Union lacks the funds to repay it. All of this possibly could necessitate the addition of renewal or other fees that the Contracting Parties at the Diplomatic Conference for the Geneva Act refused to adopt. Though this would require a deeper analysis, the text of the Geneva Act does not seem to allow this change by a simple change to Regulations. Thus the loan compromise may well represent a clever move by the U.S. to isolate the Lisbon system by making registration less attractive.

2.11 The Future of Consensus Building at WIPO

2.11.1 Implications for WIPO's Role in Norm-Setting

It remains to be seen whether the procedure leading to the adoption of the Geneva Act will serve as a precedent against multilateralism and full participation in the future revision of other WIPO treaties. The procedure provides an example that may be used to justify the exclusion of non-contracting states in future diplomatic conferences.[85] This calls into question the very role of WIPO as a forum for reaching consensus on IP-related laws in the ear of bilateral and plurilateral free trade agreements that affect IP standards (Sell 2011). If WIPO desires to keep playing a central role in the evolution of IP norm-setting through treaty development, then the organization needs to make changes.

2.12 Legal Solutions

Although we cannot go back in time and change the substance or procedure of the Geneva Act, WIPO can effect changes for future diplomatic conferences. Indeed, WIPO's system of governance, especially their rules and procedures, directly affects the ways in which substantive work is conducted and produced (Birkbeck 2014a). The unorthodox and divisive procedure behind the Geneva Act adoption could be seen as the final straw to convince the member states and the WIPO hierarchy that

[85]US Ambassador Pamela Hamamoto, in a statement discussing the closed diplomatic conference, stated, '[T]his is about more than the legitimacy of the Lisbon Agreement itself. This issue has alarming implications for the future of WIPO as an organisation.' U.S. National Statement, *All WIPO Members Should Be Able to Participate Equally in Lisbon Agreement Diplomatic Conference*, available at https://geneva.usmission.gov/2015/05/11/all-wipo-members-should-be-able-to-participate-equally-in-lisbon-agreement-diplomatic-conference/ (as consulted March 21st 2016).

now is the time to review and revise the Rules of Procedure,[86] which were last amended in 1979 (Birkbeck 2014b).

These revisions could form part of other efforts. In August 2014a, UN Joint Inspection Unit (JIU) reported numerous underlying governance issues[87] facing WIPO and recommended that the WIPO General Assembly should 'review the WIPO governance framework as well as current practices with a view to strengthen the capacity of the governing bodies to guide and monitor the work of the organisation. The report also noted that an increasing number of Member States had lately begun to repeatedly voice their dissatisfaction with regard to the current state of governance at WIPO.

This perceived increasing ineffectiveness of WIPO coincides with an era in which the opportunity for multilateralism is overtaken by the rise in bilateralism and plurilateralism (Okediji 2004). The Authors believe that multilateralism still represents the best chance for Pareto-optimal solutions between developed and developing countries, incorporating a variety of interests to impact a range of substantive economic and public policy issues including harmonization of systems, public health, education, the environment, and the protection of rights holders' interests. In the long term, plurilateral deals may fragment the international IP system. In other words, there is an urgent need for resurgence in multilateralism today but given what is at least a perception possibly reinforced by the Geneva Act kerfuffle, in order for that to occur WIPO's rules of governance and its rules and procedures must be updated.

Rule 56 of the General Rules of Procedure provides for amendments to WIPO's rules and procedures. As it currently stands, there is no provision mandating that a diplomatic conference for the adoption of a new treaty be open to the full participation of all WIPO member states.[88] Such a rule of procedure would prevent the Diplomatic Conference for the Geneva Act to the Lisbon Agreement from establishing a precedent that could be used by other systems and unions to further divide

[86]WIPO (1998) WIPO General Rules of Procedure, WIPO Doc. 339 (FE) Rev. 3., available at http://www.wipo.int/edocs/pubdocs/en/general/399/wipo_pub_399.pdf (as consulted March 21st 2016).

[87]The governance issues include, among others: a disproportionate weight that informal and undocumented meetings have on the outcome on various subjects of interest; the sheer number and variety of different WIPO bodies negatively effects the clarity of governance arrangements and, hence, their effectiveness; the lack of a comprehensive reference document detailing good practices and the roles and functions of different bodies.

[88]One could curtail worries about such a rule of procedure stifling a possibly legitimate diplomatic conference needing to be closed to certain member states by providing for the particular rule to be overridden in the case of a majority vote of WIPO member states granting permission for the diplomatic conference in question to proceed without full participation of all WIPO members. This compromise may admittedly be imperfect or undesired, but it represents one example of how WIPO could prevent future conflicting factions without sacrificing the arguably necessary flexibility to advance IP systems and policies.

WIPO and the IP community as a whole.[89] We know of no instrument administered by WIPO that is incompatible with such a rule. Correspondingly, member states could revise the Chapter regarding the conduct of business so as to give the Secretariat or Chairperson more power, including to curtail discussion on non-substantive issues. Ultimately, any revision that works towards increasing substantive member state discussions, transparency in negotiations, or clarity to the process of treaty revision and creation would be a positive step for the future of international IP norm-setting under WIPO's aegis. Taken as a whole, member states have a plethora of options at their disposal to improve both the governance framework and the rules of procedure.

2.13 Political Solutions

At the conclusion of the 2015 General Assembly a deal was reached on the Lisbon system's funding and its Chair, Vladimir Yossifov, took the opportunity to defend the Union's successive deficits. He pointed out that the Hague system on designs also loses money regularly[90] but had not been targeted for criticism by the U.S. 'What is the definition of corruption?' Yossifov questioned before answering, 'Corruption is a deal in which I don't participate' (Saez and New, 2015) Such barbed language cuts to the heart of the dispute within WIPO: the U.S. did not feel the need to instigate debate on the budget of the Lisbon system until after they had been shut out of the Geneva Act. It is also true that many Lisbon members had been increasingly frustrated by ineffective negotiations at the WTO on a registration system for wine and spirits, and did not trust non-Lisbon members to negotiate Lisbon revisions in genuine goodwill.[91]

If this cycle of distrust actually affects the normative outcomes of WIPO meetings and discussions, then all member states must recognize that such a state of affairs is suboptimal. Indeed, the U.N.'s Joint Inspection Unit noted that 'accumulated tension among Member States may adversely impact the effective

[89]There is some evidence that WIPO member states have already recognized this: the 2015 General Assembly took formal note that all diplomatic conferences in the 2016/17 budget funded by WIPO 'will be open to the full participation of the all WIPO Member States', *see* (WIPO International Bureau 2015a).

[90]While it may be easy to point out the U.S.' hypocrisy towards the Hague Union's deficit, distinctions can be found in that the Hague Union is composed of considerably more members (64) with additional members expected to join in the near future and Hague applications expected to increase rapidly (*Report adopted by PBC* 2015, 15).

[91]Neither author wishes to partition blame for the lack of progress in WTO negotiations, but do note that each side is most likely to consider the other side as the disproportionate source of the stalemate.

functioning of the organization'[92] In this regard the 2015 WIPO General Assembly is a prime example. Not only was it divisive but the debate over Lisbon financing dominated the assembly's discussions at the expense of other substantive issues. Namely, the discussion on Lisbon finances overshadowed the important work on reaching an agreement on an international legal instrument relating to the protection of traditional knowledge, the inclusion of technical assistance to developing countries within the proposed design law treaty, and the opening of new WIPO external offices (Saez and New, 2015).

2.14 Conclusion

The adoption of the Geneva Act of the Lisbon Agreement was a clear chance to update Lisbon fully and to reconcile the Lisbon system on two fronts: doctrinally, with the common law; and administratively, by allowing countries that protect GIs under trademark law (often as certification marks) to continue to do so as Lisbon members. That did not happen, at least not clearly. The absence of consensus—indeed, the deep divide—at the Diplomatic Conference, and the stated[93] incompatibility of the Geneva Act with trademark-based GI systems are unlikely to prove productive in the short- to medium-term for GI holders, authorized users of GIs, users of generic denominations, or consumers. The bridge between Lisbon and the common law system has not been built by the Geneva Act. Some useful foundations were laid, but much remains to be done. The bridge, if it is ever built, will not be the result of a normative meeting of the minds. It will be made of trade bricks, such as those that the TTIP is hoping to lay. In the end, both the U.S. and the Lisbon Union were engaged in a prisoner's dilemma in the Lisbon discussions: each party would have been better off working together, but both remained risk averse and chose to work in their self-interest. In so doing, they bucked precedent and a spirit of cooperation in favor of advancing domestic agendas. That is what they must now overcome if TTIP is going to be Lisbon-friendly. Otherwise, if applied on a broader scale, such behavior could prove toxic to ongoing and future negotiations affecting substantive issues at WIPO.

[92]From page 11 of the JIU report: '[t]he accumulated tension among Member States may adversely impact the effective functioning of the organization. A constructive and productive dialogue on the ways and means to conduct governance reform has thus become a priority for both the membership and management.'

[93]Reference is made here again to statements, notably by Australia, Japan, Korea, and the United States, during the closing plenary of the Diplomatic Conference that adopted the Geneva Act (May 20, 2015).

Bibliography

Agreement on Trade-Related Aspects of Intellectual Property Rights, Art. 22.2, Apr. 15, 1994, Marrakesh Agreement Establishing the World Trade Organization. 1869 U.N.T.S. 299. https://www.wto.org/english/docs_e/legal_e/27-trips.pdf. (as consulted 10 Feb 2016).

Belson, Jeffrey. 2002. *Special Report: Certification Marks*. London: Sweet & Maxwell.

Bird, David. 2005. Understanding Wine Technology: The Science of Wine Explained. Wine Appreciation Guild 2005) 1–16.

Birkbeck, Carolyn Deere. 2014a. Inside Views: Strengthening WIPO's Governance For The Next 50 Years: A Time For Action. Intellectual Property Watch, September 26. http://www.ip-watch.org/2014/09/26/strengthening-wipos-governance-for-the-next-50-years-a-time-for-action/. (as consulted 12 January 2016)

Birkbeck, Carolyn Deere. 2014b. The Governance of the World Intellectual Property Organization: A Reference Guide. Global Economic Governance Programme, Working Paper No. 93, 2014.

Bowen, Sarah. 2010. Development from Within? The Potential for Geographical Indications in the Global South. 13 *Journal of World Intellectual Property* 231.

Bramley, Cerkia, Estelle Biénabe, and Johann Kirsten. 2009. The Economics of Geographical Indications: Towards a Conceptual Framework for Geographical Indication Research in Developing Countries. In *The Economics of Intellectual Property: Suggestions for Further Research in Developing Countries with Economies in Transition*. 109-49. www.wipo.int/edocs/pubdocs/en/economics/1012/wipo_pub_1012.pdf (as consulted January 2016).

Calboli, Irene. 2015. Of Markets, Culture and Terroir: The Unique Economic and Culture-Related Benefits of Geographical Indications of Origin. In *International Intellectual Property: A Handbook of Contemporary Research*, ed. Daniel J Gervais, 433. Massachusetts: Edward Elgar Publishing.

Creditt, Emily C. 2009. Terroir vs. Trademarks: The Debate over Geographical Indications and Expansions to the TRIPS Agreement. 11 *Vanderbilt Journal of Entertainment and Law* 425.

Dinwoodie, Graeme B. 2009. Developing a Private International Intellectual Property Law: The Demise of Territoriality? 51 *William and Mary Law Review* 711.

Diplomatic Conference For The Adoption Of A Revised Trademark Law Treaty -- Rules Of Procedure. WIPO Doc. TLT/R/DC/2 (March 14, 2006), http://www.wipo.int/edocs/mdocs/mdocs/en/tlt_r_dc/tlt_r_dc_2.pdf. (as consulted 9 Feb 2016).

Dreyfuss, Rochelle and Susy Frankel. 2015. From Incentive To Commodity To Asset: How International Law Is Reconceptualizing Intellectual Property. 36 *Michigan Journal of International Law* 557.

EurActiv.com. 2015. EU and ASEAN to Jumpstart Trade Agreement Talks. www.euractiv.com/sections/trade-society/eu-and-asean-kick-start-free-trade-agreement-talks-314100. (as consulted 12 Jan 2016).

Gangjee, Dev. 2007. Quibbling Siblings: Conflicts Between Trademarks and Geographical Indications. 82 *Chicago-Kent Law Review* 1253.

General Report adopted by the Assemblies. 2016. WIPO Doc. A/55/13. http://www.wipo.int/edocs/mdocs/govbody/en/a_55/a_55_13-main1.pdf. (as consulted 9 Feb 2016).

Geneva Act Of The Lisbon Agreement On Appellations Of Origin And Geographical Indications And Regulations Under The Geneva Act Of The Lisbon Agreement On Appellations Of Origin And Geographical Indications. 2015. WIPO doc. LI/DC/19. http://www.wipo.int/edocs/mdocs/geoind/en/li_dc/li_dc_19.pdf. (as consulted 9 Feb 2016).

Geneva Act of the Hague Agreement Concerning the International Registration of Industrial Designs of 2 July 1999. 2279 U.N.T.S. 156. http://www.wipo.int/wipolex/en/treaties/text.jsp?file_id=285214. (as consulted 12 Jan 2016).

Gervais, Daniel J. 2010a. Reinventing Lisbon: The Case for a Protocol to the Lisbon Agreement (Geographical Indications). 11 *Chicago Journal of International Law* 67.

Gervais, Daniel J. 2010b. The Lisbon Agreement's Misunderstood Potential. 1 *World Intellectual Property Organization Journal* 87.

Gervais, Daniel J. 2012a. Traditional Innovation and the Ongoing Debate on the Protection of Geographical Indications. In *Indigenous Peoples' Innovation: Intellectual Property Pathways to Development*, eds. Peter Drahos and Susy Frankel, 121. Canberra: Australian National University Press.

Gervais, Daniel J. 2012b. Country Clubs, Empiricism, Blogs and Innovation: The Future of International Intellectual Property Norm Making in the Wake of ACTA. In *Trade Governance in the Digital Age*, eds. Mira Burri and Thomas Cottier, 323–43. Cambridge: Cambridge University Press.

Gervais, Daniel J. 2015. Irreconcilable Differences? The Geneva Act of the Lisbon Agreement and the Common Law. 53 *Houston Law Review* 339.

Guy, Kolleen M. 2009. *When Champagne Became French: Wine and the Making of a National Identity*. Baltimore: Johns Hopkins University Press.

Hughes, Justin. 2006. Champagne, Feta, and Bourbon: The Spirited Debate About Geographical Indications. 53 *Hastings Law Journal* 299.

Joint Inspection Unit. 2014. Review Of Management And Administration In The World Intellectual Property Organization (WIPO). U.N. Doc. JIU/REP/2014/2. https://www.unjiu.org/en/reports-notes/JIU%20Products/JIU_REP_2014_2_%20English.pdf. (as consulted 15 Jan 2016).

Kemp, Deborah J and Lynn M Forsythe. 2006. Trademarks and Geographical Indications: A Case of California Champagne. 10 *Chapman Law Review* 257.

Lince, Tim. 2015. TPP could aid WIPO effort to spread Madrid Protocol around Latin America, *World Trademark Review*, 13 October. http://www.worldtrademarkreview.com/Blog/detail.aspx?g=a7f018f9-c983-498f-818d-abe5d47aeeeb. (as consulted 10 Feb 2016).

Lisbon Agreement for the Protection of Appellations of Origin and Their International Registration. Oct. 31, 1958, *as revised*, July 14, 1967. 923 U.N.T.S. 205. http://www.wipo.int/wipolex/en/treaties/text.jsp?file_id=285856. (as consulted 10 Feb 2016).

McCarthy, J. Thomas. 1999. *McCarthy on Trademarks and Unfair Competition*. 2nd edn. New York: West Group.

Montén, Lina. 2006. Geographical Indications of Origin: Should They Be Protected and Why? – An Analysis of the Issue From the U.S. and EU Perspectives. 22 *Santa Clara Computer & High Technology Law Journal* 315.

Mostert, Frederick W. 2004. *Famous and Well-Known Marks: An International Analysis* (2nd edn, International Trademark Association

O'Connor, Bernard and Laura Richardson. 2012. The legal protection of Geographical Indications in the EU's Bilateral Trade Agreements: moving beyond TRIPS. *Rivista di Diritto Alimentare*, Oct.-Dec. at 39, *available at* http://www.rivistadirittoalimentare.it/rivista/2012-04/2012-04.pdf.

OAPI. 2010. États Membres. www.oapi.int/index.php/en/aipo/etats-membres. (as consulted 12 January 2016).

Okediji, Ruth L. 2004. Back to Bilateralism? Pendulum Swings in International Intellectual Property Protection. 1 *University of Ottawa Law & Technology Journal* 127.

Regulations Under The Singapore Treaty On The Law Of Trademarks. 2011. http://www.wipo.int/wipolex/en/treaties/text.jsp?file_id=290013. (as consulted 9 Feb 2016).

Report adopted by the Assembly. 2016a. WIPO Doc. PCT/A/47/.http://www.wipo.int/edocs/mdocs/govbody/en/pct_a_47/pct_a_47_9.pdf. (as consulted 3 Feb 2016).

Report adopted by the Assembly. 2016b. WIPO Doc. MM/A/49/5. http://www.wipo.int/edocs/mdocs/govbody/en/mm_a_49/mm_a_49_5.pdf. (as consulted 3 Feb 2016).

Report adopted by the Program and Budget Committee. 2015. WIPO Doc. WO/PBC/23/10. http://www.wipo.int/edocs/mdocs/govbody/en/wo_pbc_23/wo_pbc_23_10-main1.pdf. (as consulted 3 Feb 2016).

Report adopted by the Working Group. 2012a. WIPO Doc. LI/WG/DEV/4/7. http://www.wipo.int/edocs/mdocs/madrid/en/li_wg_dev_4/li_wg_dev_4_7.pdf. (as consulted 9 Feb 2016).

Report adopted by the Working Group. 2012b. WIPO Doc. LI/WG/DEV/5/7. http://www.wipo.int/edocs/mdocs/mdocs/en/li_wg_dev_5/li_wg_dev_5_7.pdf. (as consulted 9 Feb 2016).

Report adopted by the Working Group. 2013. WIPO Doc. LI/WG/DEV/6/7. http://www.wipo.int/edocs/mdocs/mdocs/en/li_wg_dev_6/li_wg_dev_6_7.pdf. (as consulted 3 Feb 2016).

Report adopted by the Working Group. 2014a. WIPO Doc. LI/WG/DEV/8/7. http://www.wipo.int/edocs/mdocs/mdocs/en/li_wg_dev_8/li_wg_dev_8_7.pdf. (as consulted 9 Feb 2016).

Report adopted by the Working Group. 2014b. WIPO Doc. LI/WG/DEV/9/8. http://www.wipo.int/edocs/mdocs/mdocs/en/li_wg_dev_9/li_wg_dev_9_8.pdf. (as consulted 9 Feb 2016).

Rhee, Chase C. *Principles of International Trade (Import-Export): The First Step Toward Globalization* (5th edn, AuthorHouse 2012).

Saez, Catherine. 2015a. 'Bound By Law', 26 Lisbon Members Say No To Fully Open Diplomatic Conference On GIs. *Intellectual Property Watch*, May 7. http://www.ip-watch.org/2015/05/07/bound-by-law-26-lisbon-members-say-no-to-open-diplomatic-conference-on-gis/. (as consulted 9 February 2016)

Saez, Catherine and William New. 2015. WIPO Assembly Agrees Budget For 2016/2017; Work Continues After Hours On Remaining Items. *Intellectual Property Watch*, 15 October. http://www.ip-watch.org/2015/10/15/wipo-assembly-agrees-budget-for-20162017-work-continues-after-hours-on-last-items/. (as consulted 20 Dec 2015).

Saez, Catherine. 2015b. United States Hopeful Lisbon Members Will Open Diplomatic Conference To All. *Intellectual Property Watch*, May 7. http://www.ip-watch.org/2015/05/07/united-states-hopeful-lisbon-members-will-open-diplomatic-conference-to-all/. (as consulted 9 February 2016).

Saez, Catherine. 2015c. WIPO Diplomatic Conference On GIs Starts With 28 Member Governments. *Intellectual Property Watch*, May 11. http://www.ip-watch.org/2015/05/11/wipo-diplomatic-conference-on-gis-proceeds-despite-protests/. (as consulted 9 February 2016)

Sell, Susan K. 2011. TRIPS Was Never Enough: Vertical Forum Shifting, FTAs, ACTA and TPP. 18 *Journal of Intellectual Property Law* 447.

US Patent and Trademark Office. Geographical Indication Protection in the United States. www.uspto.gov/sites/default/files/web/offices/dcom/olia/globalip/pdf/gi_system.pdf. (as consulted 12 January 2016).

van Caenegem, William. 2004. Registered GIs: Intellectual Property, Agricultural Policy and International Trade. 26 *European Intellectual Property Review* 170.

WIPO. 2010. Schedule of Fees. www.wipo.int/export/sites/www/hague/en/fees/pdf/sched.pdf. (as consulted 11 Jan 2016).

WIPO. 2013. Contracting Parties > Lisbon Agreement. www.wipo.int/treaties/en/ShowResults.jsp?lang=en&treaty_id=10. (as consulted 1 February 2016).

WIPO. 2015a. Individual Fees Under the Madrid Protocol. www.wipo.int/madrid/en/madridgazette/remarks/ind_taxes.html. (as consulted 11 Jan 2016).

WIPO. 2015b. PCT Fee Tables. www.wipo.int/export/sites/www/pct/en/fees.pdf. (as consulted 11 Jan 2016).

WIPO. 2016. Agenda of the Fifty-Fifth Series of Meetings of the WIPO Assemblies. http://www.wipo.int/about-wipo/en/assemblies/2015/a_55/agenda.html. (as consulted 10 Feb 2016).

WIPO. 2016. Payment of Fees—Lisbon System. www.wipo.int/about-wipo/en/finance/lisbon.html. (as consulted 11 Jan 2016).

WIPO International Bureau. 2015. Proposal To Update The Fee Schedule Under Rule 23 Of The Regulations Under The Lisbon Agreement. http://www.wipo.int/edocs/mdocs/govbody/en/li_a_32/li_a_32_2.pdf. (as consulted 10 Feb 2016).

WIPO Secretariat. 2015a. List Of Decisions. WIPO Doc. A/55/INF/11. http://www.wipo.int/edocs/mdocs/govbody/en/a_55/a_55_inf_11.pdf. (as consulted 10 Feb 2016).

WIPO Secretariat. 2015b. Options For The Financial Sustainability Of The Lisbon Union. WIPO Doc. WO/PBC/24/16. http://www.wipo.int/edocs/mdocs/govbody/en/wo_pbc_24/wo_pbc_24_16.pdf. (as consulted 10 Feb 2

Yu, Peter K. 2004. Intellectual Property At A Crossroads: The Use Of The Past In Intellectual Property Jurisprudence. 38 *Loyola of Los Angeles Law Review* 323.

Author Biographies

Daniel J. Gervais Dr. Daniel Gervais is Professor of Law at Vanderbilt University law School and Director of the Vanderbilt Intellectual Property Program. He is Editor-in-chief of the Journal of World Intellectual Property and editor of www.tripsagreement.net. Dr. Gervais is also an elected member of the American Law Institute and a member of the Academy of Europe.

Matthew Slider is a JD candidate at Vanderbilt University Law School. In 2015, he interned at the World Intellectual Property Organization (WIPO) in Geneva, Switzerland. He is B.S. Economics and Political Science at Florida State University.

Chapter 3
The Global Struggle Between Europe and United States Over Geographical Indications in South Korea and in the TPP Economies

Bernard O'Connor and Giulia de Bosio

Abstract The WTO TRIPs Agreement does not set out the means by which countries must achieve the standards laid down for GIs. Thus, the legal means for the protection of GIs can vary from country to country. The two largest trading blocks, the EU and the US, have very distinct views on the nature of GIs and the appropriate means for their protection. The different approaches taken by the EU and US are, in turn, reflected in the different FTAs that both entities have, and are currently, negotiating in Asia. This raises issues of law for any Asian country wishing to conclude an FTA with both the EU and the US. The analysis in the Chapter shows that the global arm-wrestling between the EU and the US with regard to specific GIs is coming down to a new version of the "first come first served" rule in relation to FTAs: the first to conclude a trade deal with an Asian country determines the space remaining for concluding a different deal on GIs with the other. This Chapter analyses the protection of GIs in South Korea and how that country has addressed the potentially conflicting commitments under the FTAs it has concluded with the EU and US respectively. The Chapter concludes that South Korea, in its struggle to find a balance between the different requirements under the two FTAs, risks introducing discrimination between GIs into its domestic law in order to comply with its international obligations.

3.1 Introduction

Do consumers need to know that the Feta they are purchasing is really from Greece or that the Darjeeling they are drinking effectively originates in India? This question is at the core of one of the fiercest on-going debates in the World Trade

B. O'Connor (✉) · G. de Bosio (✉)
State University of Milan, Milan, Italy
e-mail: bernard.oconnor@nctm.it

G. de Bosio
e-mail: gdebosio@trevisancuonzo.com

W. van Caenegem and J. Cleary (eds.), *The Importance of Place: Geographical Indications as a Tool for Local and Regional Development*, Ius Gentium: Comparative Perspectives on Law and Justice 58, DOI 10.1007/978-3-319-53073-4_3

Organization. It concerns the nature and the scope of the protection of Geographical Indications (GIs).

The TRIPs Agreement establishes GIs as a category of IP alongside patents, copyrights and trademarks. The Agreement defines GIs and sets out differing levels of protection. However it does not set out how the protection is to be achieved. The inclusion of GIs within TRIPs was a key demand of the EU and was resisted by the US. Thus the provisions in the TRIPS are the result of a compromise between the EU and the US. As such, the Agreement neither specified the preferred legal means for the protection of GIs nor even identified the possible options, leaving Member States considerable room to choose.

Different historical backgrounds and divergent economic interests between the "old world" and the "new world" stand out as the main sources of different positions of the EU and the US. The "old world"—Europe and its supporters - emphasises the notion of *terroir*, i.e. the idea that there is an essential nexus between a product's characteristics and its place of origin (Barham 2003), and considers that an independent and unique form of IPR protection is needed. The EU considers that GIs, unlike all other IPRs, are collective in nature and have an inherent cultural component which can include terroir, environment, climate, traditional knowledge, skills, processes and ingredients. The "new world"—the US and its supporters—on the other hand consider GIs as a subset of trademarks emphasising a utilitarian economy theory based on individual, exclusive, ownership. It is the antithesis of the idea of collectivity, culture and terroir.

Consequently, the EU and the US have adopted two very different approaches to the protection of GIs relying on different mechanisms to pursue different goals. The EU sui generis regime is strict, reflecting the belief that the link between quality or reputation and place is objective, resulting from a history of traditional practices. In the European view, when others use GIs as generic names, even if accompanied by indications of the true origin, they are unfairly usurping the value and the reputation created in that name by generations of local producers. In order to prevent fraud and ensure business fairness, under the EU regime, GIs enjoy a TRIPs-plus level of protection. The US, on the other hand, views the EU position as a trade barrier, as a protectionist attempt to control the use of names which it considers to be generic or potentially generic. The US approach to protecting GIs through trademark law seeks to categorise GIs as a sub-set of TMs. The divergence between the EU and US approaches to GIs is so preeminent that the best these parties achieved during the post-Uruguay Round discussions was to agree to disagree.

In the absence of a generally accepted and satisfactory standard, the EU and the US moved from the stalemate in the WTO, to a new field of the plurilateral and bilateral trade agreements as a mean to try and impose their concept of GIs across the world and, in particular, in Asia. The EU is aggressively pushing to enhance the level of GI protection to TRIPs plus, whereas the US seeks to ensure that names it considers generic remain so and that GIs are subject to rules taken from trade mark law and practice. This Chapter aims at shedding light on the protection GIs enjoy under the most recent FTAs the EU and the US have concluded with South Korea and other countries in the Asia-Pacific Region with particular attention paid to the level of protection granted to GIs and to the relationship between GIs and TMs.

The Chapter examines how Korea has effectively dealt with the different approaches to GIs it agreed to under the EU-Korea FTA (KOREU) on the one side, and the US-Korea FTA (KORUS) on the other side. The case study of Korea is an illustration of the difficulties created by these free trade agreements for the domestic law of the third country. Countries that, as a matter of fact, still occupy weak bargaining positions *vis-à-vis* their more powerful trading partners—i.e. the US and the EU.

3.2 The EU and the US Approach on GIs as Reflected in Recent FTAs

3.2.1 The External Actions of the EU in Relation to GIs

Rather than seeking a common understanding of the nature of GIs and the protection necessary in law, the main objective of the EU trade policy is to obtain a satisfactory level of protection of EU GIs at international level, meaning a level of protection similar to that offered by the system in the EU. Quite frequently EU GIs products with high economic value suffer from usurpation or abuse of reputation in third markets (e.g. "Feta", "Gorgonzola", "Asiago", "Parma ham", "Champagne"). This is due to the fact that often the level of protection of GIs in other WTO countries is limited to TRIPs basic provisions, i.e. Article 22, which is considered by the EU not to be sufficient to protect misuse, imitation and evocation at least for agricultural products and foodstuffs (Dg-Agri 2012: 8; Càceres 2007: 6).

Since the EU considers that the TRIPs does not offer a satisfactory protection for EU GIs other than wines and spirits, the EU has sought to achieve its objectives through a variety of bilateral and multilateral international agreements (Eugui and Spennemann 2006: 308; O'Connor and Richardson 2012: 2; Lupone 2009: 38). The "TRIPs-plus" level of protection pursued by the EU in international negotiations means, in particular: the establishment of an open list of EU names to be protected directly and indefinitely in the third country (including controversial expression such as "Feta", "Prosciutto di Parma", "Champagne") from the entry into force of the agreement; the extension of the higher protection of Article 23 to products other than wines and spirits; the co-existence with prior trademarks if they were registered in good faith; the phasing out of prior generic EU names; the attempt to limit the extensive use of Article 24 TRIPs exceptions and the obtainment of an *ex officio* protection (Dg-Agri 2012: 9; Handler and Mercurio 2015: 334).

The first WTO-plus FTA was signed with South Korea in October 2010 and entered into force in July 2011. Since it was the first of a long series of FTAs it was extremely important for the EU to reach the maximum of its negotiating objectives in the final text of the agreement. The EU managed to obtain: the mutual recognition of over 226 GIs with the possibility of adding more, the extension of Article

23 TRIPs to agricultural products and foodstuffs other than wines, the coexistence of GIs and prior TMs and the administrative enforcement of the protection (O'Connor and Richardson 2012: 11). As will be seen in this chapter, this great accomplishment was possible mainly because the EU managed to conclude the agreement before the US.

3.2.2 *The External Actions of the US in Relation to GIs*

The US also attempts to export its particular model of GI protection in its trade agreements. In particular, the US is encouraging other countries to adopt a "minimalist, trademark oriented model of legal regulation of GIs" distinct from the sui generis system in the EU. It seeks to prioritise TMs over GIs in the event of any conflict between them (Handler and Mercurio 2015: 344). The US preference for protecting GIs through the trademark system is well illustrated by the US proposal presented to the TRIPs Negotiating Group in 1990 which provided: "Contracting Parties shall protect Geographical Indications that certify regional origin by providing for their registration as certification or collective trademarks" (Eugui and Spennemann 2006: 325).

The main objective of the US external policy on GIs include: the partial extension of the TRIPs definition of GIs to come within trademark law; the retention of the scope of protection for goods other than wines and spirits to the "confusingly similar" standard; the possibility for the holder of a pre-existing trademark to prevent the later protection of GIs; the emphasis on the exceptions under Article 24 of the TRIPs and the inclusion of certain procedural safeguards such as cancellation and opposition proceedings to address concerns regarding protection of generic terms (Eugui and Spennemann 2006; O'Connor 2014a).

The divergent approaches of the US and the EU to GIs become evident in the context of the bilateral and plurilateral agreements both are currently negotiating and have recently concluded. The main issues of contention concern: the additional protection extended to goods other than wines and spirits as against the minimalistic approach; the principle of priority as against the principle of coexistence in the relationship between GIs and TMs and the problem of generic terms (USTR Special 301 Report 2015: 26; Raustiala and Munzer 2007: 350; O'Connor 2014a). This last matter, in particular, involves important economic interests for many producers in the US, particularly in relation to cheeses and cured hams, and thus animates the dispute between the US and the EU. It is a difference that will become one of the major obstacles to the conclusion of the Trans-Atlantic Trade and Investment Partnership negotiations (TTIP) (The Hagstrom Report 2014; Bella 2014; Sanburn 2014). In fact, the EU's efforts to claim exclusive rights to use names such as "Feta", "Asiago", "Fontina", "Gorgonzola", in its FTAs, dramatically affect the US dairy industry that considers these names as non-specific and uses them on products for sale around the world. The US considers that EU trade policy would prevent the import of products bearing the contested names in those third countries that have signed a bilateral agreement with the EU. The comments made by the Vice President of Trade Policy at US Dairy Export Council and National Milk Producers Federation clearly embrace the American concerns over the European action: "the EU's approach to restricting common food names through the use of GI registrations abuses a good concept in order to impose trade barriers against competitors… In forcing its trading partners to adopt the same trade-restrictive GIs in recent FTAs, the EU has turned FTAs, which are supposed to expand trade, into tools for discriminating against third countries to gain unfair market shares" (Testimony by Shawna Morris 2014). For this reason, the US tries to use its own FTAs to contain the effects of the European's push to globalise what the US sees as "restrictions" on the use of place names (Watson 2015). The consequence is that the battle over GIs is being transferred to other countries as the EU and the US both use trade agreements to influence the protection of GIs in foreign markets (Watson 2015).

3.3 The KOREU and the KORUS

3.3.1 The KOREU

On 1 July 2011, the EU-Korea FTA (KOREU) provisionally entered into force,[1] bringing to an end a process promoted five years earlier by the Commission's Communication on "Global Europe in a Competing World" which required the EU to renew its engagement in Asia. The Agreement is unprecedented both in its scope and in the speed at which trade barriers are to be removed, becoming the most ambitious trade agreement ever negotiated by the EU (European Commission 2011). With regard to Geographical Indications, a full section of the Chap. 10 of the FTA is devoted to the mutual recognition and protection of the GIs of both Parties, including, for the first time, in addition to wine and spirit GIs also agricultural products and foodstuffs GIs (Kim 2007; Trevisan and Cuonzo 2009). The Chapter reflects the EU's advocacy of a specialised system for their registration and protection, aimed at preventing their misuse in the respective markets (Wattanapruttipaisan 2009: 167).

While producers must normally register their GIs before the relevant national authorities to obtain a protection in a third country, the main advantage of the FTA signed between the EU and South Korea relies on the mutual recognition of hundreds of GIs. Indeed, KOREU includes an extensive list of GIs originating in one of the Parties that are required to receive automatic protection as GIs in the other. The GIs in question are listed in Annex 10-A (agricultural products and foodstuffs) and 10-B (wines, aromatised wines and spirits) of KOREU, and have been recognised without any need to go through the normal national procedure for recognition and

[1]Notice concerning the provisional application of the KOREU, 2011 O.J. L 168, Volume 54, 28 June 2011.

protection.[2] With the entry into force of the FTA these particular GIs have been immediately and inherently protected in both territories (Italian Intellectual Property Rights Desk—IIPRD 2010: 78). In total, the GIs listed in the Annexes are 64 for Korea and 162 for the EU. Of course, given the numbers, the list does not include all the GIs protected in the EU but only those GI's names that according to the EU are more likely to be usurped on a specific market and/or for which there is evidence of an economic interest (Trojanova 2014: 7). Among these are: Budějovické pivo, Feta, Prosciutto di Parma, Asiago, Fontina, Gorgonzola, Grana Padano, Parmigiano Reggiano, Pecorino Romano, Chianti, Porto, Korean Red Ginseng, Jeju Green Tea. Instead, the number of Korean GIs seems to include all the GIs registered in South Korea as of October 2009, when the Agreement between the EU and South Korea was initiated (Kim 2011).

The FTA also makes sure that the GI lists are flexible and may be amended if needed. Article 10.24 and 10.25(3)(b) of the FTA permit the addition of new GIs to the list and Article 10.24 and 10.25(3)(b) provide for the possibility to remove GIs if they cease to be protected in their country of origin. The Parties explicitly establish a "Working Group on Geographical Indications" that is responsible for such changes. Thus, new GIs may continue to be added for protection to the list, even after the FTA's entry into force (IIPRD 2010; 95). This approach, according to an authoritative opinion, goes beyond the TRIPs minimum standard. The latter, in fact, does not oblige Members to accord automatic protection to a foreign GI but rather allows their authorities to maintain the discretion to previously examine if any of the exceptions under Article 24 of the TRIPs applies (Eugui and Sepennemann 2006: 315).

In relation to the registration and control procedures, parties conclude that their respective laws meet all the following requirements: (a) a register listing the GIs protected in their respective territory; (b) an administrative process aimed at verifying that the GI fulfils the "essentially attributable test" as required by Article 22 of the TRIP; (c) a requirement that a registered name corresponds to a specific product for which a product specification is laid down that may be amended only through an administrative process; (d) control commitments in relation to the production process; (e) legal provisions ensuring that a registered name may be used by any operator marketing the agricultural product or foodstuff conforming to the product's specifications; (f) an objection procedure that safeguards the legitimate interests of prior users of the same or similar names.[3]

From this wording it is clear the extent to which the KOREU embraces most of the main pillars of the EU system, namely: the need for a specific registration and administrative control system; the necessity to guarantee a certain traceability of the GI product by laying down the product specifications to be respected; the assurance that a protected GI may be use by any operator marketing agricultural products, foodstuffs, wines, aromatised wines or spirits conforming to the specification.

[2]KOREU, Article 10.18(3)(4).

[3]KOREU, Article 10.18(6).

A few more words on the objection procedure prescribed by Article 10.18 paragraph 6 letter (f) are necessary. This requirement refers only to the country of origin of the GIs but not to the other Party where recognition and protection via the Agreement is sought. This is inferred by the systematic context of the paragraph. The standards required by paragraph 6 for a GI protection system are relevant only in relation to paragraph 1 and 2 of the same Article, where both Parties agree that the system of the respective contracting Party complies with the standards prescribed. This seems to confirm that the objection procedure in question only refers to the registration of domestic GIs. Consequently, under the language of the Agreement, the protection of the GIs listed in the FTA seems assured even if there are pre-existing rights that may be affected by such protection.

As to the level of protection, Article 10.21 paragraph 1 of the Agreement provides that "geographical indications referred to in Article 10.18 and 10.19 shall be protected against: (a) the use of any means in the designation or presentation of a good that indicates or suggests that the good in question originates in a geographical area other than the true place of origin in a manner which misleads the public as to the geographical origin of the good; (b) the use of a geographical indication identifying a good for a like good not originating in the place indicated by the geographical indication in question, even where the true origin of the good is indicated or the geographical indication is used in translation or transcription or accompanied by expressions such as 'kind', 'type', 'style', 'imitation' or the like; and (c) any other use which constitutes an act of unfair competition within the meaning of Article 10 bis of the Paris Convention".

Among the three types of actions mentioned above, the ones described by letter (a) and letter (c), proscribing the use of a GI in a manner that would mislead the public as to the true place of origin, resemble the level of protection granted by Article 22 of the TRIPs to all GIs. On the contrary, the misconduct illustrated by letter (b), preventing the use of a GI regardless of any misleading test, clearly reflects the type of protection provided for by Article 23 of the TRIPs in relation to GIs identifying wines or spirits. In other words, KOREU extends the additional Article 23 protection to agricultural products and other foodstuffs thus reproducing the TRIPs-plus EU approach.

However, one crucial issue to be addressed is the introductory sentence that limits its scope exclusively to "geographical indications referred to in Article 10.18 and 10.19". The wording is obscure. It is unclear if the provision refers to GIs as defined by footnote 51 (Engelhardt 2015: 799)—i.e. GIs protected under the respective legal regimes listed in that footnote - or if it is limited to those GIs listed in the Annexes 10-A and 10-B (Mandel 2011). Considering the nature of the FTA and its main goal, i.e. the mutual protection of a certain number of GIs without incurring the costs and delays of the national procedures, it seems that the obligations under this Article is limited to those GIs listed in the Annexes. Otherwise, what would have been the scope of the provisions allowing the addition to the list of specific names to the Agreement if its provisions had referred to all the GIs protected in both countries? This conclusion appears to be the one endorsed by the European Commission and by the Korean Government during the negotiation of the

Agreement. In fact, when officials in Seoul modified their national law so as to comply with the new FTA, they extended the level of protection only to those GIs already included in the FTA[4] and to those GIs that will be added to the Annexes in accordance to the procedure set out in Article 10.24.

As for the enforcement, Article 10.22 of the FTA requires the Parties to ensure protection of all GIs both on their own initiative through appropriate authorities and at the request of an interested party. This implements one of the main goals of the EU, i.e. an administrative enforcement. The FTA also provides for an obligation to implement broader measures directed against acts infringing GIs.[5] Namely, acts that mislead consumers as to the true geographical origin and acts of unfair competition in relation to the GI in question. By contrast, the Parties refrained from introducing strict obligations on criminal sanctions for GI infringements. Article 10.55 only expresses the Parties' intention to "consider adopting measures to establish the criminal liability for counterfeiting geographical indications" (Engelhardt 2015: 802).

Under KOREU, a TM, which has been applied for, registered or established by use in the territory of a party before the date of the application for the protection of the GI, can continue to be used, provided that no grounds for the TM's invalidity or revocation exist in the legislation of the Party concerned.[6] Whereas if a TM that corresponds to any of the prohibited actions under Article 10.21 is applied for on like goods after the date of application for the protection or recognition of the GI in the territory concerned, the registration of the trademark has to be refused or invalidated.[7] One more time these provisions unquestionably reflect the EU approach in the text of the FTA. In particular, Article 10.21 paragraph 5 expressly allows for the co-existence of GIs with prior TMs (Engelhardt 2015: 802) and this clearly mirrors Article 14 paragraph 2 of the Regulation (EU) 1151/2012. In fact, shifting the balance in favour of the GI when a conflict between a TM and a GI arises, suggests that KOREU endorses the European view of GIs as completely independent IPRs and as public assets with a collective nature. However, again, it is essential to underline that the above mentioned provisions only apply to those GIs already listed or that will be listed in the Annexes 10-A and 10-B of the Agreement (Mandel 2011). This is inferred by paragraph 2 of Article 10.23 which clarifies what is meant for "date of application for protection": for registered GIs already included in the FTA, the date of application for protection shall be the date of entry into force of the FTA, i.e. July 1, 2011[8]; whereas for registered GIs which are not currently on the list but could be added in the future on the basis of Article 10.24 of the Agreement, the date for protection shall be the date of a Party's receipt of a request by the other Party to protect or recognise a GI.[9]

[4]See infra par. 4.1.

[5]KOREU, Article 10.67(1), footnote 76(c)(iv).

[6]KOREU, Article 10.21(5).

[7]KOREU, Article 10.23(1).

[8]KOREU, Article 10.23(2)(a).

[9]KOREU, Article 10.23(2)(b).

There is no provision in the KOREU dealing with genericness. In the absence of a specific rule, the question is whether there is room for the exception under Article 24 paragraph 6 of the TRIPs which allow Members to exclude from the scope of protection of GIs terms customary in common language as common names for certain goods or whether the absence of such provision is to be interpreted as an implicit phasing out of the TRIPs exception. According to Eugui and Spennemann (2006), a distinction is needed. On one hand, with regard to the particular GIs enumerated in the list annexed to the Agreement it might be concluded that the exception provided for by Article 24 paragraph 6 has been implicitly eliminated. They are granted automatic protection without reference to any exception. Therefore, even if one of the GIs listed in the Annex of the Agreement had been used previously in a presumptive generic manner in South Korea (e.g. the terms: "feta", "champagne", "cognac", "scotch", bordeaux"), the latter is obliged to phase out its use locally from the date of the entry into force of the FTA. On the other hand, a different thinking might be more appropriate in relation to individual applications for the protection of GIs in Korea.[10] Here, in the absence of a specific provision in the FTA, for the principle of territoriality that governs IPRs, the national legal system applies. And according to both European and South Korean law, generic terms shall not be registered.[11] It therefore seems that for GIs individually applied for the TRIPs exception has been implicitly adopted.

Another important issue remains uncertain: after a foreign GI has been registered can it become generic? The answer again, given the silence of the FTA, varies depending on the national law of the two countries. According to EU law, GIs, once registered, cannot become generic.[12] Korean law on GIs does not have any specific provision in this regard. However subsequent genericness is not mentioned among the grounds for cancellation of GIs.[13] It seems thus that once legitimately and correctly registered, GIs do not risk to become generic either in the EU or in South Korea.

3.3.2 *The KORUS*

After many years of negotiations, on March 15, 2012, the FTA between the US and the Republic of Korea (KORUS) finally entered into force. For the US, this Agreement is one of the first FTAs agreed with a country in an advanced stage of

[10]According to Article 10.26 of the KOREU the provisions of this Sub-section are without prejudice to the right to seek protection of a geographical indication through individual applications under the relevant legislation of the European Union or Korea.

[11]See Article 6(1) of the Regulation (EU) 1151/2012 and Article 32 (9)(4) of the Agricultural and Fishery Products Quality Control Act (AFPQCA), as amended by Act No. 10885, July 21, 2011.

[12]Regulation (EU) 1151/2012, Article 13(2).

[13]AFPQCA, Article 44.

development and that includes a satisfactory, strong and comprehensive IPR chapter (ITAC 2007: 2).

With regard to GIs, KORUS reflects the US approach to GIs as a sub-set of TMs and the US interest to ensure that the protection granted to GIs in other countries does not undercut the US market access or, more specifically, does not prevent access to foreign markets for US exporters whose products are identified by names it considers to be generic such as "parmesan", "feta", or "gorgonzola" for cheeses. In fact, among the objectives of the US pursued through this Agreement can be seen: the desire to ensure that the registration of a GI does not violate prior rights (e.g. trademark rights) and does not deprive interested parties of the ability to use common names; the opposition of efforts to extend the protection given to wines and spirits to other products; and the desire to see that interested parties have the opportunity to oppose or to seek cancellation of any GI applied for or protected (USTR Special 301 Report 2015).

The US' aversion to GIs has been put on paper in a Report to the Congress from 2008, which expressly asserts that the US and South Korea in their bilateral agreement should make IPRs commitments beyond the TRIPs standard with provisions that "would facilitate the registration and protection of trademarks and established limitations on the use of geographical indications" (Cooper et al. 2008: 40). This means that the US' proposal, while expanding the scope of trademark protection, restricts the operation of GIs as a field of protection distinguishable from TMs (Flynn et al. 2011: 8).

There is no stand alone section for GIs in the text of KORUS. GIs are assimilated to TMs and therefore regulated by the same Article 18.2, the title of which provides: "Trademarks Including Geographical Indications". Even the definition of GIs contained in footnote 5 of Article 18.2 resembles more the one of TMs than the one of GIs. If on one side the first part reflects the wording of Article 22 of the TRIPs, on the other side, the second part adds something that de facto enlarges the concept of GIs as to include not only indications and names but also any sign or combination of signs, exemplified by a non-exhaustive list, which are the constitutive elements of the TMs' definition (Flynn et al. 2011: 9). This addition results in a sharp contrast not only with the idea of GIs in the EU where only a name in relation to a particular product can be protected as a GI (O'Connor 2014a: 66), but also with the idea embraced by the TRIPs that specifically establishes two different IPRs (Eugui and Spennemann 2006: 327).

Regarding the recognition procedure, Article 18.2 paragraph 14 of the KORUS enumerates a list of commitments that the US and the Republic of Korea have to respect regardless of the system of protection adopted for GIs (i.e. trademark system or otherwise). Three key words can be used to summarise the content of this Article: transparency, publicity and opposition. Both Parties are required to set out clearly the procedures to file and to process applications and petitions, to make them readily available to the public, to ensure that applications and petitions for GIs are published for the purpose of initiating opposition procedures and to lay down unambiguous procedures to oppose or cancel a registration resulting from an application or a petition.[14] What is clear from the Article in question is that KORUS does not provide automatic protection for any GI, but rather, it emphasis the important role of the opposition process and the possibility for any interested party to defend their rights before granting protection.

Article 18.2 paragraph 2 requires the Parties to provide that GIs be eligible for protection as TMs. It seems thus that the level of protection granted to GIs in the KORUS, mirroring the one required for TMs, be limited to confusingly similar signs and be based on the risk of misleading the public.[15] In relation to the enforcement procedure, even if KORUS does not say anything in this regard, it is clear that, in light of the inclusion of GIs in the trademark system, which does not benefit from any administrative procedure, their enforcement is entirely up to the owner of the trademark GI who has the duty to monitor continually the market against potential infringing uses or against the subsequent registration request of a potential conflicting trademark.

Concerning the relationship between prior TMs and GIs, KORUS clearly shifts the balance in favour of TMs (O'Connor 2014a: 67). This is made clear by two provisions: Article 18.2 paragraph 15 and Article 18.2 paragraph 4. The first provision provides for the refusal, opposition or cancellation of GI protection or recognition if the designation of geographical origin is: (i) likely to cause confusion with a prior trademark that is the subject of a good faith pending application or registration in the Party's territory; (ii) likely to cause confusion with prior trademark acquired through use in good faith; (iii) likely to cause confusion with a prior well-known trademark. The second provision provides as follow: "Each Party shall provide that the owner of a registered trademark shall have the exclusive right to prevent all third parties not having the owner's consent from using in the course of trade identical or similar signs, including geographical indications, at least for goods or services that are identical or similar to those goods or services in respect of which the owner's trademark is registered, where such use would result in a likelihood of confusion. In the case of the use of an identical sign, including a geographical indication, for identical goods or services, a likelihood of confusion shall be presumed".

This Article reflects the terms of Article 16 of TRIPs, however, by adding an explicit reference to geographical indications it stresses the idea of superiority of TMs over GIs. In fact, these two Articles incorporate one of the core elements of the US Lanham Act: "the first in time first in right principle", also said "principle of priority". This principle, in theory, only governs the relationship between TMs themselves. However, since GIs can be protected as TMs both under US domestic law and KORUS, this basic rule applies without any distinction to the relationship between TMs and GIs. This means that a latter GI which is confusingly similar to a prior TM could never be protected under KORUS. This provision represents an

[14]KORUS, Article 18.2(14).

[15]KORUS, Article 18.2(4), gives the owner of a TM the exclusive right to prevent all third parties from using confusingly similar signs for like goods or services.

instance of the US approach of making GI protection similar to that of TMs (Flynn et al. 2011: 9).

Surprisingly, in KORUS, unlike in the more recent Trans Pacific Partnership Agreement (TPP), a provision addressing the idea of generic terms is missing. In the absence of any regulation, Article 24 paragraph 6 of the TRIPs applies. Therefore neither the US nor the Republic of Korea are obliged to apply its provisions in respect of a GI of the other Party in relation to goods or services for which the indication concerned is identical with the term customary in common language as the common name for such goods or services in the territory of that Party.[16] This solution is implicitly suggested by the reference to the TRIPs in KORUS and, if this view is shared, it can be concluded that the exception of genericness may always be claimed by any interested party as an opposing ground for GI registration. In fact, in this Agreement, unlike in the EU-Korea FTA, there is no list of GIs to be mutually recognised and granted automatic protection: national authorities maintain their discretion to evaluate, for each individual application, the existence of genericness. Rather, KORUS emphasis the important role of the opposition process and the possibility for any interested party to challenge every application for protection of GIs.[17]

3.3.3 A Brief Comparison Between KOREU and KORUS

KORUS and KOREU are far from each other on several important issues: the scope, structure and definition of GIs, the registration and the enforcement, the level of protection, the relationship with trademarks and the issue of genericness. As already mentioned, this is the result of different economic interests (Lee 2014). The EU considers GIs as part of its cultural heritage as well as a tool to consolidate the reputation and market niche of certain agricultural products, thus as a tool to increase, in quantity and in quality, its level of exports. The US considers the protection of GIs as a potential barrier to the export of its processed agricultural products (Eugui and Spennemann 2006: 335).

One of the most debated issue at the international level is precisely what type of protection should be given to GIs. According to KOREU the level of protection should at least match that provided for by Article 23 of the TRIPs and be applicable to all those GIs mutually recognised through an agreement, including GIs for agricultural products and other foodstuffs, while under KORUS the protection should be limited to confusingly similar signs or at most to the basic level of protection as accorded by Article 22 of the TRIPs to products other than wines and spirits. In short, while the US requires proof that consumers are potentially misled in order to protect a certain GI, the EU protects GIs as such, even in the absence of

[16]TRIPs, Article 24(6).

[17]KORUS, Article 18.2(14).

misleading practices, and even if the true origin is indicated, if it is used in translation or with expressions such "like", "type", "style".[18]

The real question is whether South Korea is able to combine these conflicting provisions so as to be compliant with both Agreements. On a very practical level, a label such as "Feta made in the US" would be fully admissible under the KORUS while prohibited under the KOREU. Another difficult debate concerns the relationship between GIs and TMs previously registered in good faith. This conflict is solved, according to the KORUS, in light of the principle of priority,[19] and according to the KOREU, in light of the principle of co-existence.[20] How is Korea going to manage this evident incompatibility? Can a GI be registered even if there is an identical or similar TM previously registered in good faith for similar goods?

Article 24.6 of the TRIPs Agreement allows Members to grandfather the use of a GI which has been legally deemed to be generic in their territory. Both the Agreements, in principle, recognise this exception. However the EU, in Annex 10-A and Annex 10-B, listed a number of GIs to be granted automatic and absolute protection in South Korea thus taking away the opportunity for stakeholders with an interest in the Korean market to present arguments that the GIs at issue were in fact generic names. As a result of the entrance into force of the KOREU certain US dairy products normally sold around the world under names identical to the ones listed in the Annexes (such as "gorgonzola", "feta", "asiago" and "fontina"), on the basis that those names are considered generic in the US, have been prevented from being imported in South Korea. This, not surprisingly, caused furious protests from the US dairy industry (Trojanova 2014: 7; Cooper et al. 2011). The Consortium for Common Food Names (CCFN), representing mainly the dairy industry, has accused the EU of de facto using its FTAs as a trade barrier and as a mean to claw back the use of common names of certain dairy products for the sole use of EU producers (Testimony of Shawna Morris Testimony by Shawna Morris 2014; O'Connor 2014a, b: 68).

Illustrative of the US concerns over this issue is the exchange of letters between the USTR Ambassador, Mr. Kirk, and the Korean Minister of Trade, Mr. Kim, just before the entrance into force of the KOREU (Trojanova 2014: 7). Mr. Kirk, expressing the US industry's worries, asked for clarifications on the GIs provisions in the FTA with the EU (Official Letter Kirk). Mr. Kim, in his answer, tried to reassure the US Government that most of the GI terms set forth in Annex 10-A are compound terms, i.e. "Grana Padano", "Parmigiano Reggiano", and "Provolone Valpadana", and therefore any restriction imposed on the use of these indications would pertain only to the protection of the compound terms in their entirety and not of the individual components which would be freely usable by anybody, i.e. "grana", "parmigiano" or "provolone" (Official Letter Kim). However, the complaint advanced by the Consortium for Common Food Names and the clarifications

---

[18]KOREU, Article 10.21(1)(b).

[19]KORUS, Article 10.2(4)(15).

[20]KOREU, Article 10.21(5).

required by the US Trade Representatives seem to be based on an erroneous interpretation of the TRIPs exception. The fact that certain names are deemed generic in the US in no way implies that in South Korea these terms have to be considered generic as well since the entire concept of genericness is limited by the principle of territoriality. And this is clear from the wording of Article 24.6 of the TRIPs which provides that: "nothing in this Section shall require a Member to apply its provisions in respect of a geographical indication of any other Member with respect to goods or services for which the relevant indication is identical with the term customary in common language as the common name for such goods or services in the territory of that Member". Therefore, in the absence of any judicial decision or public consultation aimed at demonstrating the genericness of certain GIs in South Korea, the indications listed in the Annexes of the KOREU seems to be fully respectful of the exception under Article 24.6 of the TRIPs. Rather, it would be in breach of the TRIPs Agreement if a GI not generic in Korea would not be protected because it is considered generic in the US.

The most debated issue remains the inclusion of a list of names, which are deemed generic in the US, but are required to be protected in KOREU (Viju 2013). How has South Korea actually implemented the KOREU prohibiting these contentious indications from entering the Korean market?

3.4 The Korean Compromise

3.4.1 Amendments to South Korean Law with the Entry into Force of KOREU and KORUS

How has KOREU and KORUS been implemented by the Korean Government and to what extent have they impacted on Korean law in this regard? South Korea is the only Asian country where this phenomenon can be properly examined since both the FTAs have been ratified and entered into force. Whether Korea was able or not to comply with the conflicting commitments in both agreements is the focus of this section of the chapter.

In Korea, there are two types of protection of GIs: passive and active (Jung 2013). Passive protection means preventing third parties from registering or using a name of a place they have no right to monopolise. The major laws providing for passive protection include: the Trademark Act, the Unfair Competition Prevention and Trade Secret Protection Act and the Act on the Investigation of Unfair International Trade Practices and Remedy Against Injury to Industry. Active protection means the formal recognition of GIs through their registration and enforcement. Seoul adopted this positive protection system in 1999 only with the Agricultural and Fishery Products Quality Control Act (AFPQCA) and added to it in 2005 with the incorporation of GI collective marks into the Trademark Act (IIPRD 2010).

More specifically, Korea enacted AFPQCA, in 1999, to deal with the new protection obligations imposed by the TRIPs Agreement of 1994 and by the Korea-EU Framework Agreement on Trade and Cooperation signed in 1996. The Republic of Korea, as a member of the WTO, had the duty to make its domestic regulations related to IPRs conform with the International Treaty by January 1, 2000 and as a contracting Party of the Framework Agreement, had the obligation to establish a system of GI registration.[21] However, lacking any provision granting remedies against infringement of GIs, AFQCA turned out to be insufficient to protect GIs as IPRs, and, in this situation the Korean Intellectual Property Office decided to amend its Trademark Act as to allow GIs to be registered as collective marks. Nevertheless, years later, with an amendment that took effect in December 2009, AFQCA introduced civil remedy procedures against unjust uses of GI rights clearly stating that GIs are independent IPRs with an exclusive right of use (IIPRD 2010).

So, as of now, Korea has two different systems for the protection of GIs by registration: the GI registration system of the Ministry for Food, Agriculture, Forestry and Fisheries (MFAFF) and the GI collective mark registration system of the Korean Intellectual Property Office (KIPO). Therefore, South Korea is one of those countries, together with China, which provides for a dualism of protection of GIs in its national system allowing their registration as trademarks and at the same time having in place a sui generis system (Kireeva and O'Connor 2010). That being said, the fact that the protection of GIs has been incorporated into several domestic laws creates confusion and legal uncertainty (Yook 2014).

In this context, the KOREU and the KORUS brought important changes, in particular with regard to the level of protection and the relationship between GIs and TMs. In KOREU, the Parties recognised that each has legislation in place for the registration and control of GIs that meet the criteria set out in the FTA[22]; they agreed to list a number of GIs annexed to the Agreement that are to be protected at a TRIPs-plus level[23]; they inserted a "grandfather" clause which allows continued use of TMs already in existence before the date of the application for protection or registration of a GI that is included in the annexes[24]; they agreed on an enforcement procedure at the initiative of the respective authorities in each Party or at the request of an interested party[25]; and finally they established a Working Group on GIs with the aim of adding new GIs to the annexes.[26] These achievements made KOREU one of the most successful examples of the EU external action in relation to GIs within a preferential trade agreement. This successful result from an EU perspective

---

[21]Article 25 of the Framework Agreement between the EU and the Republic of Korea, signed in Luxembourg on 28 October 1996 and entered into force on 1 April 2001.

[22]KOREU, Article 10.18(1)(2).

[23]KOREU, Article 10.18(3)(4), Article 10.19 and Article 10.21.

[24]KOREU, Article 10.21(5).

[25]KOREU, Article 10.22.

[26]KOREU, Article 10.24 and 10. 25.

has probably been favoured by two main factors: first, by the fact that Korea already had in place a sui generis system of GI protection and second, by the fact that KOREU entered into force before KORUS. This priority was essential in order to avoid the detrimental effects for EU GIs contained in the US agreement.

KORUS, on the other hand, expressly required GIs to be protected as TMs, and, in particular, certification and collective marks,[27] required procedures for opposing geographical indications that are subject of applications and petitions,[28] and mandated the refusal and cancellation of those GIs that are likely to cause confusion with prior TMs.[29] So that begs two questions, first, did the Korean dual system of protection, as described above, have to be modified to be able to satisfy the provisions of the US Agreement? Second, will the amended system be capable of complying with possibly conflicting commitments in the two Agreements?

Article 10.18 paragraph 6 of KOREU requires Korea to have in place an ad hoc register listing GIs, an administrative process aimed at verifying that GIs respect the "essentially attributable test" with their product specifications and at continually monitoring the production process and, finally, an objection procedure to safeguard legitimate interests of prior users. In the same Article, at paragraph 1, the EU has been able to get Korea agree on all these requirements. In fact, AFPQCA mandates the establishment of a specific subcommittee constituted in the Council on Quality Control of Agricultural and Fishery products with the task to deliberate on the registration of GIs[30] and establishes ad hoc procedures for the registration and the post management of GIs.[31] Under Article 10.22 of the agreement, Parties shall enforce the protection of GIs on their own initiative by appropriate intervention of their authorities and at the request of any interested party. This commitment finds its transposition into Korean law through Article 43 of AFPQCA which enumerates, among the persons entitled to file a request for the invalidation or the cancellation of a certain GI, both any interested party and the subcommittee. Thus the EU has been able to get agreement on a system that appears to replicate the EU's domestic system.

Article 10.21 of the FTA deals with the scope of protection and requires all GIs listed in the FTA to be protected at a TRIPs-plus level. The highest protection granted to GIs under the Korean system before the implementation of the KOREU was established by Article 2 of the Unfair Competition and Trade Secret Protection Act which ensured to GIs a level of protection similar to the one required by Article 22 of the TRIPs Agreement. Hence, a level of protection based on the misleading test.[32]

---

[27]KORUS, Article 18.2(2).

[28]KORUS, Article 18.2(14)(e).

[29]KORUS, Article 18.2(15).

[30]AFPQCA, Article 3.

[31]AFPQCA, Article 32, Article 33 and Article 39.

[32]Unfair Competition Prevention and Trade Secret Protection Act, Article 2(1)(d)(e).

As a consequence of the entrance into force of the KOREU, the biggest change that the Korean Government had to carry out was the inclusion of an additional level of protection for GIs related to products other than wines or spirits. In fact, according to the common understanding of the treaty, the absolute protection provided for by Article 23 of the TRIPs Agreement was extended to all those GIs listed in the FTA. For this purpose, by virtue of the Amendment of June 30, 2011[33] the Korean Government inserted a new Article in the Unfair Competition Prevention and Trade Secret Protection Act, namely Article 3-2, which provides as follow: "as to geographical indications protected under a free trade agreement which is concluded bilaterally or mutually and takes effect between the Republic of Korea and a foreign country, or foreign countries […].in addition to the act of unfair competition under subparagraphs 1 (d) and (e) of Article 2, any person who does not have a legitimate source of right shall not conduct any of the following acts with respect to the goods whose place of origin is not the one indicated in the geographic mark concerned (limited to goods that are identical to or recognised to be identical to the goods with the relevant geographic mark): using a geographic mark separately, in addition to the authentic place of origin; using a geographic mark which is translated or transliterated; using a geographic mark with the expression of 'kind', 'type', 'mode', 'counterfeit' or other expressions".[34] However this rule does not apply if the prior use of the mark began before July 1, 2011 and the mark has become recognised by consumers.[35] This provision increases the level of protection in relation to all those GIs listed in an FTA where South Korea is party, regardless of the type of goods on which such GIs are used. However, while on the one hand this rule brings Korea into compliance with the requirements of the agreement with the EU, on the other hand, it generates discrimination between those foreign and national GIs listed in the FTA and those registered Korean or foreign GIs not listed in the FTA. Is this discrimination admissible under Korean law?

Another amendment that the Korean Government had to make in order to satisfy the commitments agreed upon in the KOREU pertained to the relationship between TMs and GIs. In fact, under Article 10.23 of the FTA, if a TM, that may infringe the scope of protection granted to a GI, is applied for on similar goods, its registration shall be refused or invalidated, provided that the TM application was submitted after the date of application for the protection or recognition of the GI in the territory concerned. By contrast, the Korean TMA only prevented a TM, identical or similar to another person's GI which was filed earlier for identical goods, to be refused or invalidated. As such this provision was amended[36] to expand its scope as to include TM applications that are filed for goods identical or recognised as identical to the designated goods. Despite this modification, however, doubts persist on whether this change may really fulfil the concept of similar or like goods.

---

[33]Act No. 10810, June 30, 2011.

[34]Unfair Competition Prevention and Trade Secret Protection Act, Article 3-2(1).

[35]Unfair Competition Prevention and Trade Secret Protection Act, Article 3-2(3).

[36]Act No. 10810, June 30, 2011.

Under Article 10.21 the protection of GIs is without prejudice of the continued use of prior TMs.[37] This is the famous principle of co-existence between prior TMs and GIs. Korean law does not provide anything in this regard, however, the fact that the list of GIs annexed to the Agreement has been automatically recognised with the entrance into force of the FTA notwithstanding the existence of prior or similar TMs may lead to the conclusion that this ruling has also been respected.

In relation to KORUS, the obligations related to GIs are less demanding than those in KOREU. However four provisions deserve attention and seem to be directly targeted to undermine the approach in KOREU. Article 18.2 paragraph 2 requires Parties to provide that GIs be eligible for protection as TMs. Footnote 5 of the same Article contains a definition of GIs which includes the wording: "any sign or combination of signs, in any form whatsoever". Article 18.2 paragraph 14 demands the guarantee of a procedure for opposing GIs subject of applications and Article 18.2 paragraph 15 mandates the refusal or the cancellation of the GI that is likely to cause confusion with a prior TM.

Concerning the first commitment, since 2005, the TMA provides for the protection of GIs as collective marks and, since 2011, also as certification marks. In this regard thus, the Korean system can be considered compliant with Article 18.2 of KORUS. In contrast, the definition of GIs as provided for by footnote 5 is a more contentious issue. The TMA defines GIs as mere "indications"[38] excluding "signs" or "combination of signs" from being eligible for registration as geographical collective or geographical certification marks. This may be due to the fact that GIs are viewed by Koreans as a separate form of IPR (Jeon and Lee 2010), different from traditional TMs.

Additionally, the opposition phase required by the bilateral agreement in question in relation to each GI application has been guaranteed by Korean law. The relevant provisions in this regard can be found in Article 32 paragraph 5 and 6 of AFPQCA and in Article 25 of TMA. However, if on one hand this may satisfy the KORUS obligation, on the other hand, it may contrast with the automatic recognition of certain GIs under the KOREU. In this respect, it should be noted that the direct protection of listed GIs has been possible, without infringing any international obligations, only because KOREU entered into force prior to KORUS.

Finally, with regard to the principle of priority as established under Article 18.2 paragraph 15, its equivalent in the Korean system can be recognised in Article 32 paragraph 9(2) of AFPQCA according to which an application for the registration of the GI shall be rejected where the GI in question is identical or similar to a third party's TM for which an application has already been filed or registered or to a third person's TM widely known in the Republic of Korea. Thus, except for those GIs listed in the KOREU that have been automatically recognised and that are governed by the principle of co-existence with prior similar TMs, the general rule that

[37]KOREU, Article 10.21(5).

[38]Trademark Act, Article 2(3-2).

regulates the relationship between TMs and GIs is the rule of priority in conformity with the US requests.

Last but not least, the level of protection granted to GIs under KORUS seems to be limited to confusingly similar signs. However, in light of the reference, made in the same Agreement, to the TRIPs obligations, the basic level of protection provided for by Article 22 of the TRIPs has to be considered as the minimum standard that both contracting Parties have to ensure. South Korea complies with this requirement thanks to Article 2 of the Unfair Competition Prevention and Trade Secret Protection Act.[39]

3.4.2 Is South Korean Law Capable of an Interpretation Compliant with the Provisions on GIs Contained in Both FTAs?

It seems that, overall, South Korea, with its dual system of protection, has implemented its GI law in a manner that is compliant with most of the obligations agreed under the FTAs with the EU and with the US. The only important exception regards the definition of GIs, which, according to the KORUS includes "any sign or combination of signs" while under the KOREU is limited to "indications". However, with respect to the two main contentious issues in the Trans-Atlantic debate, namely the level of protection and the relationship between GIs and TMs, the Republic of Korea has been able to balance those different approaches in its domestic law and to enforce all the commitments established under KORUS and KOREU in this regard. This has been possible for two reasons.

First because the KOREU just required the extension of the level of protection and the imposition of the co-existence rule only to those GIs listed in the FTA and not to all GIs protected in the contracting Parties. The key provision in this context is Article 3-2 of the Unfair Competition Prevention and Trade Secret Protection Act which solves, in a clever way, the problem of the US and EU conflicting agendas over GIs. It grants absolute protection only to those GIs directly recognised through FTAs as required by KOREU and at the same time it implicitly confirms that all the other agricultural and foodstuffs GIs shall be protected at a lower level, namely the one provided for by Article 22 of the TRIPs as desired by the US.

Second, because KOREU entered into force before KORUS, the automatic recognition of some GIs without a mandatory public opposition phase was still possible. However problems may arise in the future when GIs are added to the annexes according to the procedure laid down in Article 10.25 of the KOREU. In fact, under the terms of the KOREU, new GIs may be included in the FTA upon a simple decision of the Working Group. As a consequence, the US may contend that Korea, by enlarging the list of GIs which benefit from automatic and direct

[39]Unfair Competition Prevention and Trade Secret Protection Act, Article 2(d)(e).

protection in South Korea, infringes the obligation to always ensure an opposition or cancellation procedure in relation to each GI application.

That being said, although to date Korea seems not to be in a position of incompatibility with regard to the majority of the international obligations assumed, the ratification of the KOREU and the KORUS affected the legal coherence of Korean law.

The ratification of the two FTAs has increased the complexity of an already intricate system, where provisions on GIs are placed in several different regulations and where it is hard to understand the applicable rules (Yook 2014). The complexities are in part due to the inadequate information provided by the Korean Government in this regard (Jung 2013). According to the TMA and to AFPQCA, for any person who produces agricultural or fishery products the characteristics, the quality or the reputation of which is essentially attributable to its place of origin and who wants to obtain a GI registration in Korea, has two options: register such indication as a TM or as sui generis GI. Both regimes give the registrant the following advantages: the exclusive right to use the GI, the right of priority against subsequent identical or similar TMs applications in relation to goods identical or recognised as identical and the right to prevent confusingly similar uses of such registered indication. TMA and AFPQCA require similar documents for the filing of a GI application and both demand the proof of the causal link between the characteristics of the good and its geographical origin. Finally, GI collective or certification marks and sui generis GIs, as well as unregistered indication of geographical origin, are all protected under the Unfair Competition Act, which ensures TRIPs 22 level of protection. The only main difference concerns the duration of the rights, perpetual for sui generis GIs and limited for GI collective or certification marks and the administrative control to which only the sui generis GI is subject. This overlap between the two systems generates confusion as well as high costs for the maintenance of different procedures and administrations (Jung 2013). For all these reasons, some distinguished authors have envisaged the need for reform creating a single system (Yook 2014). In this respect, the entrance into force of the two FTAs has removed any chance of a reorganisation. In fact, having in place a TM as well as a GI regime for the protection of GIs is necessary in order to comply with the different commitments established under the KOREU and the KORUS.

Beside the obstacles posed by the bilateral agreements in question to the reform of the system, the implementation of the FTAS has led to the establishment of a domestic system that may conflict with fundamental principles of domestic law. The rule of protecting GIs listed in FTAs at a higher level than all the other national registered GIs as a consequence of the KOREU may violate the non-discrimination principle as stated in the Korean Constitution. Comparing Article 3-2 of the Unfair Competition Prevention and Trade Secret Protection Act with Article 2 letter (d) and (e) of the same Act it is clear that GIs protected under FTAs benefit from a higher level of protection, namely from Article 23 TRIPs level of protection, than all the other national or foreign GIs not listed in the agreement. Neither the TMA nor AFPQCA grant to GIs such a high and comprehensive protection. Is this compliant with Korean law? There is as yet no case law on the issue. The Korean Constitution calls for the principle of equality among Korean citizens. Under Article 11: "(1) all citizens shall be equal before the law, and there shall be no discrimination in political, economic, social or cultural life on account of sex, religion or social status. (2) No privileged caste shall be recognised or ever established in any form".[40] This fundamental principle of the law, according to the common understanding of the Korean Supreme Court, requires that similar situations shall not be treated differently unless differentiation is reasonably justified[41] (Kim and Jeong 2015). Thus, if foreign GIs or a certain number of national GIs—i.e. Korean GIs registered before 2008—enjoy better rights under Korean law, there might be room for alleging a violation of the equality principle. Hence, if a violation was contended by an unlisted GI holder based on such inequality, the Korean Supreme Court would have to consider, as a decisive factor, whether such inequality was the result of Government's arbitrariness or of Government's justifiable discretion in support of a legitimate position.

The new Unfair Competition Act created two different types of rights, one for listed GIs and one for unlisted GIs and it is conceivable that this distinction was made on the belief that listed and unlisted GIs are not to be considered absolutely identical. This view would justify the different treatment. However, what would make those GIs different? The mere fact that some GIs have been inserted in a free trade agreement does not make them per se different from all the other GIs registered through national procedure. In GI law, unlike TM law, a separate category of "well-known GIs" does not exist and, in theory, all GIs should be treated equally. The only legitimate distinction, according to TRIPs, concerns wines and spirits GIs as opposed to agricultural products GIs. In support of this allegation it should be noted that, in the KOREU, Korea listed basically all its GIs registered at the time the negotiating talks were closed, and therefore the choice was not based on a supposed prevalence of certain Korean GIs over others. If all GIs were inserted in the agreement, why did Korea not change its internal law as to give every sui generis GI the same level of protection? According to the current regime, dozens of national GIs registered under AFPQCA after 2008 and foreign GIs registered via national procedure in Korea enjoy a lower protection. Following this argument, it is reasonable to say that no justification exists for such a different treatment and it is conceivable to believe that the Government adopted this halfway solution as to undermine the least possible the US' commercial interests.

[40]The Constitution of the Republic of Korea, as amended on 29 October 1987.

[41]In the case law see: KCCR May 29, 2008, 2007Hun-Ma1105; KCCR February 3, 2005, 2001Hun-Ga9; KCCR August 30, 2007, 2004 Hun-Ma670; KCCR June 28, 2007, 2004 Hun-Ma 644.

3.4.3 Korea at the Crossroads of Economic Interest and Legal Coherence

A country's primary aim, when entering into FTA negotiations, is to obtain economic advantages, to reduce tariff and non-tariff barriers and to increase trade relationships with the negotiating partner (Feridhanusetyawan 2005: 14). However, all new generation FTAs concluded by the EU and the US go well beyond the elimination of tariffs and incorporate, to varying degrees, WTO provisions (Semertzi 2014) including detailed obligations with regard to IPRs. In this framework, if a contracting party wants to achieve the privileges and the economic advantages of a free trade area with the EU or the US, it also has to deal with the legal consequences of such an agreement that affect its domestic law. New agreements now require new IP standards so as to align the level of protection desired by the EU and the US, the two "regulators of the world" (Horn et al. 2009: 3). However, problems arise when the commitments required by either the EU or the US are different. When this happens, the present Chapter shows that the first agreement has an impact on the room for manoeuvre left to the parties in relation to the second (Garcia 2014: 6). In the case of Korea, the EU succeeded in closing its Agreement before the US. This allowed the EU to achieve the objectives without making compromises, in particular the EU obtained the mutual recognition of 164 GIs (among which, e.g., "Gorgonzola", "Fontina", "Asiago"). The US was not in a position, on the basis of its later FTA, to prevent such automatic protection. This had the potential for negative consequences for US exports to Korea. In fact, American producers and suppliers, since the entrance into force of the Korea-EU FTA, have been prevented from selling, in Korea, products bearing names identical to those listed in the Agreement even if the true origin is indicated, or if they are accompanied with expressions such as "like", "style" or the like or if they are used in translation. US stakeholders have also been blocked from registering such indications as TMs.[42]

In this battle over conflicting interests Korea found itself in a difficult position. In order to benefit from the trade advantages arising out of the FTAs, Korea had to accept the EU hard stance over GIs and to modify its internal law in order to give European GIs the level of protection sought.[43] Likewise, US trade representatives disclosed their concerns over the GI system outlined by the EU-Korea FTA and investigated to what extent the KOREU would have affected the successful conclusion of KORUS (Official Letter Kirk). Korean Ministers, in an exchange of letters, had to reassure the American partners that the KOREU would not have effectively undermined the US export market (Official Letter Kim). Officials in Seoul knew that the problem of GIs had to be treated with caution as it was a burning issue for both trade partners.

---

[42]Information on the refusal of US TMs' applications in relation to "gorgonzola", "fontina" and "asiago" can be accessed at: http://engportal.kipris.or.kr/engportal/search/total_search.do.

[43]Unfair Competition Prevention and Trade Secret Protection Act, Article 3-2.

As a result, the current GI regime in South Korea seems to be the outcome of a compromise, of an attempt to mutually please. Both the KORUS and the KOREU were too important for Korean exports to the EU and the US to be held up by GIs. In this respect, however, Korea's efforts to develop a GI system capable of complying with the different obligations required by the KOREU and the KORUS should be acknowledged and admired. What this Chapter underlines are the hurdles caused by the entrance into force of such comprehensive FTAs as a consequence of the divergent approaches adopted in relation to GIs in the EU and the US. In the absence of a common strategy, these bilateral agreements may jeopardise the coherence of a country's domestic system with respect to basic principles of the law.

3.5 New Agreements on the Way: The TPP

3.5.1 The TPP

The TPP is an agreement between twelve countries: the US, Australia, Canada, Japan, Brunei, Chile, Malaysia, Mexico, New Zealand, Peru, Singapore and Vietnam that seeks to expand the US approach to GIs in Asia. This Agreement, unlike all other US FTAs, contains detailed provisions in relation to GIs and its provisions seem to be specifically targeted against the EU policy on GIs in Asia (O'Connor 2014a: 67). On October 4, 2015 an agreement in principle has been reached and Ministers of the twelve countries announced the conclusion of the negotiations. More recently, on February 4, the Agreement has been signed by the TPP Ministers in Auckland, New Zealand. The result is an ambitious, comprehensive and high-standard trade deal and an important platform for regional integration across the Asia-Pacific region.

The main purpose of TPP in relation to GIs is well stated in a recently published overview of the IP Chapter of the Agreement which expressly affirms that the FTA aims at keeping generic terms available for US producers: "the chapter helps address the potential for inappropriately 'overprotecting' geographical indications in ways that shut the U.S. agricultural and food producers, including by providing opportunities for due process and requiring guidelines on how TPP partners should determine whether a term is generic in its market, as well as safeguards for owners of pre-existing trademarks" (Overview on chapter 18 2015).

Four provisions in the TPP reflect the US approach: Article 18.32 "Grounds of Opposition and Cancellation", Article 18.33 "Guidelines for Determining Whether a Term is the Term Customary in the Common Language", Article 18.34 "Multi-Component Terms" and Article 18.36 "International Agreements". According to Article 18.32 paragraph 1, among the grounds of opposition and cancellation of a GI, beside the usual circumstance of the GI being confusingly similar to a pre-existing TM, letter (c) adds the case of the GI being a term customary in common language as the common name for the relevant good in the territory of the Party. Again, the same Article in paragraph 2 expressly affirms the possibility for a regularly registered and protected GI to be cancelled on the basis of a potential subsequent genericness (Stewart and Stewart 2015: 20). In other worlds, under the TPP, a registered GI can become generic and be cancelled on that basis.

On the other hand, Article 18.33 gives some guidelines on how to determine whether a term is generic in the territory of a Party: "Party's authorities shall have the authority to take into account how consumers understand the term in the territory of that Party. Factors relevant to such consumer understanding may include: (a) whether the term is used to refer to the type of good in question, as indicated by competent sources such as dictionaries, newspapers and relevant websites; and (b) how the good referenced by the term is marketed and used in trade in the territory of that Party". Moreover, Party's authorities may also take into account how the term is used in relevant international standards.[44] This last addition, referring to international standards as a criterion for the evaluation of genericness, seems to open the possibility for an extra-territoriality effect of this exception: a GI term may be considered generic in one country (e.g. Japan) because it has been used, in relevant international standards recognised by the Parties, to refer to a type or class of goods. Examples of relevant international food standards might be the once established under the Codex Alimentarius, a code adopted in 1961 and jointly administered by the Food and Agriculture Organization of the United Nations (FAO) and by the World Health Organization (WHO) (MacMaolain 2007: 151). The Codex establishes a collection of standards and product descriptions for a variety of foods with the aim to ensure food security and consumer protection (MacMaolain 2007: 152). Among others it lists the standards required for a cheese to be qualified as "Brie", "Provolone", "Cheddar", "Camembert", "Emmental" or "Mozzarella". Thus implicitly affirming that those terms are considered as common food names by the international community.[45]

Article 18.34 requires Parties not to prohibit a third party's use of any individual component of a multicomponent protected GI if that component is the term customary in the common language as the common name for the associated good. This Article expressly declares, that a multicomponent term, e.g. "Brie de Meaux", is protected only in its entirety while the individual term "Brie" would be freely available.

---

[44]TPP, Article 18.33, footnote 24.

[45]However, questions may arise as to whether *Codex* Standards can be taken as an "enforceable" point of reference with regard to IPRs. In fact, it should be kept in mind that *Codex* Standards can be adopted by non-consensual proceedings and therefore a certain name may be inserted in the list even without the approval of those CAC Member States where IPRs corresponding to these standards exist. According to an opinion that can be shared, non-unanimous approval of *Codex* Standards, with no formal implementation in domestic law, cannot "overrule" basic principles of international IP and GI law, namely the principle of *pacta sunt servanda* (Vienna Convention on the Law of Treaties, Article 26) in relation to existing bilateral and multilateral treaties and the principle of territoriality of IPRs (Simon 2009: 323).

Finally, Article 18.36 is specifically designed to prevent future international agreements from granting automatic protection to a pre-established list of GIs without giving the opportunity for interested parties to oppose those applications. Exempted from this provision are only GIs for wines and spirits[46] and those GIs listed in international agreements that: (a) have already been concluded or agreed in principle by that Party prior to the date of conclusion or agreement in principle of TPP; (b) have been ratified by that Party prior to the date of ratification of TPP; (c) or have entered into force for that Party prior to the date of entry into force of TPP[47] (Di Mambro 2015). With regard to those GIs, Parties shall only provide an opportunity for interested persons to make comments but not to oppose their recognition.

In line with this clarification, Article 18.19 of TPP, reflecting Article 18.2 paragraph 2 of KORUS, requires each party to provide that GIs be capable of protection under trademark law. As a consequence, the level of protection granted to GIs under TPP seems to be limited to confusingly similar signs.[48] TPP also reaffirms the principle of priority as the leading rule governing the relationship between TMs and GIs.[49] However, compared to Article 18.2 paragraph 4 of KORUS, this provision goes a little further. While KORUS prohibits the use of GIs for identical or similar goods or services benefiting from a registered TM if such a use would result in a likelihood of confusion, the TPP replace the expression "identical or similar" with "related to", thus apparently enlarging the benchmark to evaluate the likelihood of confusion (O'Connor 2014a: 67).

3.5.2 The TPP and the EU-Vietnam FTA

The Free Trade Agreement between Vietnam and the EU (EUVFTA) was agreed in principle on 4 August 2015 (European Commission, Press Release 2015) and on 2 December 2015 the Parties announced the conclusion of the negotiations. This FTA, the negotiations for which started in October 2012, is the most ambitious and comprehensive Agreement that the EU has ever concluded with a developing country, the second in the ASEAN region after Singapore and a further stepping stone towards the EU's objective of a region to region EU-ASEAN FTA (European Commission Memo 2015).

Nevertheless, probably as a consequence of the TPP's negotiations, the EU was not able to achieve the same successful results with regard to GIs as it was with South Korea. In fact, next to the general provisions—i.e. the agreement to maintain a sui generis system of GI protection for agricultural GIs, the mutual recognition of

[46]TPP, Article 18.36(4).

[47]TPP, Article 18.36(6).

[48]TPP, Article 18.20.

[49]TPP, Article 18.20 combined with Article 18.32(a)(b).

hundreds of GIs listed in Annex GI-I, the establishment of a Working Group responsible for the amendment of the lists of GIs the decision to grant to listed GIs the absolute protection as provided for by Article 23 of the TRIPs Agreement and to allow them to co-existence with prior TMs[50]—the EU-Vietnam FTA contains some specifications that seem to be the result of a concession with the TPP's obligations. Namely, four main provisions weaken the EU objectives to ensuring a strong and unassailable protection to its most famous GIs in Vietnam.

According to Article 6.2, footnote 8, each Party shall provide the legal means for the invalidation of registered GIs. This provision goes beyond the terms of KOREU, which only requires an opposition phase to be guaranteed. Instead, under EUVFTA, the Parties commit to ensuring that a registered GI may always be invalidated taking into account the legitimate interests of third parties and of the right holders of the GI in question. This means that a legally registered GI may be challenged at any time on the grounds of subsequent genericiy or previous rights.

Under the terms of Article 6.3, the EU and Vietnam undertake to protect the GIs listed in Annex GI-I only after the completion of a specific objection procedure. Interestingly, unlike in KOREU, EUVFTA makes the objection proceeding an explicit requirement probably as to avoid the hurdles posed by Article 18.36 of the TPP which prevent International Agreements from requiring the automatic protection of certain GIs without ensuring an objection procedure.

Again, the FTA allows for the amendment of the lists of GIs by removing those GIs that ceased to be protected in their country of origin and by adding, after the completion of the objection and the examination procedure, new GIs.[51] The Parties explicitly establish a "Working Group on Intellectual Property Rights, including Geographical Indications" that is responsible for such changes. However, unlike in KOREU, this Article contains a second paragraph that de facto denies all those GIs already protected in the EU and in Vietnam on the date of signing of the Agreement to be added to the list, thus preventing thousands of EU registered GIs to enjoy direct recognition and enhanced protection in Vietnam[52] (O'Connor 2014b).

Finally, Article 6.5 provides that listed GIs benefit from the absolute protection as provided for by Article 23 of the TRIPs Agreement. Instead, unlisted GIs, according to letter (b) and (c) of the same Article are granted TRIPs Article 22 level. Nevertheless, unlike in KOREU, the level of protection granted to listed GIs according to the EU-Vietnam FTA is not without exceptions. Under Article 6.5a the protection of the GIs "Asiago", "Fontina", "Gorgonzola" and "Feta" shall not prevent the use in the territory of Vietnam of any of these indications by any individuals who made actual commercial use in good faith of those indications with regard to products in the class of "cheeses" prior to 1 January 2017.[53] A similar provision applies to the GI "Champagne". According to Article 6.5a paragraph 3, the use of such GI, its

[50]EUVFTA, Article 6.2 and Article 6.7.

[51]EUVFTA, Article 6.4.

[52]EUVFTA, Article 6.4(2).

[53]EUVFTA, Article 6.5a(1)(2).

translation or transliteration by any person who made actual commercial use in good faith of this indication with regard to products in the class of "wines" shall not be prevented for at least ten years from the entry into force of this Agreement.

As already pointed out in this Chapter, those specific GIs are among the most contested indications worldwide and their recognition stands at the core of the battle between the US and the EU. While they are strictly protected in the EU they are considered generic in the US. The compromise solution found in this Agreement reflects, to a large extent, the solution adopted in the Comprehensive Economic and Trade Agreement (CETA) concluded between the EU and Canada. However, while for Canada the need to find a middle way with regard to such terms was quite comprehensible because of Canada's strong commercial relationship with the US, the solution adopted by the Vietnamese Government can only be explained by the influence exercised by the US and Australia with the Trans-Pacific Partnership. Those names are not deemed as generic under the Agreement but their protection, in comparison to that granted to all the other listed GIs, is lower. In addition, not only any person who made, in the past, commercial use of such indications in Vietnam, will be allowed to use these terms forever notwithstanding their protection as GIs but also any person who will make commercial use of such indications in the next months until 1 January 2017 will have the same benefit.

Following this analysis, it is undeniable that the TPP has exercised a certain influence over the provisions contained in the EU-Vietnam FTA. This latter Agreement, unlike KOREU, expressly requires an objection proceeding to be guaranteed before a GI is added to the list, it requires the Parties to always ensure the possibility to invalidate a registered GI, and, like CETA, it addresses specific contentious GIs by lowering their level of protection compared to all the other listed GIs. This is probably due to the fact that TPP negotiations started almost two years before the negotiations with the EU were launched. This compromise confirms the "first come first served" rule.

As proof of this, it should be recorded that Vietnam protects GIs by means of sui generis rules, and therefore, the less satisfactory protection of GIs under EUVFTA if compared to KOREU cannot be considered the result of a different legal tradition in Vietnam, but it can only be regarded as the effect of the TPP Agreement. As one author has pointed out, Vietnam will be the largest beneficiary of the TPP from an economic perspective: "according to the World Economic Forum, Vietnam is predicted to have the most significant change in GDP in 2025 (i.e., 28.2%) compared with other TPP economies" (Massmann 2016). This prediction helps in understanding the importance of the TPP for the ASEAN country in question. That said, one question remains open: will Vietnam be able to balance the need to implement different commitments in its domestic law?

It is not clear whether the provision of Article 18.36 of the TPP which prevents the EU from imposing a list of GIs to be directly protected in a contracting party which is also party to the Trans-Pacific Partnership, applies or not to the EU-Vietnam Agreement. Hopefully this uncertainty will be clarified during the ratification process. But one thing remains certain: the Trans-Pacific Partnership, once it enters into force, will prevent the EU from signing FTAs of the scope of

KOREU with TPP economies. In particular, the model that the EU will probably have to embrace with future trade partners in Asia, i.e. with Malaysia Brunei and Japan, will be similar to the one adopted in the EU-Singapore FTA, even if the countries in question already have in place a sui generis system of GI protection. In other words what will be required is a model that does not allow for the mutual recognition of GIs as such but that requires a national opposition procedure with respect to each GI for which protection is sought. A model, hence, that does not properly satisfy the needs envisaged by the "old continent" (Letter Kiang).

3.6 Conclusion

The analysis shows that the arm-wrestling between the EU and the US with regard to GIs in Asia is basically governed by a new version of the old "first come first served" rule: the country which concludes a deal first gets to set the scene for the other. Korea and Vietnam are the perfect examples. It has been possible to study the impact of the different FTAs on the domestic law of Korea. South Korea, in its struggle to strike a balance between the two systems of GI protection imposed in the two FTAs it has signed with the EU and the US, changed its domestic law which, even if it managed to balance quite different obligations in the two FTAs, may have had to infringe its own constitutional obligations not to discriminate in the treatment of different local GIs. Vietnam will probably find itself in the same difficult position when it will implement its obligations on GIs as imposed by the TPP and the EU-Vietnam FTA.

Some conclusions can be drawn. The competition between the EU and the US to impose their concept of GIs across the world by means of bilateral and plurilateral trade agreements with the same third country has two consequences: first, it is threatening the legal coherence of national laws for trading partners, and second it is clearly undermining the possibility to create a uniform approach to the protection of all GIs throughout the world.

The protection of GIs should not be dependent on the patchwork of deals being concluded by the EU and the US. This is especially true given that Asian countries are still the economically weaker bargaining partners *vis-à-vis* their more powerful trading partners and seem willing to accept whatever is required on GIs just to get the deal done. The fact that GIs are regulated and protected differently in each country may prevent equal market access undermining some of the main objectives of the TRIPs itself, i.e. to set up a uniform and fair international intellectual property law for all Member States in order to reduce distortion and impediments to international trade.

Rather than locking themselves into a mercantilist battle over a limited number of specific GIs by means of overlapping and conflicting bilateral agreements with the same Asian partners, the EU and the US should take a step back from Asia and focus on achieving a coherent compromise that could provide the basis for a multilateral agreement, possibly in a revived Doha Round or in the TTIP. The new "first come first served" rule plays a pivotal role for the successful conclusion of these FTAs. However this race to conclude deals cannot mask the failure to address the underlying need to get the formal recognition of GIs as a separate form of IPRs. The small victories that the EU has achieved in ensuring some form of protection for some of Europe's most famous GIs in some of the agreements is not necessarily a reason for the EU or EU GI producers to celebrate. The object of any international negotiation should rather be to set out a series of basic and commonly agreed principles to address the "essentially attributable test" required by TRIPs Article 22, the need for co-existence between GI and TM law and the problems surrounding a limited number of specific names considered both generic and specific by consumers in different markets.

The difficulties that the current bilateral rather than multilateral approach is creating should not be underestimated. In this respect, it seems appropriate to recall a statement made by the first WTO Director General, Mr. Ruggiero, on the occasion of the WTO symposium held in Geneva in April 2005: "it seems that preferential agreements are no more considered as an exception to the non-discrimination rule, but as an instrument of normal competition with the multilateral system. If the Doha Rounds fails, the movement towards the consolidation and the creation of new vast preferential areas, sometimes covering a whole continent, will increase. This will not only change the trade geography, but also influence political relations. The risk is an international trade system with no more rules agreed by everyone, where the poor and the weak will have to fear 'a return to the law of the jungle'[…] Are we 'deglobalizing' the international trade system? The rigidities of the system will increase and the disputes between these vast regional preferential areas could become dangerous. This is, I believe, the most important challenge in today's international trading system".

Director General Ruggiero was addressing the general phenomenon of FTAs. But his concerns can easily describe the current situation for the harmonised protection of GIs. Having different GI systems and attempting to impose those conflicting systems on third countries through FTAs is undermining the main WTO principles of non-discrimination and harmonised trading rules. More efforts must be made by the US and the EU to reopen a constructive dialogue and to find a balanced solution at the multilateral level. The EU-US Trans-Atlantic Trade and Investment Partnership (TTIP) currently under negotiation should be seen as an opportunity to redefine how GIs should be protected around the world and not a mean to continue this battle of ideas.

Bibliography

Books and Articles

BARHAM E., 2003, Translating Terroir: the global challange of french AOC labeling, in Journal of Rural Studies, Vol. 19.

BELLA P., 2014, European Union declares cheese war on America, available at: http://www.chicagonow.com/cooking-cop/2014/03/european-union-declares-cheese-war-on-america/.

CÀCERES E.O., 2007, Perspectives for Geographical Indications, paper presented at the International Symposium on Geographical Indications, jointly organised by the World Intellectual property Organization (WIPO) and the State Administration for Industry and Commerce (SAIC) of the People's Republic of China, Beijing, June 26-28, 2007, WIPO/GEO/BEI/07/13.

COOPER W.H. et al., 2008, The Proposed U.S-South Korea Free Trade Agreement (KORUS FTA): Provisions and Implications, Report for Congress, Congressional Research Service (CRS).

COOPER W.H., et al., 2011, The EU-South Korea Free Trade Agreement and Its Implications for the United States, Congressional Research Service (CRS).

DI MAMBRO A., 2015, Così il TPP affossa Dop e Igp: l'accordo USA-Pacifico tutela solo i marchi registrati, ItaliaOggi.

ENGELHARDT T., 2015, Geographical Indications Under Recent EU Trade Agreements, in IIC - International Review on Intellectual Property Law, Vol. 46: 781–818.

EUGUI D.V. – SPENNEMANN C., 2006, The treatment of geographical indications in recent regional and bilateral free trade agreements in M.P. Pugatch (ed.), The Intellectual Property Debate, Cheltenham, Edward Elgar.

FERIDHANUSETYAWAN T., 2005, Preferential Trade Agreements in the Asia-Pacific Region, International Monetary Fund, Working Paper WP/05/149.

FLYNN S. et al., 2011, Public Interest Analysis of the US TPP Proposal for an IP Chapter, Program on Information Justice and Intellectual Property, American University Washington College of Law, available at: http://infojustice.org/wp-content/uploads/2011/12/TPP-Analysis-12062011.pdf.

GARCIA M., 2014, Squaring the Circle? Approaches to Intellectual Property Rights and the TTIP, UACES Conference, Cork, Ireland, available at: http://www.uaces.org/documents/papers/1401/garcia.pdf.

HANDLER M. – MERCURIO B., 2015, Intellectual Property, in S. Lester – B. Mercurio (eds.), Bilateral and regional trade agreements: commentary and analysis, New York, Cambridge University Press, 2nd edit.

HORN et al., 2009, Beyond WTO? An Anatomy of EU and US preferential trade agreements, Brussels, Bruegel Blueprint Series.

JEON H.J. - LEE J.H., 2010, A study on the Problems of Geographical Indications and a Regional Development Plan in Korea, Kyobo Book Center.

C.K. JUNG, 2013, A Study on the Protection of Geographical Indications, Master Thesis, Korea University.

KIM D., 2007, Geographical Indications Surfacing as Obstacle to Korea-EU FTA Talks, The Korea Times.

KIM H., 2011, Future Prospects of Korea-Latvia Trade Relations in the Framework of the Korea-EU FTA, Korea Institute for International Economic Policy, available at: http://www.mfa.gov.lv/data/Prezentacijas/Korea_EU_FTA.pdf.

KIM Y.H. - JEONG H.C., 2015, Basic Classes of Constitution, Seoul, Willbes. (Korean title: 정회철, 기본강의 헌법 2015,사법시험 및 변호사시험 대비, 저자 김유향, |윌비스).

I. KIREEVA - B. O'Connor, 2010, Geographical Indications and the TRIPS Agreement: What Protection is Provided to Geographical Indications in WTO Members?, in The Journal of World Intellectual Property, Vol. 13, No. 2.

LEE J., 2014, Korea's Intellectual Property Law Strategies in the Korea-China FTA Negotiations in Comparison with Other International FTAs, Master Thesis, Seoul National University.

LUPONE A., 2009, Il dibattito sulle indicazioni geografiche nel sistema multilaterale degli scambi: dal Doha Round dell'organizzazione mondiale del commercio alla protezione TRIPs plus, in B. Ubertazzi - E.M. Espada (eds.), Le indicazioni di qualità degli alimenti, Milano, Giuffrè Editore.

MACMAOLAIN C., 2007, EU Food Law: Protecting Consumer and Health in Common Market, Portland, Hart Publishing.
MANDEL O., 2011, The Recognition and Protection of key EU geographical indications in South Korea following the adoption of the EU - South Korea Free Trade Agreement, available at: http://www.mandel-office.com/the-recognition-and-protection-of-key-eu-geographical-indications-in-south-korea-following-the-adoption-of-the-eu-south-korea-free-trade-agreement/.
MASSMANN O., 2016, Overview on the Trans Pacific Partnership Agreement - Commitments above WTO Level - An Analysis, available at: http://www.lexology.com/library/detail.aspx?g=d69941cf-2f7b-4ea2-a36e-a278d81e7174.
O'CONNOR B., 2014a, The European Union and the United States: Conflicting Agendas on Geographical Indications – What's happening in Asia?, in Global Trade and Customs Journal, Vol. 9, Issue 2.
O'CONNOR B., 2014b, Geographical Indications in CETA, the Comprehensive Economic and Trade Agreement between Canada and the EU, at: http://www.origingi.com/images/stories/PDFs/English/14.11.24_GIs_in_the_CETA_English_copy.pdf.
O'CONNOR B. – RICHARDSON L., 2012, The Legal Protection of Geographical Indications in the EU's Bilateral Trade Agreements: moving beyond TRIPS, in Rivista di Diritto Alimentare, Anno VI, No. 4.
RAUSTIALA K. - MUNZER S.R., 2007, The Global Struggle over Geographic Indications, in The European Journal of International Law, Vol. 18, Issue 2.
SANBURN J., 2014, Europe's War on American Cheese, at: http://time.com/2201l/europes-war-on-american-cheese/.
SEMERTZI, 2014, The Preclusion of Direct Effect in the Recently Concluded EU Free Trade Agreements, in Common Market Law Review, Vol. 51: 1125–1158.
SIMON J., 2009, Geographical Indications (GIs), Trademarks and International Standards (e.g. Codex Alimentarius), in B. Ubertazzi - E. M. Espada (eds.), Le Indicazioni di Qualità degli Alimenti, Milano, Giuffrè Editore.
TROJANOVÀ K., 2014, Intellectual Property Rights in Preferential Trade Agreements: The Comparison of KORUS FTA and EU-South Korea FTA, Milan, E-Leader.
WATSON K.W., 2015, U.S: Trying (and Failing) to Contain the Spread of European Geographical Indications, Cato at Liberty, at: http://www.cato.org/blog/us-trying-failing-contain-spread-european-gis.
VIJU, 2013, CETA and Geographical Indicators: Why a Sensitive Issue?, Canada-Europe Transatlantic Dialogue: Seeking Transnational Solutions to 21st Century Problems, CETA policy Briefs Series, at: http://carleton.ca/ces/wp-content/uploads/CETD_CETA-policy-brief_GIs_Viju.pdf.
YOOK S.Y., 2014, A study on Geographical Indication under FTAs and Its Domestic Implementation, Korea Legislation Research Institute.

Others

Dg-Agri, 2012, working document on international protection of EU geographical indications: objectives, outcome and challenges, Ref. Ares(2012)669394–06/06/2012, Advisory Group International Aspects of Agriculture, Meeting of 25 June 2012.
European Commission, 2011, The EU-Korea Free Trade Agreement in practice, European Union Publication Office, Luxembourg.
European Commission, 2015, EU and Vietnam reach agreement on free trade deal, Press Release, IP/15/5467.
European Commission, 2015, Facts and Figures: Free Trade Agreement between EU and Vietnam, MEMO/15/5468.
ITAC, 2007, The U.S. - Korea Free Trade Agreement (FTA): The Intellectual Property Provisions, Report of the Industry Trade Advisory Committee on Intellectual Property Rights (ITAC - 15).

Italian Intellectual Property Rights DESK (IIPRD), Ufficio ICE di Seoul, 2010, Protection of Geographical Indications in the Republic of Korea, Ministero dello Sviluppo Economico, Italia in Corea, Italian Trade Commission, Seoul.

Letter sent by the Singaporean Minister for Trade and Industry L.H. Kiang to the Trade Commissioner Mr. Karel De Gucht concerning Geographical Indications in the EU-Singapore Free Trade Agreement on 21 January 2013.

Official Letter sent by Ambassador Ron Kirk to the Minister of Trade in the Republic of Korea Jong-hoon Kim on June 09, 2011.

Official Letter sent by the Minister of Trade in the Republic of Korea Jonh-hoon Kim to Ambassador Ron Kirk on June 20, 2011.

Stewart and Stewart, 2015, The Trans-Pacific Partnership, A Side-by-Side Comparison with: The United States - Colombia Trade Promotion Agreement of 2012; The United States - Korea Free Trade Agreement of 2012; the United States - Peru Free trade Agreement of 2009, Comparison, Vol. 3, TPP Chapter 18: Intellectual Property.

Testimony by Shawna Morris, Vice President of Trade Policy, U.S. Dairy Export Council & National Milk Producers Federation to the United Stated Senate Committee on Finance, Subcommittee on International Trade, Customs and Global Competitiveness, 2014, The U.S. – Korea Free Trade Agreement: Lessons Learned Two Years Later, at: http://www.finance.senate.gov/imo/media/doc/Testimony%20to%20Senate%20Trade%20Subcommittee%20on%20Korus_July%2029%2014_FINAL.pdf.

The Hagstrom Report, 2014, GI supporters express puzzlement at U.S. dairy position, Vol. 4, No. 217.

Trevisan&Cuonzo, 2009, Free Trade Agreement EU-South Korea: for the first time in a Bilateral Agreement between the EU and a Third Country, the EU provides for the Protection of European GIs for Agricultural Products and Foodstuffs, at: http://www.trevisancuonzo.com/pubblicazioni-e-legal-update/legal-update/index.html?page=20.

USTR, 2015, Overview on Chapter 18: Intellectual Property, at: https://medium.com/the-trans-pacific-partnership/intellectual-property-3479efdc7adf#.jcqjgds85.

USTR, 2015, Special 301 Report.

Free Trade Agreements

Framework Agreement between the European Union and its Member States, on the one part, and the Republic of Korea, on the other part, signed in Luxembourg on 28 October 1996 and entered into force on 1 April 2001.

United States - Singapore Free Trade Agreement, with annexes and related exchange of letters, signed at Washington on May 6, 2003 and entered into force on January 1 2004, available at: https://ustr.gov/trade-agreements/free-trade-agreements/singapore-fta/final-text.

Free Trade Agreement between the European Union and its Member States, of the one part , and the Republic of Korea, of the other part, 2011 O.J. L 127/6.

Free Trade Agreement between the United States of America and the Republic of Korea, entered into force on March, 15, 2012, available at: https://ustr.gov/trade-agreements/free-trade-agreements/korus-fta/final-text.

Comprehensive Economic and Trade Agreement between the European Union and Canada, 2014 (not yet ratified), available at: http://trade.ec.europa.eu/doclib/docs/2014/september/tradoc_152806.pdf.

Free Trade Agreement between the European Union and Vietnam, 2015 (not yet ratified), available at: http://trade.ec.europa.eu/doclib/press/index.cfm?id=1437.

Trans-Pacific Partnership Agreement, signed in Auckland, New Zealand, on February 4, 2016 (not yet ratified), available at: https://ustr.gov/tpp/.

Websites

https://ustr.gov/about-us/policy-offices/press-office/press-releases/2015/october/summary-trans-pacific-partnership
https://ustr.gov/tpp/
http://ec.europa.eu/trade/policy/countries-and-regions/countries/canada/;
http://ec.europa.eu/trade/policy/countries-and-regions/countries/singapore/;
http://ec.europa.eu/trade/policy/countries-and-regions/countries/vietnam/.
http://trade.ec.europa.eu/doclib/press/index.cfm?id=1358.
https://ustr.gov/about-us/policy-offices/press-office/press-releases/2016/February/TPP-Ministers-Statement.
http://www.international.gc.ca/trade-agreements-accords-commerciaux/agr-acc/tpp-ptp/index.aspx?lang=eng
http://www.codexalimentarius.org/about-codex/codex-timeline/en/.
http://www.codexalimentarius.org/standards/list-standards/en/?no_cache=1
http://wtocenter.vn/content/facts-and-figures-free-trade-agreement-between-eu-and-vietnam-0
http://wtocenter.vn/content/eu-and-vietnam-reach-agreement-free-trade-deal.
http://www.naqs.go.kr/eng/contents/Agrifood/Agrifood/H_01.naqs
http://www.mofa.go.kr/ENG/policy/fta/status/effect/eu/index.jsp?menu=m_20_80_10&tabmenu=t_2&.

Author Biographies

Bernard O'Connor is a partner in the Italian law firm NCTM and teaches trade law at MILE (Bern) and IELPO (Barcelona) and Food Law at the State University of Milan.

Giulia de Bosio is a Law graduate from the State University of Milan and a trainee lawyer at Trevisan & Cuonzo Avvocati.

Chapter 4
The Aroma of Opportunity: The Potential of Wine Geographical Indications in the Australia-India Comprehensive Economic Cooperation Agreement

Susanne Taylor and Madeline Taylor

Abstract As the Australian federal government seeks greater economic and trade integration with its regional neighbours, this chapter reviews the potential to include Australian wine Geographical Indications (GIs) in the current Free Trade Agreement (FTA) negotiations with India in the Comprehensive Economic Cooperation Agreement (CECA). FTA's have been characterised as a 'spaghetti bowl' of bilateral and multilateral agreements in the Asia Pacific region, leading to a debate about harmonisation, integration and codification of GIs in international trade negotiations. There are two distinct methods of including Intellectual Property (IP) content within FTA's: the minimum level of regulation, under the Agreement on Trade-Related Aspects of Intellectual Property Rights (TRIPS), termed 'TRIP-minus' and a comprehensive regulatory and protection regime, termed 'TRIPS-plus'. Both Australia and India have adopted a TRIPS-minus approach to FTA negotiations, reiterating the minimum GI standards under Section 22 of the TRIPS agreement. This is despite the high export value of Australia's wine GIs to India and India's extensive and stringent national GI system. Australian wine industry associations continue to lobby the federal government to include wine GIs in Australia's FTA negotiations asserting that a TRIPS-plus trade position will boost Australia's wine exports, meet consumer demand for wine and support local producer communities. Using the Trans Pacific Partnership (TPP) as an example of heightened TRIPS-plus GI protection and development in regional FTA's, this chapter concludes with a recommendation to include an annexure in the current CECA negotiations in recognition of the value GIs represent in the Australian agricultural sector.

S. Taylor · M. Taylor (✉)
Bond University, 14 University Drive, Robina, QLD 4226, Australia
e-mail: mataylor@bond.edu.au

S. Taylor
e-mail: sustaylo@bond.edu.au

W. van Caenegem and J. Cleary (eds.), *The Importance of Place: Geographical Indications as a Tool for Local and Regional Development*, Ius Gentium: Comparative Perspectives on Law and Justice 58, DOI 10.1007/978-3-319-53073-4_4

4.1 Introduction

Geographical indications (GIs) are tools to identify a good as originating in the territory, or a region or a locality in that territory, where a given quality, reputation or other characteristic of the good is essentially attributable to its geographical origin [Article 22 Agreement on Trade-Related Aspects of Intellectual Property Rights Agreement (TRIPS)]. Efficient GI protection fosters international trade for the benefits of its producers in commercially utilising a GI name to ensure high quality and fight counterfeiting in trade. Over the last few decades, Australia's engagement with Free Trade Agreements (FTAs) with its Asian trading partners has paid scant attention to the inclusion GIs. In late 2015, the China-Australia Free Trade Agreement (ChAFTA) was concluded and contains limited references to GIs. Article 11.15 of the ChAFTA (DFAT 2015) does not create any particular obligation on either party to protect GIs nationally, rather it merely confirms that GIs may be protected through a trade mark or sui generis system or other legal means. Nor does Article 11.15 provide for the acceptance by China to protect and recognise Australian wine GIs domestically. This is despite the fact that the ChAFTA will reduce the current wine import tariff to China of 14% for bottled wine and 20% for bulk wine to a zero tariff rate over four years (Rabobank 2014).

The absence of a provision protecting Australian wine GIs in the ChAFTA is perceived by the Australian wine industry as a "lost opportunity" (ABC Rural 2015) to protect and extend the regional identity of Australian wines in China. The lack of an Australian wine GI provision within the ChAFTA is viewed by the wine industry as an omission-while Australia's burgeoning export market continues to grow with wine exports rising 66% to total a value of $172 million in China alone (Worthington 2016). As stated by Weatherall, "the Australian wine industry might legitimately query why there has been no attempt to negotiate a list of protected GIs, especially in a growing wine market" (Weatherall 2015, p. 2).

India is currently Australia's twelfth largest trading partner and seventh largest export market, with trade between both countries valued at $15 billion. According to Alam et al., Australia's trade with India has expanded faster than with any of its other major trading partners, with both goods and services exports increasing by an annual average of over 32% in the last five years (Alam et al. 2013). Consequently, the CECA can be viewed as a platform into an increasingly significant consumer market within India, as stated by Jayasuriya and Panza, India has "a sizeable and assertive middle class [...] providing an expanding market for a wide range of services and western-style consumer goods" (Jayasuriya and Panza 2011, p. 450).

Within the current CECA discussions, the Australian Trade Mission to India has been focussed on securing tariff liberalisation and market access policies in a variety of sectors including; dairy, fresh fruit, pharmaceuticals, meats and wines (Cho and Yoon 2014). As stated by Deputy Minister Barnaby Joyce, "deepening our economic relationship with India is important to Australia, as we seek to diversify our economy in this critical post-mining boom period" (North Queensland Register 2015).

India has emphasised promoting foreign investment in its manufacturing and services sector. This policy standpoint is consistent with the Modi administration's 'Make in India' manufacturing policy, favouring zero duty on service industries such as auto machinery and textiles (Make in India 2016). Consequently, CECA talks have consistently stalled over the past five years due to differing trade policy priorities amongst the parties. India's CECA position is based upon its desire to attract investment and become an international manufacturing hub (Make in India 2016). In contrast, Australia's CECA policy standpoint aims to reduce and eventually eliminate tariffs for its strongest economic sectors, including its premium agricultural products. In order to access India's growing wine market, India's stringent wine tariff and regulatory burdens require negotiation to eliminate high tariffs allowing Australia's wine sector to gain greater market access as, "wine exports from Australia to India are increasing, and accounted for 15% of India's total wine imports in 2008" (DFAT 2010, p. 38).

Part I of this chapter examines the international standard for GIs found within the TRIPS and the recent treatment of intellectual property (IP) provisions in trade negotiations within the Asia-Pacific region. Part II provides a comparative analysis of both the Australian and Indian regulatory positions of their respective domestic GI legislation. Part III analyses the application of the virtuous economy theory illustrating the socio-economic benefits of GIs and the economic potential of the Indian wine market to Australian producers. Finally, this chapter outlines recommendations to include Australian wine GIs in the CECA to safeguard Australian wine producers and promote rural and regional development in Australian wine regions.

4.2 International Regulation of Geographical Indications

4.2.1 The TRIPS Agreement

Currently, the international standard of GI protection is found within in Article 22.1 of TRIPS, and is defined as follows:

> Indications which identify a good as originating in the territory of a member, or a region or locality in that territory, where a given quality, reputation or other characteristic of the good is essentially attributable to its geographical origin.

Article 22 of TRIPS provides the general level of protection applicable to all GI products and prohibits the use of misleading GIs or indications which constitutes an act of unfair competition (Bramley et al. 2014). Article 24 of TRIPS provides exceptions to the broad Article 22 definition, allowing precedent for member states to deny generic terms in GI protection within their domestic market that indicates a type of good rather than the goods geographical origin. Article 23 (1) of TRIPS prescribes a "heightened level" of GI protection for wines and spirits and prohibits the incorrect use of GIs and additionally prohibits the use of an indication of a region in conjunction with words such as 'like', 'style' etc. Accordingly, it stipulates:

> Each Member shall provide the legal means for interested parties to prevent use of the geographical indication identifying wines for wines not originating in the place indicated by the geographical indication in question, even where the true origin of the goods is indicated or the geographical indication is used in translation or accompanied by expressions such as 'kind', 'type', 'style', 'imitation' or the like (TRIPS, Article 23.1).

However, it must be noted that pursuant to article 1.1 TRIPS, WTO member countries may determine the appropriate method of implementing these GI provisions within their own legal framework. The recognition of wines and spirits requiring 'heightened' GI legal protection is partly due to the policy impetus of enacting the Law of 1905 in France based on the need to geographically identify wines to curtail the significant misuse of appellations. As stated by Marie-Vivien, 'these misuses were due to the redevelopment of the wine growing areas or "terroirs", after the Phylloxèra disease which destroyed most of the French vineyards in the mid-1800s. The misuses of the appellations also led to the poisoning of consumers by fake products leading to a public health threat' (Marie-Vivien 2008, p. 123).

The Doha Ministerial Declaration in 2001 placed on the negotiation agenda a provision to extend the international GI system in allowing 'heightened' Article 23 (1) protection to products other than wine and spirits. The Doha debate regarding the legitimacy of GIs has caused contention between two opposing ideologies led by the EU and US. The EU trade bloc member states (and India), has argued for the granting of extended protection of TRIPS Article 23 for products other than wines and spirits in order to localise economic control, promote rural socio-economic development and ensure higher economic returns for producers (Rangnekar 2004). The opposing US-led trade bloc, including Australia, considers extended Article 23 protection for products other than wines and spirits is not necessary as trademark laws provide adequate brand equity protection. Van Caenegem summarises the debate as follows:

> The current international debate revolves around the European Union's (EU) proposals for the expansion of a hybrid system, that of "registered geographical indications (GIs)", which already exists under municipal and EU laws. Such registered GIs fulfil two functions: they guarantee both the geographical origins of goods and certain qualities or characteristics of goods in relation to which the GI is used (van Caenegem 2003, p. 3).

The 'heightened protection' debate illustrates the crossroads between three major policies found within GI regulation namely: intellectual property, trade, and agricultural standpoints. Differing cultural perceptions, trade practices, national regulation and rural sectors have inevitably affected how GIs are interpreted and enacted at the local and national regulatory level (Audier 2008). These different national and trans-national approaches have been identified as 'problematic', particularly for the enactment of GI provisions in FTAs (Blakeney 2001).

The World Trade Organization (WTO) Secretariat (WTO document IP/C/W/253, 4 April 2000) has classified the diverse means of protection available for GIs in different WTO member countries into three broad categories: (1) laws focusing on business practices (the common law tort of "passing-off" and laws relating to the repression of unfair competition); (2) trademark law; and (3) Sui Generis protection (such as laws specifically dedicated to the protection of GIs).

According to Blakeney, sui generis systems 'provide for substantial examination by states of the product's link to its origin, embodied in the GI specification' (Blakeney 2001). Since 1999, India has embraced the economic and cultural potential of GIs by enacting a stringent domestic sui generis regime comprising two-tier protection for GIs. The first level is based on provisions in Article 22, with a further opportunity for producers to apply for extended protection under Article 23 for goods, textiles and agricultural foodstuffs. As yet, the national government has not used the extended provisions in FTA negotiations. The strict domestic regulation is partly due to the threat of misappropriation of GIs that India has faced with products such as Basmati rice and Darjeeling tea, although it is unclear why the national government has shied away from enacting GI regulation and protection in international trade agreements. Since 2004, Australia has enacted a more limited domestic sui generis regulatory regime by restraining GI protection to wines within protected GI regions and internationally agreed upon GIs within the EU-*Australia Wine Trade Agreement* only.

4.2.2 The Asia-Pacific FTA 'Spaghetti Bowl'

The CECA trade negotiations between Australia and India have been slow and multifaceted (Earl 2016). In part, this is due to an apparent contradiction in India's negotiating stance. There is a reluctance to decrease its high tariff in key agricultural sectors as a compromise to attract increased foreign investment in manufacturing to cement its place as a rising regional and global power. As stated by Naidu et al., India has, 'shown tremendous interest (and confidence) in becoming a new global player and aspires to be treated as an equal partner with the US' (Naidu and Mumin 2015, p. 3). Understanding the lack of progress in the CECA negotiations requires a broader analysis and policy context of the inclusion of GI and IP provisions in Asia-Pacific regional and bi-lateral FTA's.

There has been a tidal wave of new FTAs in the Asian region in the past 20 years, specifically stemming from the Asian Financial Crisis of 1997–8, with varying degrees of success in introducing trade liberalisation to multilateral trade negotiations amongst the 'Tiger economies' (Antons and Hilty 2014). According to The Asian Development Bank, Singapore has the most FTA's in the region with 32 FTA's, followed by India (amongst other countries) with 28 FTA's (ARIC 2015). Many of these FTA's have been signed within the context of larger regional groupings and forums for example, ASEAN (Antons and Thampapillai 2014), and with neighbouring countries in a specific region. For example, the Bangladesh-China-India-Myanmar Forum for Regional Cooperation (BCIM) and the Bay of Bengal Initiative of neighbouring South and Southeast Asian countries (BIMSTEC) which incorporates Bangladesh, India, Sri Lanka, Thailand, Myanmar, Bhutan and Nepal.

The drive towards greater regional economic integration in Asia has created a 'spaghetti bowl' effect (Low 2003) of differing FTAs, bilateral, multilateral and regional economic agreements encompassing countries with differing levels of economic development. This, in turn, has produced FTA's in the region with diverging levels of IP standards with the distinction being drawn between TRIPS-minus and TRIPS-plus trade agreements (Antons and Hilty 2014). TRIPS-minus FTA agreements take a minimalistic approach and are either absent of specific IP content altogether; include general provisions in agreements to achieve trade 'collaboration and cooperation', or contain single sector protection clauses (Antons and Hilty 2014, p. 37).

FTA's incorporating TRIPS-plus provisions are generally based upon three principles as stated by Ono: (i) to simplify and increase the transparency of patenting and other administrative procedures; (ii) to strengthen IP protection and (iii) to strengthen IP enforcement (Ono 2014, p. 239). TRIPS-plus agreements traditionally tend to be linked to the more developed Asian-Pacific economies who already have in place strong national IP laws (e.g. Singapore, New Zealand and Australia) or to regional blocs such as the AANZFTA agreement between the ASEAN trading bloc (Ruse-Khan 2009).

However, there are indications that the Asian region is starting to recognise and protect technical, industrial and value-added industries by protecting and promoting IP in their agreements. For example, Japan has concluded an Economic Partnership Agreement with TRIPS-plus content with Malaysia, Thailand and the Philippines. However, it is important to note that the TRIPS-plus provisions negotiated by Japan did not require changes to Japan's IP regulations; therefore Japan seeks to disseminate the TRIPS-plus provisions in Japanese national IP laws to negotiating parties (Ono 2014). The Japan- Peru FTA (JPEPA) provides an illustration of the inclusion of IP provisions for a developing economy, which emphasizes the general streamlining of the administrative procedures with respect to the acquisition and maintenance of IP rights between the countries (Article 171 JPEPA). A list of GIs for wines and spirits, subject to Article 22 (1) of TRIPS, is contained in Annex 10 (Article 177) of the JPEPA (Antons and Hilty 2014, p. 237). Although, brief, the list outlines GIs for both countries as follows:

With respect to Japan: Iki, Kuma, Ryukyu and Satsuma for spirits, as specified in accordance with the laws and regulations of Japan. With respect to Peru: Pisco Perú (Pisco Peru) for spirits, as specified in accordance with the laws and regulations of Peru (Annex 10).

An example of a TRIPS-minus agreement is the 1998 FTA between India and Sri Lanka that stresses the general exceptions (rather than the inclusion of specific GI provisions), in accordance with Article XX and XXI of the General Agreement on Trade and Tariffs (GATT 1994) which allows: 'either party to the agreement to take measures necessary to protect national security, public morals, animal or plant life and health, and articles of artistic, historic and archaeological value'. Similarly, the Eurasian Economic Community Customs Union (ACUSEA) authorises the negotiating parties, 'to take measures to prevent the misuse of a dominant position by one or several business entities in order to hinder, among other things, use of unfair competition methods' (Article 33(1) ACUSEA).

Another limited form of IP content in FTA's is the single permissive provision, modelled on the provision of Article XX GATT, stating that the agreement 'shall not preclude the adoption or enforcement of measures necessary for the protection of internationally property rights such as in trademarks, patents and copyrights, to protect a specific item, product or service from 'unfair or deceptive' conduct' (Antons and Hilty 2014, p. 44). This is commonly used in agreements between developing countries such as the Pacific Island Countries Trade Agreement (PICTA) and in early examples of FTA's in the region, for example the Australia-New Zealand Closer Economic Relations Trade Agreement (1983) (Antons and Hilty 2014).

An example of a TRIPS-plus approach to regional multilateral agreement is the development of IP provisions in Trans-Pacific Partnership Agreement (TPP). Chapter 18 of the Draft TPP (MFAT 2016) includes expanded IP provisions in Section E: Geographical Indications (MFAT 2016). Although not formally signed by all 12 participating members and still in a draft format, the TRIPS-plus approach to heightened IP provisions in the TPP includes acknowledgement of: Administrative Procedures; Opposition or Cancellation of GIs; Guidelines to determine use of GI terms; (including multi-component terms), and GIs in International Agreements (Articles 18.31–18.36, MFAT 2016). Article 18.31 *Administrative Procedures for the Protection or Recognition of Geographical Indications* (MFAT 2016 P18-14), states:

If a Party provides administrative procedures for the protection or recognition of geographical indications, whether through a trademark or a Sui Generis system, that Party shall with respect to applications for that protection or petitions for that recognition:

(a) accept those applications or petitions without requiring intercession by a Party on behalf of its nationals;
(b) process those applications or petitions without imposition of overly burdensome formalities;
(c) ensure that its laws and regulations governing the filing of those applications or petitions are readily available to the public and clearly set out the procedures for these actions;

(d) make available information sufficient to allow the general public to obtain guidance concerning the procedures for filing applications or petitions and the processing of those applications or petitions in general; and allow an applicant, a petitioner, or their representative to ascertain the status of specific applications and petitions;
(e) ensure that those applications or petitions are published for opposition and provide procedures for opposing geographical indications that are the subject of applications or petitions; and
(f) provide for cancellation of the protection or recognition afforded to a geographical indication (MFAT 2016, 18–14).

The inclusion of this provision thus provides a 'baseline' recognition that member States will ensure their GI systems are transparent, clearly set out and be able to be regulated with a minimum of formalities. Further, the grounds for GI opposition and cancellation by member countries is also stated and must include the fact that the GI is likely to cause confusion with a trade mark or GI that was either the subject of an earlier application, or an earlier good faith use (Article 2.15 Draft TPP). Another important provision is the International Agreements section found in Article 18.36 of the TPP which states:

> If a Party protects or recognizes a geographical indication pursuant to an international agreement, as of the applicable date under paragraph 6, involving a Party or a non-Party and that geographical indication is not protected through the procedures referred to in Article 18.31 (Administrative Procedures for the Protection or Recognition of Geographical Indications) or Article 18.32.4 (Grounds of Opposition and Cancellation), that Party shall apply at least procedures and grounds that are equivalent to those in Article 18.31(e) (Administrative Procedures for the Protection or Recognition of Geographical Indications) and Article 18.32.1 (Grounds of Opposition and Cancellation) (MFAT 2016, p. 18).

This Article aims towards a multinational consensus to facilitate compliance with international agreements on GI provisions by imposing the minimum standard found in Article 18.31 to harmonize and administer GI protection and applications for TPP member states. If signed in its current format, the IP chapter of the TPP will create a new negotiating environment whereby an increasing number of countries will pursue GI based trade strategies which sets the groundwork for revision of a TRIPS-minus stance on IP protection in the region.

The TPP parties currently includes; US, Japan, Malaysia, Vietnam, Singapore, Brunei, Australia, New Zealand, Canada, Mexico, Chile and Peru. India is not included as a TPP member state, and has made no indication that it will join, however the new IP standards set within the TPP will likely produce a lasting impact on future trade multinational fora, including future FTA and multilateral negotiations. To allow for this change to occur, the inclusion of IP provisions in FTA's has also been recognized by the (WTO) as an issue of harmonization, particularly across less developed nations. Hence, the transition period to introduce TRIPS IP standards was extended in 2013–2021 (WTO 2010). This extension will allow smaller Asian economies to incorporate longer transitional periods for the harmonization of complex international trade standards like GIs into FTA's and other trade agreements.

India has enthusiastically embraced regional, bilateral FTAs, however only its agreements with Japan, Korea, Singapore, Thailand and ASEAN mention IP (ARIC

2015). With the exception of the Japan-Republic of India Comprehensive Economic Partnership Agreement (JICEPA), the other four agreements aim at cooperation only and/or confirm the minimum respective rights and obligations under TRIPS (Antons and Hilty 2014). Hence, India takes a TRIPS-minus approach in FTAs, which creates a curious distinction between its own stringent national GI regulatory system that promotes and protects over 227 domestically registered GIs. This is despite the fact that India has recognised a number of foreign GIs in compliance with its domestic GI system since 2009 including: Champagne; Cognac; Scotch Whisky; Napa Valley wines; Douro wine; Peruvian Pisco and Prosciutto di Parma (Nanda et al. 2014, p. 2). This has created a dichotomy between its domestic and international policy stance on GIs which will require harmonization of the differing GI standpoints. As Sethi suggests, 'If India is to attract large investments, it is also vital to provide stability and certainty on the area of regulation and policy' (Sethi 2015).

4.3 A Comparative Analysis of Geographical Indications in Australia and India

4.3.1 A Case of Wine: Australia's GI Regulatory System

On 1 December 2008, negotiations between Australia and the European Community (EC) resulted in the signing of a revised agreement entitled the, *Agreement Between the European Community and Australia on Trade in Wine, European Community-Australia*, Jan. 30, 2009, 2009 O.J. (L 28) 3 (EC) (Wine Treaty) which entered into force on 1 September 2010. The *Australian Wine and Brandy Corporation Act 1980* (Cth), now the *Australian Grape and Wine Authority Act 2013* (Cth), was amended to give effect to the introduction of the new EU-Australia Wine Treaty in legislating for the protection of Australian wine GIs. The agreement provided a transition period of 12 months until 1 September 2011 for the use of certain sensitive GIs (Article 15), traditional expressions (TEs) (Article 17), vine variety names [Article 22(3) and (4)] for Australian produced wines. GIs, TEs and vine varieties (see Appendix 1 and Appendix 2). The Australian sui generis registration system for wine GIs created by the EU Agreement closely resembles the EU system, which was used as a model in its design given the need to align local regulations with EU standards for export purposes.

The *Australian Grape and Wine Authority Act 2013* (Cth) in accordance with the EC Wine treaty defines a GI in s 4(1) as, 'an indication that identifies the goods as originating in a country, or in a region or locality in that country, where a given quality, reputation or other characteristic of the goods is essentially attributable to their geographical origin'. The *Australian Grape and Wine Authority Regulations 1981* (Cth) (AGWAR) specifies the legal definitions of zones, regions and subregions protected by Australian wine GIs. According to s 25 of the AGWAR, the Committee may have regard to the following criteria when determining whether to grant a GI for a wine area including, but not limited to: (a) whether the area falls within the definition of a sub-region, a region, a zone or any other area; (b) the history of the founding and development of the area; (c) the existence in relation to the area of natural features, including rivers, contour lines and other topographical features; (d) the existence in relation to the area of constructed features, including roads, railways, towns and buildings; (h) the existence in relation to the area of a word or expression to indicate that area, including: (i) any history relating to the word or expression; and (ii) whether, and to what extent, the word or expression is known to wine retailers beyond the boundaries of the area; and the degree of discreteness and homogeneity of the proposed GI. Furthermore, an Australian GI registered product must ensure at least 85% of the fruit comes from the specified region (Crawford 2010). Where a region has developed a reputation for particular products, it will be a misrepresentation amounting to 'passing off' for a trader to falsely indicate that his or her products come from, or have a connection with that region.

The surge in demand for Australian wine from the mid-1980s onwards has continued and can be partly attributed to the protection of wine GIs in international trade. The Australian wine market had total revenue of $7829.1 million in 2014, representing a compound annual growth rate (CAGR) of 3.1% between 2010 and 2014. Market consumption volume increased with a CAGR of 2% between 2010–2014, to reach a total of 536.4 million litres in 2014 (MarketLine 2015). The Australian wine market's volume is expected to rise to 584.5 million litres by the end of 2019, representing a CAGR of 1.7% for the 2014–2019 period. In 2015, Australian wine exports accounted for 15.4% of the Asia-Pacific region (including Australia, China, Hong Kong, India, Indonesia, Japan, Malaysia, New Zealand, Philippines, Singapore, South Korea, Taiwan, Thailand, and Vietnam) (MarketLine 2014, 2). Additionally, in 2019, the Australian wine market is forecast to have a value of $8777.1 million, an increase of 12.1% since 2014.

A study by Verdonk et al. (2015) analysed wine producers from five zones in South Australia, Australia's dominant wine-producing state, including: The Barossa Valley; Fleurieu; Limestone Coast; Lower Murray and Mount Lofty Ranges, to investigate the importance and use of GIs. Of the producers surveyed, approximately half (49%) crushed less than 100 tonnes per annum and the majority had been operating for more than 10 years. 66% of wine producers indicated they regard GIs as extremely important and 85% indicating they always use their GI designation in marketing. Further, consumer levels of awareness of Australian wine regions was high for top-tier regions (i.e. Barossa Valley, Hunter Valley and Margaret River) with a high level of unprompted awareness of the regions for the quality and reputation of their GI protected wines (53–73%). This study validates wine producers' use of GIs to establish their brand identity and the associations consumers make between regions and grape cultivars.

Schamel and Anderson (2003) suggest that the Australian wine GI regulatory system has resulted in a greater investment in marketing wine regions by producers who receive higher pricing for their GI protected product. In McCutcheon et al.'s study (2009), it was found 72% (1672) of Australian wineries crushed less than 100 tonnes in 2007 and were therefore boutique or 'small-to-medium size enterprises' (SMEs), who ordinarily do not have the marketing budgets to spend on national magazine advertising campaigns. These regional boutique wineries struggle to contend with market forces such as an increasing downstream consolidation (KPMG 2003). Therefore, regional branding provides a clear point of differentiation from competitors, both foreign and domestic, that assists small wineries to compete (Tolley 2005; Bruce-Gardyne 2005). GIs are also presented in the literature as possible tools for upgrading quality in food systems, (Ponte and Ewert 2009) in food producing regions, or for creating new market segments (Cusmano et al. 2010). GIs are a distinctive mark of quality and origin for Australian wine regions understood by consumers as a value proposition, which confirms the products link to its region, unique geographical factors and other artisan skills that contribute to the products novel characteristics. Therefore, the benefit of a well-defined and well-protected GI sui generis national system allows 'the producers in the region to act as a monopoly and hence, command a premium price due to their exclusive use of the GI' (Agarwal and Barone 2005, p. 2).

4.3.2 *Travels Along the Silk Road: The Indian GI Regulatory System*

The enactment of the *Geographical Indication of Goods (Registration and Protection) Act 1999* (GI Act), symbolises India's move to safeguard the commercial status and to maintain the quality of their iconic goods, by establishing a GI registry for local products made by GI protected producers. Section 2(1)(e) of the GI Act defines a GI as:

> in relation to goods, means an indication which identifies such goods as agricultural goods, natural goods or manufactured goods as originating, or manufactured in the territory of a country, or a region or locality in that territory, where a given quality, reputation or other characteristic of such goods is essentially attributable to its geographical origin and in case where such goods are manufactured goods one of the activities of either the production or of processing or preparation of the goods concerned takes place in such territory, region or locality, as the case may be.

Therefore, to qualify for protection an indication must:

- Identify the good and its area of geographical origin; and
- Possess a given quality, reputation or other characteristics which is essentially attributable to its area of geographic origin.

A novel feature of the GI Act is the variety of product categories that may be registered as a GI in India. These include textiles, handicrafts, paintings, agricultural products, horticultural products and beverages, among others. This is in sharp contrast with the Australian regulatory regime, whereby GI protection is afforded to wines only and not other goods or foodstuffs. Despite India being an agrarian country, handicrafts lead when it comes to GI registration representing 66% of all registered GIs followed by agricultural GIs with 22% of all registered GIs. Southern regions of India are more active with regard to the registration of their GIs. For example, Karnataka has the maximum registered number of GIs at 41, followed by Andhra with 26 and Tamilnadu with 24 (Raju and Tiwari 2015).

There is a unique two-tier system in India for the protection of GIs. The first tier provides general protection required for all GIs, commensurate with Article 22 of TRIPS. The second tier of GI protection extends a 'heightened' level of protection, (granted to wine and spirits GIs, pursuant to Article 23 of TRIPS), to any domestic product via government notification (Article 84.2 of the GI Act). Although at this stage the national government of India has not invoked the second tier of GI protection, this two-tier GI system represents the policy framework that would allow India to pursue heightened protection for its domestic GIs at the international level.

India has pursued a rigorous domestic GI regime as a testament to India's strong international trade history, particularly in tea, textiles and spices, dating back to the fourth century B.C. According to Jena and Grote:

> Separate legislation for GIs in the form of the GI Act was considered to be necessary in India for two reasons. First, there was the controversial patenting of Basmati rice by a US-based company, Rice Tec, and the widespread report of tea from other countries being passed off as Darjeeling tea. The Indian government realised that if it wanted to protect its own geographical indications globally, it needed to protect them at the national level to begin with. The second reason relates to institutional commitment. As a signatory to the TRIPs agreement, India is obliged to put in place national intellectual property laws which also include GI laws (2010, p. 224).

The Indian GI registry maintains a split-level system Register of GIs providing for both a 'registered proprietor and an 'authorised user'. A 'registered proprietor' is defined as any association of persons, or of producers or any organisation entered in the register as proprietor of the GI. Conversely, an 'authorised user' is the authorised user of a GI registered under the GI Act (Das 2010). Any person claiming to be the producer of the goods in respect to a registered GI may apply in writing to the Registrar to become an 'authorised user' of the GI concerned. This split-level approach works successfully for producers, who for socio-economic reasons, may not have had the opportunity initially to be part of the collective to apply to be a registered producer (Shodhanga 2010). In this regard, an 'authorised user' does not enjoy the right to a GI as a matter of personal entitlement, but as a member of the community which imposes restrictions as to the scope of the GI use. Upon registration, the GI Registrar is required to issue a certificate of registration to the applicant and the 'authorised users', if both parties are registered with the GI. Importantly, it is the 'authorised users' and not the 'registered proprietor' who hold the exclusive right to utilise a GI. However, it is noted that both categories of GI users can take infringement actions (Shodhanga 2010).

A GI application must fulfill a number of requirements to be assessed by the GI Registry in Chennai as outlined by Marie-Vivien (2015):

- A statement as to how the GI serves to designate the goods as originating from the concerned geographical territory in respect of specific quality, reputation or other characteristics that are due exclusively or essentially to the geographical environment, with its inherent natural and human factors; and the production, processing or preparation of which takes place in such geographical location;
- The geographical map of the territory concerned;
- The particulars regarding the appearance of the GI as to whether it is comprised of the words or figurative elements or both;
- An affidavit as to how the applicant claims to represent the interest of the association of persons or producers or any organization or authority established by or under any law;
- The standards benchmark for the use of the GI or the industry standard as regards the production, exploitation, making or manufacture of the goods having specific quality, reputation or other characteristic of such goods that is essentially attributable to its geographical origin with the detailed description of the human creativity involved, if any, or other characteristics from the definite geographical territory;
- The particulars of the mechanism to ensure that the standards, quality, integrity and consistency or other special characteristic in respect of the goods to which the GI relates, which are maintained by the producers, makers or manufacturers of the goods as the case may be;
- The particulars of special human skill involved or the uniqueness of the geographical environment or other inherent characteristics associated with the GI to which the application relates; and
- The particulars of the inspection structure, if any, to regulate the use of the GI in respect of the goods for which application is made in the definite territory, region or locality mentioned in the application (p. 201–202).

While TRIPS Article 22.1 requires that ‘a given quality, reputation or other characteristic of the good should essentially be attributable to its geographical origin’, the Indian GI Act, in the case of manufactured goods, includes the additional condition that, one of the activities of either the production or of processing or preparation of the goods concerned must take place in the place of its geographical origin. This requirement is thus more expansive than that under Article 22.1 of TRIPS (Jena and Grote 2010).

For example, in the case of the manufacture of a wine in the Maharashtra GI region, the GI Act will only allow the product to be labelled as GI protected ‘Maharashtra wine’ if the grapes have been harvested and the manufacturing process completed within the specified Maharashtra GI wine registered region. However, according to the more lenient TRIPS Article 22.1 definition, the final product may be designated as ‘Maharashtra wine’ if the processing takes place outside Maharashtra given the quality or characteristics of the final product will be essentially attributable to its GI protected region. Further, unlike Article 22.1 of TRIPS that does not imply whether socio-environmental qualities and characteristics of the good are attributable to natural or human factors including artisanal skill, climate and topography, the GI Act Section 11(2)(a) 'stipulates what an application for registration should contain, refers to the geographical environment, with its inherent natural and human factors' (Gupta and Maiti 2008). Moreover, pursuant to rule 32(1) of the GI Rules, an application for a GI is required to be supported by 'the detailed description of the human creativity involved and the particulars of special human skill involved'. Therefore, according to Section 11(2) of the GI Act, all applications must include a technical specification in the form of a code of practice, which includes a description of the product, its history, its geographical environment comprising natural and/or human factors (production method) and all other elements which justify its uniqueness.

GI protection in India is very much a State-driven process as the state or its agencies generally apply for GI registration (Marie-Vivien 2010). Furthermore, 'both the GI registry and the Agricultural and Processed Food Products Export Development Authority (APEDA) are under the authority of the Ministry of Commerce, which conducts a proactive policy in favour of GI registration' (Biènabe and Marie-Vivien 2015, p. 5). As pointed out by Biènabe and Marie-Vivien:

> The interventionist State most often acts directly as the applicant with the stated objectives of preserving national heritage and supporting underprivileged farmers and artisans (2015, p. 6).

Therefore, the concept of GI as a *public right* is understood in India through the ownership of GIs conferred by the State. This is in contrast to the EU GI system, whereby GI applications may not be brought in the name of the state pursuant to EU Regulation 510/2006. The desire to protect Indian cultural heritage coupled with the large number of underprivileged members of the agrarian community provides justification of state intervention in its domestic GI regime.

Interestingly, the GI Act sets out more rigid prohibitions than both TRIPS and the Australian GI system against the infringement of registered GIs, over and above the requirements pursuant to TRIPS Articles 22.4 and 24.9. A prohibition exists for the registration of a GI which would 'be likely to deceive or cause confusion, be contrary to law, are determined generic names of goods or would otherwise be disentitled to protection in a court', as outlined in Section 9 of the GI Act. Specific prohibitions are outlined subjectively with reference to Indian cultural and religious heritage. Sections 9(c–d) specifies the prohibition of registration of certain GIs:

> (c) Which comprises or contains scandalous or obscene matter; or (d) which comprises or contains any matter likely to hurt the religious susceptibilities of any class or section of the citizens of India (Section 9, GI Act).

Section 39 and 40 of the GI Act provides that:

> In the case of infringement/false indication of a GI, involves punishment or imprisonment for a term which shall not be less than 6 months and may extend to 3 years and with fine which shall not be less than 50 thousand rupees which may extend up to 2 lakh rupees.

As previously discussed, the Indian national GI regime provides a regulatory system whereby the state and its agencies are often themselves the applicant and the proprietor of GIs (Marie-Vivien 2010). Therefore, the GI application process is a State-driven, top-down process, in contrast to the Australian producer-driven bottom-up process to obtain GI protection. Consequently, the registration of producers in India is an a posteriori procedure after the GI has been applied and registered (Biènabe and Marie-Vivien 2015). This contrasts Australia's position whereby producers must be members of a recognised GI regional group in order to gain GI protection.

Further, in comparison to the TRIPS and Australian GI regulatory regime, the GI Act in India provides for more encompassing principles, beyond extrinsic factors such as climate and topography, when considering GI applications. For example, the emphasis on the importance of 'special human skills', provides evidence of a regime acknowledging the importance of the socio-economic aspects inherent within GIs. The acknowledgement that GIs will not be granted that may offend any religious principles is an indication of the acknowledgement of social equity and good faith to be applied when granting GI applications in India. GIs in the Indian scenario provides an illustration of multifaceted GI protection, not only to preserve the national identity of Indian products and protect its cultural heritage, but as a tool of public policy to increase the welfare of its producers, often in low socio-economic environments, through the ownership of GI rights conferred in some instances to the state itself.

4.4 The Socio-economic Importance of Geographical Indications

4.4.1 The 'Virtuous Economy'

The pillars of GI policy are founded upon four key principles: (1) protection of consumers against fraud; (2) protection of the producer of the good; territorial; (3) local regional and rural development; and (4) conservation of biological resources, biodiversity and cultural diversity (Sylvander et al. 2006). Objectives three and four are notable as they signal GIs unique purpose in preserving social and environmental sustainability, absent from traditional intellectual property rights. GIs are thus built on a philosophical platform of collective entitlement connected to the artisan skills of producers and locally created public goods as an asset encompassing history, customs and cultural of local communities (Rangnekar 2003). As stated by Biènabe and Marie-Vivien, 'what makes GI a peculiar and valuable instrument socially is the capacity to embody the link to the origin on which the product reputation is established and to account for public considerations both from quality and intellectual property perspectives' (2015, p. 2).

As previously discussed, the Indian GI Act includes a unique recognition of social factors when considering GI applications. For example, India has pursued the recognition of iconic brands, not only to create market share, but also to recognise the value of the GI system to encourage development in poor, rural regions with high unemployment rates. This is consistent with the broad WTO objective to encourage trade liberalisation in developing countries to reduce poverty. As an example, the Darjeeling tea industry employs women as 60% of the tea plantation labour force (Das 2006), who rank amongst the poorest and most disadvantaged groups in rural, India. The Darjeeling area has a history of ethnic violence and a dependence on two key industries: tea production and tourism to create employment (Darjeeling Tourism 2015). This remote region offers few employment options for workers and the Darjeeling tea industry is critical to economic survival. In this respect, '(GIs) provides an original scheme of governance…due to their collective and public dimensions' (Marie-Vivien 2010, p. 121). This dimension of the GI debate has been supported by international development agencies including The Food and Agriculture Organisation of the United Nations (FAO). The FAO argues that origin-linked food quality schemes are an appropriate rural development tool outside of Europe and a way to minimise the socio-economic risks linked to globalisation that small producers face (Barham 2003).

These features of GIs have been termed by Belletti et al. (2015) as a 'virtuous economic model' incorporating social, economic and environmental components. Consequently, local skills and sustainable production techniques which use traditional farming or production processes carry an inherent economic and social 'value' based on traditional knowledge, know-how and collectivisation under GI regimes. This 'ideal' economic model emphasises three areas of GI product value: development of farms and small food companies; collective action on product production and control and specific local source management. Taken together these elements create a 'virtuous circle' (Belletti and Marescotti 2011), working to unite sustainable outcomes in the supply chain by:

> Enhancing three principal pillars: the economic viability of firms in the value chain, consumer knowledge and confidence in the ties between product quality and origin, and the preservation of the locally specific resources that determine the territory – specific quality of the GI product (Belletti et al. 2015, p. 2).

In this manner, it is argued that GIs play a role in evolving local community identity and social cohesion centred around a 'public good' which is of benefit to the producer group and by extension, the community. As reiterated by Belletti et al.:

> GI products are not only tied to a local supply chain, but also to the whole local community, as demonstrated by the product's ties to consumption habits, local gastronomy, tourism, fairs and festivals, daily life and traditions stemming from production methods. As a consequence, they possess a particular status related to their heritage (Belletti et al. 2015, p. 2).

4.4.2 The Growing Indian Wine Market

According to MarketLine (2015), The Indian wine market grew by 4.6% in 2014 to reach a value of $314.6 million. The compound annual growth rate of the market in the period 2010–14 was 8.8%. According to a 2014 Vinexpo survey, wine consumption in India will rise to 2.1 million cases by 2017, an increase of 73% from the 1.21 million cases in 2013 (Pai 2015). Australia increased its sales of wine to India by 71.7% between 2008 and 2012 and has become the country's leading supplier of imported wines. Nashik Vintners Pvt. Ltd. is the leading player in the Indian wine market, generating a 30.6% share of the market's volume and United Spirits Ltd. accounts for a further 17% of the market. Although Nashik Valley wine in the Maharashtra region is the sole wine GI registered in India currently, the rising Indian consumer demand for wine will likely stimulate local production and increase applications for wine GIs in India. Further, in 2001, implementation of the 'Maharashtra Grape Processing Policy' stimulated a boom in the State's wine production from 712 kl in 2002 to over 20 million liters in 2008–09. In 2008, the 'Karnataka Wine Policy' simplified the process for, and reduced the cost of obtaining a winery license to significantly expand the market (Marie-Vivien 2010).

Religious observance is a major part of cultural life in India and over 80% of the population is Hindu, and the religion does not prohibit the use of wine of alcohol, leaving alcohol consumption to personal preference (JBC 2008). However, it is noted of 20% of other religions in India, there are a number of religions that prohibit alcohol usage including, Buddhists, Islam, Jains and Sikhs. The highest consumption of alcohol occurs in many coastal Indian states with a large tourist industry; Maharashtra, Goa, Karnataka (George 2007).

There is evidence that wine drinking is seen as a more palatable choice for youth (where data correlates wine consumption to college level education) and women as, 'Indian women are beginning to prefer wine as a more socially acceptable form of drinking for females[…] it is seen as more feminine to consume wine'(JBC 2008, p. 10). This view of wine as more socially acceptable in a traditional Indian culture suggests that the market for wine is an alternative to the hard liquor culture that has tended to dominate Indian life, as 'Wine has a softer tone and connotation which is seen as more acceptable to consume in view of the public' (JBC 2008, p. 10). Further, market segmentation which takes into consideration religious, geographic, gender and income related data suggests the total figure of potential wine consumers in Indian is 24 million out of a total population of 1.25 billion (JBC 2008).

However, the rigorous regulatory landscape for wine production and imports provides a barrier to growth for the Indian wine market. For example, an excise tax of 150% on imports of alcohol from all countries to India is currently imposed on all alcoholic beverages (Wine Australia 2012, p. 7). The nine wine import tariffs existing in India apply to: sparkling wine, port and other still reds, sherry, port, still red wine (greater than 2 l), still white wine (greater than 2 l), vermouth and other fresh wine varieties (WFA 2015, p. 2). This stringent regulatory regime has been identified as the trade barrier for Australia to export wines. This has been noted in wine industry submissions to DFAT concerning the India-Australia FTA negotiations. As acknowledged by ABARES, 'In India, the greatest barrier to trade is the significant import tariff on wine of around 150% on imports of alcohol from all countries' (ABARES 2012, p. 3). Further, as recognised by Wine Australia, 'India raises an "extra additional duty" of 4% on imported wine. Despite theoretically being refundable once the domestic taxes have been collected, this impost is, in practice almost impossible to recover' (Wine Australia 2011, p. 3).

Australia is consequently seeking to reduce tariffs on its wine imports to India, in a similar fashion to the tariff reduction clause found in the ChAFTA, to create a more liberalised wine market in India. The addition of including an annexure of Australian wine GIs, similar to annexure in the Japan-Peru FTA, in combination with a wine tariff reduction strategy within the CECA, is essential to allow for the protection and development of Australia's wine GIs and to reap the economic benefit of an expanding Indian consumer base.

4.5 Conclusion

The trade negotiations between Australia and India are occurring during a period of rapid change to IP provisions in the FTA policy framework. The most obvious sign of future trade trends is the TPP trade negotiations, with early indications suggesting that it will include landmark IP provisions that may lead to greater harmonisation of GIs amongst member countries. Although the final effects of the TPP on IP provisions is yet unclear, it is evident that cooperation in the field of IP, harmonisation of standards, technical regulations and conformity assessments will be become prominent matters in future FTA agendas.

Although India is not a member State of the TPP, its 'TRIP-minus' approach to trade negotiations may no longer be tenable for a number of reasons: firstly, the trend towards national economic policy favouring value-added and technical industries to drive the economy; secondly, the size of the SME sector (60%) in India involved in traditional industries providing employment and growth in regional and rural areas; and finally, the opportunity to harmonise the rigorous national GI system at the international level. India has consistently ignored the opportunity to include GIs in 27 of its 28 FTA's. This is despite its acknowledgement by its former Minister of Commerce, Murasoli Maran, that 'GIs (are) vectors of national, regional and local cultural identities providing value addition to the products. In a globalising world, GIs represent more than a simple category of intellectual property rights' (Marie-Vivien 2010, p. 141). Failure to adopt GI provisions in future trade negotiations may find India increasingly isolated in future trade talks, unless it is willing to engage with IP provisions and adopt a 'TRIP-plus' position. A focus on IP protection within future trade strategies will indicate India's willingness to prevent the abuse of IP rights and signal a step towards the modernisation of its international trade position.

For Australia, the growth of the wine consumer market in India cannot be ignored. The wine industry in Australia has been strident in its demands for the federal government to include wine GIs in trade negotiations. The Australian wine sector represents a high proportion of small producers who have created an extremely valuable export sector based on quality, regionality, *terroir* and high value production techniques protected by its GI system. Consequently, including wine GIs in the next round of the CECA FTA negotiations will signal that both India and Australia acknowledge the unique characteristics and export value of their respective GI systems.

One negotiation option available to CECA Australian trade negotiators is to accept and protect India's wine GI in the Maharashtra region in exchange for recognition of Australia's most prominent wine GIs. This strategy has precedence in the Japan-Peru FTA annexure, which protected wines and spirits as part of broader protection for multiple GIs. Such an annexure in the CECA negotiations would represent the first step towards a TRIP-plus regulatory regime for both parties and would offer the opportunity to build market share for Australian wine producers in an increasingly important export sector.

Appendix 1

State/Zone	Region	Subregion
Australia		
South Eastern Australia[a]		
South Australia		
Adelaide (Super zone, includes Mount Lofty Ranges, Fleurieu and Barossa)		
Barossa	Barossa Valley	
	Eden Valley	High Eden
Far North	Southern Flinders Ranges	
Fleurieu	Currency Creek	
	Kangaroo Island	
	Langhorne Creek	
	McLaren Vale	
	Southern Fleurieu	
Limestone Coast	Coonawarra	
	Mount Benson	
	Mount Gambier	
	Padthaway	
	Robe	
	Wrattonbully	

(continued)

(continued)

Lower Murray	Riverland	
Mount Lofty Ranges	Adelaide Hills	Lenswood
		Piccadilly Valley
	Adelaide Plains	
	Clare Valley	
The Peninsulas		
New South Wales		
Big Rivers	Murray Darling[b]	
	Perricoota	
	Riverina	
	Swan Hill[b]	
Central Ranges	Cowra	
	Mudgee	
	Orange	
Hunter Valley	Hunter	Broke Fordwich
		Pokolbin
		Upper Hunter Valley
Northern Rivers	Hastings River	
Northern Slopes	Nevi England Australia	
South Coast	Shoalhaven Coast	
	Southern Highlands	
Southern New South Wales	Canberra District	
	Gundagai	
	Hilltops	
	Tumbarumba	
Western Plains		
Western Australia		
Central Western Australia		
Eastern Plains—Inland and North of Western Australia		
Greater Perth	Peel	
	Perth Hills	
	Swan District	Swan Valley
South West Australia	Blackwood Valley	
	Geographe	
	Great Southern	Albany
		Denmark
		Frankland River
		Mount Barker
		Porongurup
	Manjimup	
	Margaret River	

(continued)

(continued)

	Pemberton	
West Australian South East Coastal		
Queensland		
	Granite belt	
	South Burnett	
Victoria		
Central Victoria	Bendigo	
	Goulburn Valley	Nagambie Lakes
	Heathcote	
	Strathbogie Ranges	
	Upper Goulburn	
Gippsland		
North East Victoria	Alpine Valleys	
	Beechworth	
	Glenrowan	
	King Valley	
	Rutherglen	
North West Victoria	Murray Darling[b]	
	Swan Hill[b]	
Port Phillip	Geelong	
	Macedon Ranges	
	Mornington Peninsula	
	Sunbury	
	Yarra Valley	
Western Victoria	Grampians	Great Western[c]
	Henty	
	Pyrenees	
Tasmania		
Northern Territory		
Australian Capital Territory		

[a]The zone South Eastern Australia incorporates the whole of the states of NSW, Victoria and Tasmania and only part of Queensland and South Australia.
[b]Murray Darling and Swan Hill are contained within the zones of Big Rivers (New South Wales) and North West Victoria.
[c]The use of Great Western is subject to legally enforceable conditions of use.

Appendix 2

Geographical indications	Traditional expressions	Vine variety
Burgundy	Amontillado[a]	Hermitage
Chablis	Auslese	Lambrusco
Champagne (inc "Methode Champenoise")	Claret	
Graves	Fino[a]	
Manzanilla[a]	Oloroso[a]	
Marsala[a]	Spatlese	
Moselle		
Port[a]		
Sauterne		
Sherry[a]		
White Burgundy		

[a]The protection of these names only entered into force 12 months after the Agreement between Australia and the European Community on Trade in Wine.

Bibliography

ABARES. (2006). *Australian Commodities*, ABARES: Canberra.

ABARES. (2002). *World wine market Barriers to increasing trade*. ABARE Research report 02.6. On behalf of The Grape and Wine research and Development Corporation.

ABARES. (2012). *Consultation with Australian wine sector on barriers to trade in wine markets*. ABARES Report on behalf of The Grape and Wine research and Development Corporation.

Agarwal, S and Barone, M. (2005). *Emerging Issues for Geographical Indication Branding Strategies*. Retrieved from http://www.card.iastate.edu/publications/dbs/pdffiles/05mrp9.pdf.

Agreement on Trade-Related Aspects of Intellectual Property Rights, Apr. 15, 1994, Marrakesh Agreement Establishing the World Trade Organization, Annex 1C, THE LEGAL TEXTS: THE RESULTS OF THE URUGUAY ROUND OF MULTILATERAL TRADE NEGOTIATIONS 320 (1999), 1869 U.N.T.S. 299, 33 I.L.M. 1197 (1994).

Alam, S. Mukhopadhya, P and Islam, R. (2013). The Australia India Proposed Free Trade Agreement and Trade in Agriculture: Opportunities and Challenges. *Journal of World Investment and Trade*, 14:1 167–197.

Anthons, C and Thampapillai, D. (2014). An overview of free trade agreements in the Asia-Pacific region with a particular focus on Intellectual Property. In C, Antons and R Hilty, (eds.) *Intellectual Property and Free Trade Agreements in the Asia-Pacific Region*. Berlin: Springer.

Antons, C and Hilty, R. (2014). Introduction: IP and the Asia-Pacific 'Spaghetti Bowl' of Free Trade Agreements'. In Antons, C and Hilty, R. (eds.) *Intellectual Property and Free Trade Agreements in the Asia-Pacific Region*. Berlin: Springer.

ARIC. (2015). *Free Trade Agreements*. Retrieved from: https://aric.adb.org/fta.

Asia Regional Integration Centre. (2015). *Free Trade Agreements*. Retrieved from https://aric.adb.org/fta-country.

Audier, J. (1999). L'application de l'Accord sur les ADPIC par les membres de l'OMC. *Bulletin de l'O.I.V*, (821–822), 533–549.

Audier, J. (2008). Mondialisation et indications géographiques: Applications nationales de l'Accord ADPIC. *Indications géographiques. Propriétés Intellectuelles.* 26, (3).

Australian Government. (2010). *India and Australia Joint Free Trade Agreement Feasibility Study*. Retrieved from http://commerce.nic.in/WhatsNew/Final_JSG_Report_as_printed_and_released_4thMay_2010.pdf.

Balganesh, S. (2003). Systems of Protection for Geographical Indications of Origin: A Review of the India Regulatory Framework. *Journal of World Intellectual Property*, 6(1), 191–205.

Banga, R. (2008) 'Geographical Indications: UNCTAD's Initiatives'. Presentation delivered in the 'Regional Conference on IPR Protection through Geographical Indications', co- organized by the UNCTAD India Programme and the Textiles Committee, Lucknow, India, 4–5 September.

Barham, E. (2003). Translating terroir: The global challenge of French AOC labelling. *Journal of Rural Studies*, 19(1), 127–138.

Barham, E., & Sylvander, B. (2011). *Labels of origin for food: local development, global recognition*. CABI.

Belletti, G., & Marescotti, A. (2011). Origin products, geographical indications and rural development. In E. Barham, & B. Sylvander (Eds.), Labels of origin for food: Local development, global recognition (pp. 75–91). *Wallingford: CAB International.*

Belletti, G., Marescotti, A., and Touzard, J. M. (2015). *Geographical Indications, Pubic Goods and Sustainable Development: The Roles of Actor's Strategies and Public Policies, World Development*. Retrieved from: http://dx.doi.org/10.1016/j.worlddev.2015.05.004.

Biagio Famularo, Johan Bruwer, Elton Li. (2010). Region of origin as choice factor: wine knowledge and wine tourism involvement influence. *International Journal of Wine Business Research*. 22 (4), 362–385.

Blakeney, M. (2001). Proposals for the international regulation of geographical indications. *The Journal of World Intellectual Property*, 4(5), 629–652.

Bombrun, H. and Sumner, D. A. (2003). *What Determines the Price of Wine?*. Retrieved from https://www.researchgate.net/publication/248419304_What_Determines_the_Price_of_Wine_The_Value_of_Grape_Characteristics_and_Wine_Quality_Assessments.

Bramley, C and Kristen, J. F., 'Exploring the Economic Rationale for Protecting Geographical Indicators in Agriculture' (2007) 46 Agrekon 69.

Bramley, C., Bienabe, E., and Kirsten, J. (2014). The Economics of geographical indication: towards a conceptual framework for geographical indications research in development. In *Developing Geographical Indications in the South: The Southern African Experience.* edited by Bramley, C, Bienabe, E, Johann, K (eds). Springer: Berlin.

Aggarwal, H. Singh, S. Prashar. (2014). Branding of geographical indications in India: A paradigm to sustain its premium value. *International Journal of Law and Management* 56.6 (2014): 431–442.

Bruce-Gardyne, T. (2005). Heard on the Grapevine: Regionality, Rose′ and Riesling are just some of the buzzwords in Australia right now. *Wine and Spirit International*, July, 44–5.

CB Capital Research. (2015). *Three key policies to a successful 'make in India' initiative.* Retrieved from http://cbcapitalresearch.com/2015/06/17/three-key-policies-to-a-successful-make-in-india-initiative/.

Cho, S and Yoonb, G. (2014). Sectoral analysis of an Australia–India free trade agreement. *Journal of the Asia Pacific Economy*. 19 (2) 205–229.

Committee for Agriculture and the Trade Committee. (2013). COM/AGR/APM/TD/WP.

Council for Trade-Related Aspects of Intellectual Property Rights, Extension of the Transition Period Under Article 66.1 for Least Developed Country Members, Decision for the Council for TRIPS of 11 June 2013, WTO Doc. IP/IC/64 of 12 June 2013. Countries: Economic and Legal Implications.

Crawford, M. (2010). *Australia: Wine Labels - Regulatory Requirements For Domestic And Export Markets.* Retrieved from http://www.mondaq.com/australia/x/109112/Product+Liability+Safety/Wine+Labels+Regulatory+Requirements+For+Domestic+And+Export+Markets.

Cusmano, L., A. Morrison and R. Rabellotti. (2010). 'Catching-up Trajectories in the Wine Industry', Ch. 1 in *Innovation and Technological Catch-Up: The Changing Geography of Wine Production,* Giuliani, E. Morrison, A and Rabellotti, R (eds) London: Edward Elgar.

Dagne, T. (2015). Beyond Economic Considerations: (Re)Conceptualizing Geographical Indications for Protecting Traditional Agricultural Products. *IIC - International Review of Intellectual Property and Competition Law*, 46, 6, 682.

Darjeeling Tourism. (2015). *Darjeeling Tea Industry: Present and Future*. Retrieved from: http://www.darjeeling-tourism.com/darj_0000bd.htm.

Das, K. (2006). International protection of India's geographical indications with special reference to "Darjeeling" tea. *The Journal of World Intellectual Property*, 9(5), 459–495.

Das, K. (2010). Prospects and Challenges of Geographical Indications in India. *The Journal of World Intellectual Property*, 13: 148–201.

Das, K. (2007). *Protection of Geographical Indications: An Overview of Select Issues with Particular Reference to India*, Working Paper 8, Centre for Trade & Development (CENTAD): New Delhi.

Das, K. (2008). Select Issues and Debates around Geographical Indications with Particular Reference to India. *Journal of World Trade*, 42(3), 461–507.

DFAT. (2010). *Joint Free Trade Agreement (FTA) Feasibility Study*. Retrieved from https://dfat.gov.au/trade/agreements/aifta/Documents/Australia-India-Joint-FTA-Feasibility-Study.pdf.

DFAT. (2015). *Intellectual Property Chapter 11*. Retrieved from https://dfat.gov.au/trade/agreements/chafta/official-documents/Documents/chafta-chapter-11-intellectual-property.pdf.

Earl, G. (2016). *India worried Australia trade deal will drive imports*. Retrieved from: http://www.afr.com/news/world/asia/india-worried-australia-trade-deal-will-drive-imports-20160302-gn8zby.

George, B. P. (2007). *Wine Tourist Motivation and the perceived Importance of Servicescape: A study conducted in Goa, India*. Emerald.

Grosse Ruse-Khan, H. (2009). Time for a Paradigm Shift-Exploring Maximum Standards in International Intellectual Property Protection. *Trade L. & Dev.* 1: 56.

Gupta, E and Maiti, P. (2008). *Rural Development in India*. Atlantic: New Delhi. https://www.wilsoncenter.org/sites/default/files/ImpactofTrade_low.pdf.

Jayasuriya, S and Panza, L. (2011). *Will India Be the Next China? Challenges, Prospects and Implications for Australia*. Retrieved from http://papers.ssrn.com/sol3/papers.cfm?abstract_id=1967336.

JBC International (2008). *Comprehensive Study of the Indian Wine Market*. Retrieved from http://www.indianwineacademy.com/Comprehensive_Study_IWM_Reference_Section.pdf.

Jena, P. R. and Grote, U. (2010). Changing Institutions to Protect Regional Heritage: A Case for Geographical Indications in the Indian Agrifood Sector. *Development Policy Review*, 28: 217–236.

Jena, P. R. and Grote, U. (2010). Impact Evaluation of Traditional Basmati Rice Cultivation in Uttarakhand State of Northern India: What Impactions Does it hold for Geographical Indications? *World development*, 40(9) 1895–1907.

KPMG. (2003). *Shelf space is there room for me?* KPMG Publications: London.

Kulkarni V, Konde V. (2011). Pre- and post-geographical indications registration measures for handicrafts in India. *J Intellect Prop Rights* 16:463–469.

Long, W. (2015). *Farmers discuss the pros and cons of a China Australia free trade agreement ABC Rural*. Retrieved from http://www.abc.net.au/news/2015-09-04/farmer-forum-chafta-china-free-trade/6747224.

Loureiro, M. L. and McCluskey, J. J. (2000). Assessing Consumer Response to Protected geographical identification labelling. *Agribusiness*, 16 (3), 309–320.

Low, L. (2003). Multilateralism, Regionalism, Bilateral and Crossregional Free Trade Arrangements: All Paved with Good Intentions for ASEAN?. *Asian Economic Journal*. 17: 65–86.

Make in India. (2016). *The Next Manufacturing Destination*. Retrieved from: http://www.makeinindia.com/article/-/v/direct-foreign-investment-towards-india-s-growth.

Marie-Vivien, D. (2008). From Plant Variety Definition to Geographical Indication Protection: A Search for the Link Between Basmati Rice and India/Pakistan. *The Journal of World Intellectual Property*, 11: 321–344.

Marie-Vivien, D. (2010). The Role of the State in the Protection of Geographical Indications: From Disengagement in France/Europe to Significant Involvement in India. *The Journal of World Intellectual Property*, 13: 121–147.

Marie-Vivien, D. (2015). The Protection of Geographical Indications in India: A New Perspective on the French and European Experience. Sage Publications: India.

Biènabe, E and Marie-Vivien, D. (2015). Institutionalizing Geographical Indications in Southern Countries: Lessons Learned from Basmati and Rooibos. *World Development* (In Press).

Marketline. (2014). *Australia- Wine.* Retrieved from http://store.marketline.com/Product/australia_wine?productid=MLIP1635-0002.

Marketline. (2015). *India- Wine.* Retrieved from http://store.marketline.com/Product/india_wine?productid=MLIP1635-0010.

McCutcheon, E., Bruwer, J. and Li, E. (2009). Region of origin and its importance among choice factors in the wine-buying decision making of consumers. *International Journal of Wine Business Research* 21, 212–234.

Ministry for Foreign Affairs and Trade (NZ). (2016). *Chapter 18: Intellectual Property*. Retrieved from: https://www.mfat.govt.nz/assets/_securedfiles/Trans-Pacific-Partnership/Text/18.-Intellectual-Property.pdf.

Naidu, G. Mumin, C, 'Introduction' in Naidu, G, Chen, M., & Narayanan, R. (Eds.). (2015). *India and China in the Emerging Dynamics of East Asia.* Springer: Berlin.

Nanda, N, Barpujari, I and Srivastava, N. (2014). *Briefing Paper: The Protection of Geographical Indications in India: Issues and Challenges.* Retrieved from http://www.teriin.org/div/briefing_paper_GI.pdf.

OECD. (2000). *Appellations of Origin and Geographical Indications in OECD Member States.* Retrieved from http://www.oecd.org/officialdocuments/publicdisplaydocumentpdf/?cote=COM/AGR/APM/TD/WP(2000)15/FINAL&docLanguage=En.

Ono, N. (2014). Japan's Approach to FTAs and IP. In Antons, C and Hilty, R. (eds.) *Intellectual Property and Free Trade Agreements in the Asia-Pacific Region.* Berlin: Springer.

Pai, A. (2015). *The Grape Escape.* Retrieved from http://www.businesstoday.in/magazine/bt-more/food/wine-consumption-grape-varietals-vineyards-wines-in-urban-india/story/213338.html.

Ponte, S. and J. Ewert. (2009). Which Way is "Up" in upgrading? Trajectories of Change in the Value Chain for South African Wine. *World Development,* 37(10), 1637–1650.

Power, B. (2014). *What the FTA? Who is benefiting from free trade agreements? Global Accounting Alliance.* Retrieved from: http://www.gaaaccounting.com/what-the-fta-who-is-benefiting-from-free-trade-agreements/.

Queensland Country Life. (2015). *India Next on Trade Deal List.* Retrieved from http://www.northqueenslandregister.com.au/story/3499877/india-next-on-trade-deal-list/?cs=4784.

Rabobank. (2014). *China- Australia FTA.* Retrieved from http://www.agriworldsa.com/article-archive/economic-environment/china%20australia%20FTA%20dec%202014.pdf.

Raju, K and Tiwari, S. (2015). The management of geographical indications: post registration challenges and opportunities. *DECISION*, 42 (3), 293–306.

Rangnekar, D. (2003). *The socio-economics of geographical indications- A review of empirical evidence from Europe*, Centre for the Study of Globalisation and Regionalisation, Warwick University Coventry. Retrieved from http://www.iprsonline.org/resources/docs/Ragnekar%20-%20Socio%20Economics%20of%20GIs%20-%20Blue%208.pdf.

Rangnekar, D. (2004). *Demanding stronger protection for Geographical Indications: the relationship between local information and reputation.* Retrieved from http://www.intech.unu.edu/publications/discussion-papers/2004-11.pdf.

Raustiala, K., & Munzer, S. R. (2007). The global struggle over geographical indications. *The European Journal of International Law.* 18(2), 337–365.

Johnson, R and Bruwer, J. (2007). Regional brand image and perceived wine quality: the consumer perspective. *International Journal of Wine Business Research.* 19 (4), 276–297.

Schamel, G., & Anderson, K. (2003). Wine quality and varietal, regional and winery reputations: hedonic prices for Australia and New Zealand. *Economic Record*, 79(246), 357–369.

Sethi, A. (2015). *The myth of Modi's Make in India Campaign*. Retrieved from http://america.aljazeera.com/multimedia/2015/3/The-myth-of-modis-make-in-india-campaign.html.

Shodhanga. (2010). *Legal Protection of Geographical Indication in India*. Retrieved from http://shodhganga.inflibnet.ac.in/bitstream/10603/14514/15/14_chapter%207.pdf.

Skuras, D. & Vakrou, A. (2002). Consumers willingness to pay for origin labeled wine: A Greek case study. *British Food Journal*, 104 (11), 898–912.

Stoler, A. *Australia- US Free Trade: Benefits and Costs of an Agreement*. Institute for International Economics. Retrieved from http://www.iie.com/publications/chapters_preview/375/05iie3616.pdf, March 7, 2015.

Sylvander, B., Allaire, G., Belletti, G., Marescotti, A., Thevenod-Mottet, E., Barjolle, D. (2006). Les dispositifs franc ̦ais et européens de protection de la qualité et de l'origine dans le contexte de l'OMC : justifications générales et contextes nationaux. *Revue canadienne des sciences regionales*, 29(1), 43–54.

Tolley, S. (2005). Regionality issue crops up again. *Australian and New Zealand Wine Industry Journal*. 20 (4).

van Caenegem, W. (2003). *Registered Geographical Indications: Between Intellectual Property and Rural Policy – Part I*. Retrieved from: http://works.bepress.com/w_v_caen/30/.

Verdonk, N Wilkinson, K, Bruwer, J. (2015). Importance, use and awareness of South Australian geographical indications. *Australian Journal of Grape and Wine Research* 21, 361–366.

Wall Street Journal. (2015). *Five things that shows Modis Make in India Campaign in Working*. Retrieved from http://blogs.wsj.com/briefly/2015/08/12/five-things-that-show-modis-make-in-india-campaign-is-working/.

Weatherall, K. (2015). *Submission of Associate Professor Kimberlee Weatherall to the Joint Standing Committee on Treaties China-Australia Free Trade Agreement (ChAFTA)*. Retrieved from http://www.aph.gov.au/DocumentStore.ashx?id=8f528336-bc06-475a-8bd7-c427ca891785&subId=399528.

WFA. (2015). Australia-Indian Free Trade Agreement Wine Sector Issues. Retrieved from https://dfat.gov.au/trade/agreements/aifta/Documents/winemakers-federation-of-australia-submission.pdf.

Wikileaks. (2015). *TPP Treaty: Intellectual Property Rights Chapter – 5 October 2015*. Retrieved from: https://wikileaks.org/tpp-ip3/.

Wine Australia. (2011). *Submission to the Australian Government Department of Foreign Affairs and Trade Australia- India Free Trade Agreement*. Retrieved from https://dfat.gov.au/trade/agreements/aifta/Documents/wine-australia.pdf.

Wine Australia. (2012). *Consultation with the Australian wine sector on barriers to trade in wine markets*. Retrieved from http://research.wineaustralia.com/wp-content/uploads/2012/09/ABA-0901-Final-report.pdf.

World Intellectual Property Organisation (WIPO). (2006). *Agreement between the Government of Japan and the Government of Malaysia for an Economic Partnership*. Retrieved from: http://www.wipo.int/wipolex/en/other_treaties/details.jsp?treaty_id=405.

World Trade Organisation (WTO). (2010). *Agreement on Trade-Related Aspects of Intellectual Property Rights: Module IV Geographic Indication A*. Retrieved from: https://www.wto.org/english/tratop_e/trips_e/t_agm0_e.htm.

Worthington, B. (2016). *Growth in China drives Australian wine exports to highest value in almost a decade*. Retrieved from http://www.abc.net.au/news/2016-01-21/australian-wine-exports-2015/7104022.

WTO. (1994). *Analytical Index, GATT 1994. General Agreement Tariffs and Trade. Article XX General Exceptions*. Retrieved from: https://www.wto.org/english/res_e/booksp_e/analytic_index_e/gatt1994_07_e.htm#article20.

Author Biographies

Susanne Taylor holds an Honorary Adjunct Teaching Fellow position at Bond University at the Gold Coast, Queensland Australia. Currently, she is working on bus rapid transit project in Dhaka, Bangladesh.

Madeline Taylor (LLB Hons.) holds an Australian Postgraduate Award Scholarship and has recently completed a Rural Industry Development Corporation funded study for the Commonwealth Government.

Part II
Old and New World Development Perspectives

representing approximately $46 billion (ABS 4102.0, 2012). The major preoccupation of agricultural policy post-war has been creating greater production efficiency and scaling up and consolidation with its primary focus on commodity production. Australian agricultural policy has been driven by the geography of the nation and thin domestic markets associated with our small population. These conditions have given rise to the current domestic retail agrifood market. There is a very high risk for many producers to try and change their production regimes, where the current production model is the most rational economic response to the market environment. Undifferentiated products enable supermarkets to maintain continuity of supply by allowing substitution from among producers, which in turn creates competition, deters cooperation and guarantees lowest prices paid for commodity products.

In the 2011 Census, the majority of Australian farming occurred in commodity production including beef cattle (28%); mixed grain-sheep or grain-beef cattle farming (9%); other grains (9%); or specialized sheep farming (8%) (ABS 5204.0, 2012). Restructuring has significantly impacted the workforce needed to support agricultural production and agricultural jobs decline at an average rate of 2.3% p.a. (Bureau of Rural Sciences 2008).

The transformation of agricultural production has seen the previously taken for granted 'country-mindedness' supported by agrarian ideals disappear from the Australian policy landscape in consideration of RRR Australia. This radical shift away from agrarian and rural social ideals informing the operating environment has resulted in an environment where farms are considered businesses like any other, and therefore subject to the same competitive forces, and considered from the singular perspective of their overall contribution to the nation's wealth. Rather than cooperation, competition between farmers becomes the guiding norm, with declining collective organization resulting, particularly in the cooperatives sector (Van Caenegem et al. Collective Bargaining Report 2014).

Many of the processes and programs associated with the reforms of deregulation to achieve these policy goals have been developed and driven from within Treasury and the Productivity Commission, rather than the 'rural heartlands' of farming and farmers. This is an important point, as it cuts to the very nature of how farming and rural production systems are understood or not understood in the Australian policy context, and in how any factors beyond contribution to national wealth, such as potential benefits that GIs might bestow locally in RRR places, might be perceived.

Yet, despite the consolidation of agriculture in Australia where there is increasingly concentrated production in large farms, the majority of Australian farms are (comparatively) small, with 55% having an estimated value of agricultural operations of less than $100,000 (BRS 2008). An important question to be asked here is: for those farmers who may continue to work within their frame of reference of 'country-mindedness' and the ideals associated with 'rural heartland', what is their future?

There is a high risk for many producers when they attempt to change their production regimes, where the current production model appears to be the most rational economic response to the regulatory and market environment. The paradox of course is that structural change has created the situation where GIs may well be a useful tool to improve farmer conditions, at the same time as sometimes making it difficult to do so. GIs by their essence are collective efforts to produce differentiated and qualitatively superior consumer-oriented products. They are not private intellectual property rights which one local farmer can deploy against another in a manner that advantages actors higher up the value chain.

One important reason for considering a deeper policy engagement with GIs in Australia then, lies in their potential to support smaller farms and farmers, when considered from the perspective of burgeoning interest in local products, in the origin of products, and in local, organic and 'farmers' markets. Consumers want to know more about the production chain from actual producer to consumer and in direct connections between these two. They are also prepared to pay more for products emanating from specialist niche sectors (Hurst 2007). Smaller farms that do not have the capacity singly to support quality product development and a private branding strategy might benefit from collectively supporting the development and registration of a GI.

5.3 Making a Case for Australian GIs

Although our perspective up to this point has been encouragement of rural development, there is also a strong connection between food, place and culture. In the heartland of GIs, in France, for example, consumers are encouraged to visit specific regions and even to get to know producers themselves. The GI serves to organize local producers, to focus the forging of local identity and to maintain and broadly promote local character over time, encouraging tourism. GIs are concerned with the maintenance and survival of local traditions and the cultural life that surrounds and supports them.

The economic impact of GI regulation potentially thus extends beyond the farm sector itself into tourism, travel and associated activities such as artisanal production more generally. We can see parallels in the Australian context, in 'cellar door' wine sales, for example. GIs by virtue of their legal character remain embedded in a region and so have the potential to function as a regional asset, and this is further discussed later in this chapter. By virtue of this character they form a more natural starting point for a dialogue involving various stakeholders about a region's agricultural advantages and how to develop a cooperative approach to the branding of those advantages. However, for farmers locked into production and market systems that favor commodity producers, any thoughts about seeking to move from commodity based production to premium, differentiated production are often restricted by the twin forces of geography and demography. Farmers also tend to be risk averse in terms of abandoning traditional trading channels and engaging in new investment to produce varied products with direct consumer appeal that may result in potentially higher value but lower volume.

For Australian agricultural production then, the evidence from our research (van Caenegem et al. 2014) suggests that an effective GI system could only work for premium and value-added products. Geography and demography are the most significant limiters of specialist production because transaction and promotion costs are simply too high, unless the price premium on the value-added product is high enough, as we have seen in Tasmania, for example. Comparatively high value products are therefore likely to be the main products that might carry a brand and be differentiated by place under a GI regime. GI products must normally be locally produced so that they reflect the character of the region—they will inevitable be consumer oriented, low volume and high margin products. They will better withstand competitive forces as they are differentiated and unique and not easily substituted. However, this dynamic crucially depends on an efficient method of protecting the provenance brand value of the GI. If free riding on the reputation built around local quality is too easy or inefficiently combatted by the legal system, there is little incentive to invest in high quality products and provenance brands—the benefits will substantially flow to free riders.

For those producers we interviewed across agricultural value chains in operation in Australia (reported in full in van Caenegem et al. 2014), we consistently heard that differentiation and its attendant marketing and promotional activities imposes a burden of cost and effort that is too great for individual producers within the current operating environment. Where place-based marketing initiatives were evident, they generally relied on the efforts of a ‘champion’. In these instances, the initiatives were quite tenuous, reliant as they were on the efforts of concerned individuals. Frequently, the reliance of those benefitting from the place-based branding cited a lack of succession planning and the ‘fragility’ of the role-champion as a common reason for the lack of continuity of such initiatives.

For branding at the regional scale, often the challenge lay in deciding upon which product or products to focus when developing the region brand. Often, such initiatives failed, either because there were too many products across too many industries involved e.g. tourism product including experiences and accommodation, together with wine and/or food. In these instances, the brand became ‘watered down’ and it was unclear what rules were maintained for the use of the brand. In other instances, winning the trust of multiple actors, defining the brand and developing a fee structure that was acceptable hindered region-branding initiatives.

Interestingly, we did find multiple examples where an individual entrepreneur had garnered a reputation for high-value products and had created a supply chain from other producers in their locality. This entrepreneur would take on the role of ‘agent’ and the burden of negotiation and risk in supply contracts and in turn, his or her suppliers would benefit from higher prices. Examples of this included local dairy farmers supplying to a local cheesemaker or growers supplying a company that was building a particular brand in a particular product, e.g. chia, and in another instance, premium beef.

For many farmers supplying supermarkets, we heard of the benefits of regular and reliable payments and the positive impacts of such contracts upon longer-term sustainability of their businesses. However, this was often tempered by comments about the economic path-dependency that resulted from 'having all one's eggs in one basket'. There was a sense that taking up this kind of contract with a supermarket was a one-way street that required committing to the kind of volume demanded by supermarkets and made any thoughts of switching at a later date very difficult. For many, this created a 'lock-in' situation that invariably saw their volumes grow, but margins decrease over time.

Compliance and an increasing burden of regulation associated with both quality, environmental management and food safety are also having an impact on producer efforts at differentiation through branding. Farmers may find themselves dealing with multiple regulators, e.g. supermarkets may impose their own auditing regimes including environmental management systems (EMS). Under such regimes, premiums may initially be offered, but may be relatively short-lived. A case in point involved Tasmanian onion farmers supplying Field Fresh Tasmania who obtained certification under Nature's Choice, a code of practice required of UK supermarket Tesco. Under the pilot project, farmers were paid premiums by Field Fresh in order to encourage participation. These were eventually dropped however, and such certifications are simply a condition of doing business with Tesco. EUREPGAP (European Good Agricultural Practice) and later GlobalGAP are other examples. Such phenomena are increasingly affecting farmers, where powerful actors in the supply chains set standards and create codes of practice around food safety, quality, animal welfare and environmental management that result in certification obligations. Many of our interviewees reported that supermarkets had become the 'new regulators' (van Caenegem et al. 2014). For many, the investment in such systems in order to meet contractual obligations and regulations in force in the broader market often resulted in a level of 'sunk cost' which saw them reluctant to change their business models. For example, to change from commodity production to producing branded, high-value (and often lower volume) goods. We found that the compliance and cost burden of various regulatory systems that exist in both public and private spaces has increased, but farm-gate returns have not, while simultaneously making it ever more difficult for producers to switch. Against this backdrop of increasing commodification; an ever more complex public and private regulatory environment; complex market structures and decreasing farmgate returns, how might GIs improve the current situation for Australian producers and the regions in which they are located?

In the course of our research, we conducted a number of case studies to better understand the role GIs might potentially play as tools for regional development. It is worth reproducing here (in part) two of these for the purpose of illustrating the context in which GIs are currently deployed in Australia (i.e. only in the wine industry), and, in the second case study (King Island), to illustrate what may have occurred had they been deployed more broadly.[2]

[2]Both case studies are reproduced (with permission from RIRDC) from van Caenegem et al. (2014), *Provenance of Australian food products: is there a place for Geographical Indications?* RIRDC, Canberra.

5.3.1 *Case Study 1: The Australian Experience with GIs—the Wine and Grape Industry*

The wine industry in Australia is the only industry that has the option of registering a GI under legislation specially dedicated to that purpose, namely Part VIB of the *Australian Grape and Wine Authority Act 2013* (AGWA Act). Without going into the detailed legal history, GI protection for wine in Australia goes back to agreements on trade in wine between the EU and Australia, the first of these coming into operation in 1994 (1994 Agreement)[3] and a replacement agreement that was signed by the EU and Australia in 2008, coming into force in 2010 (2010 Agreement).[4] The *Australian Wine and Brandy Corporation Act 1980* was amended in 1993 to provide a system of protection for GIs (see *Australian Wine and Brandy Corporation Amendment Act 1993*).

The trade incentive for Australia to continue negotiating with the EU over wine is significant. In 2008 the then Minister for Agriculture Tony Burke, pointed out that the European Community had in 2007–08 accounted for about 50% of Australia's wine exports, worth about $1.3 billion (Burke 2008). Moreover, the 2010 Agreement also saw Australia come to the table with more than 100 of its own registered GIs for which it was able to obtain protection in the European market.[5] The 2010 Agreement offers Australia the strongest protection for its wine GIs of any of the international markets to which Australian wine is exported. Protecting Australian wine GIs in emerging markets such as China is for the time being more difficult, a point made by a number of interviewees.[6]

If Australia wishes to obtain better protection for its GIs in markets such as China then a special bilateral agreement of the kind it negotiated with the EU is one proven pathway to reciprocal protection. Such an agreement provides a framework for an ongoing dialogue between the industries and officials of the countries party to the agreement. The EU is not the only partner with which Australia has negotiated provisions relating to GIs. As countries have intensified their bilateral trade negotiations Australia has found itself having to negotiate provisions on GIs with other trading partners, the *Australia-United States Free Trade Agreement* (AUSFTA) being an important example.

Australia's experience with wine GIs shows that for an industry to change the way it labels its products requires careful management and a long implementation

---

[3]*Agreement between Australia and the European Community on Trade in Wine, and Protocol*, signed 26–31 January 1994, [ATS] 6 (entered into force 1 March 1994).

[4]*Agreement between Australia and the European Community on Trade in Wine*, opened for signature 1 December 2008, [2010] ATS 19 (entered into force 1 September 2010). A copy is available at http://www.agriculture.gov.au/SiteCollectionDocuments/ag-food/wine/wine-agreement.pdf as consulted May 17th 2016.

[5]*See* Annex II, Part B of the 2010 Agreement.

[6]There are reported to be provisions on GIs in the *China—Australia Free Trade Agreement* (ChAFTA) but at the time of writing these have not been made public.

period. The 1994 Agreement with the European Community was a case of where Australia saw GIs as a means to an end, the end being access to the European market. However, the focus of our study is not on whether GIs are a useful bargaining chip in trade negotiations, but rather on whether they have some of the inherent benefits claimed for them in terms of aiding regional development. In the questions we put to our wine industry interviewees we were aiming to find out whether they believed there were benefits to GIs beyond the bargaining chip story.

The industry's experience with GIs can be thought of as forming an important case. If the benefits of GIs are not be to be found in this industry one might ask whether they are likely to eventuate for any other Australian agricultural product. However, the Australian wine GI system has atypical features, in particular its limitation to origin requirements only.

5.3.1.1 The GI System for Wine

The GI system is a specific system that needs to be distinguished from other laws that affect what may or may not be put on the label of a wine bottle. For example, a country of origin claim on a wine bottle is mandatory under the Australia New Zealand Food Standards Code, as are other claims relating to alcohol content or the name of the food. Similarly there are consumer protection laws that set limits upon what winemakers may claim about their product and its origins (see Sect. 5.3).

A principal regulatory contribution of the Australian GI system is to provide legal certainty over the boundary of a given wine region. When the GI system was introduced into Australia, the then Minister Simon Crean in his second reading speech asked: 'where does the Coonawarra end and the Riverland start? This question is just as relevant for Australian wine consumers as it is for the EC requirements'.

For present purposes, the key features of the wine GI system are that the Act establishes a Geographical Indications Committee (GIC) with powers to determine GIs. The Act lists organisations that may apply for a GI as well as specifying that a winemaker or grower of grapes may make an application. The GIC may determine a GI of its own initiative. Detailed criteria for determining a GI are prescribed in regulations. These include geological, climatological, cultural and historical factors (see regulation 25 of the *Australian Grape and Wine Authority Regulations 1981* (Cth)).

Lying at the heart of a GI system is the delineation of the boundaries of an area and deciding on the indication to be used in relation to that area. Boundaries have to be described as accurately as possible. This description is included in the Register of Protected GIs and becomes the legal basis for deciding whether, for example, someone has made a false claim on a wine label. By way of example, the lengthy description for the Barossa GI begins in the following way:

> The beginning point of the boundary is identified on the map: TRURO 6729-3, and is located at the intersection of the Hundreds of Belvidere, Julia Creek and Dutton at a topographical feature identified as "Mount Rufus", and proceeds thence in a generally south-west direction and thence in a generally south direction along the eastern boundary of the Hundred of Belvidere to its intersection with the Hundred of Jellicoe.

The rules that govern the use of a GI registered in Australia are straightforward compared to those found in many European systems, especially the French appellation system (*Appellation d'Origine Contrôlée* or AOC). The French system allows mandatory rules to be prescribed for many things including allowable grape varieties for a region, minimum levels of alcohol, production methods, maximum yield levels and methods of harvesting (van Caenegem and william 2003). The purpose behind this detailed prescription or 'specification' is to ensure the production of a wine of good quality from grapes that best fit the terroir of the area. The Australian wine GI system can be said to represent the other end of the spectrum. It is a simple system concerned with the regional designation of grapes and is silent on prescribed processes.

In Australia there is no obligation to use a GI, but where a single GI such as 'McClaren Vale' is used then at least 85% of the wine must have come from grapes grown in the region. Multiple GIs can be claimed up to a maximum of three. In the case of multiple claims the basic rule is that 95% of the wine must have come from grapes grown in those regions with at least 5% of the wine coming from each region.

The Australian GI system takes a strict bright line approach to GIs, minimising through its rules game-playing opportunities and ensuring that truth in labelling remains a genuinely paramount value. For example, the guide to labelling with GIs makes clear that:

> Statements such as "our winery is situated in McLaren Vale" when the particular wine in question is not from McLaren Vale are not permitted, even if true and even if supplemented by clarifying information.[7]

The 85% rule reveals a certain flexibility in terms of sourcing grapes, since if required, up to 15% can come from outside the region. If so there is no requirement for this to be mentioned on the label. Similarly, there is no obligation under GI rules to process the grapes in the region in which they are grown. The advantage of a GI registration regime over other forms of legal protection against the misuse of brand marks is that there is no requirement to establish that the use of the GI, in relation to wine that does not meet the 85% standard, misleads consumers. There is no possibility of using the GI while making it clear that the wine does not originate in the region, nor is there a good faith defence as exists in relation to registered trade marks (for example, the use of a mark in good faith to indicate the origin of goods). It is also important to note that the particular Australian wine GI system, because it relies only on rules of origin and imposes no other standards, does set up a potential free-rider problem. Winemakers within a region may work hard to create a

[7]For this and other examples, see Wine Australia (2014).

reputation for quality through the adoption of best practice methods and drawing on the strong innovation base of the industry. In the absence of detailed prescribed standards there is nothing to stop someone from buying into the region, making a lower quality wine, but being able to use the GI to market the wine because they have sourced 85% of the grapes from the region. Some of the views we heard suggest that some winemakers see the reputation of a region as depending on a small pool of excellence: 'six out of 50 wineries are excellent in [region], there are about 10 other wineries that are good'.

However, as we will see later, informal social mechanisms have developed to try and counter this potential free riding problem. At the time the rules of the Australian GI system were being debated, the industry produced wines of varying quality and price. A number of interviewees pointed out that in order to accommodate this diversity, a 'keep it simple' approach was adopted. The GI system was not intended to be a barrier to entry into the industry. Rather than using formal rules and sanctions to set a standard of wine production in a region, the industry has pursued quality in other ways. Important to the drive for quality at the regional level have been informal social mechanisms such as tasting groups organised by winemakers. These provide new entrants into the region with information, constructive feedback, support and the opportunity to win recognition through competing for prizes. The aim has been to socialise those who are interested into a culture of quality production rather than prescribe it through law. According to our interviewees, the Australian approach to GIs has helped the industry to grow. A highly prescriptive system might have functioned as a deterrent. In the words of one interviewee:

> How would it have been if people were allowed to say "mate you're not good enough to get into this even though you grow your fruit here"?

While these rules seem clear enough it should be noted that the process of adopting rules for an entire industry may not be straightforward. Obtaining an agreement amongst producers may prove difficult. Moreover, while a simple GI system might be desired by an industry it seems that in practice this might be difficult to achieve. If one examines the laws that define Australia's wine GI system there is a degree of complexity, especially in procedural terms. Indeed, the current system took considerable time to organise, gain agreement within the industry and to implement—a process first mooted and begun in the late 1980s, according to one interviewee who was prominent in the negotiations and establishment of the system.

5.3.1.2 Procedural Complexity, Trade Marks and Public Assets

One source of procedural complexity is the need to address the relationship between trademarks and GIs. Australia's law on this issue has been affected by the AUSFTA, this agreement leading to provisions that allow for objections to the determination of a GI based on pre-existing trade mark rights.[8] For example, Rothbury Wines were able to use these provisions to object successfully to the determination of a 'Rothbury' GI, citing the existence of a suite of registered, pending and common law trade marks.[9] The 2010 Agreement led to further changes to Australia's trademark and GI laws.

If Australia did decide to have a GI system for food, a likely cost would be the integration of that system into the existing trade mark landscape. One experienced interviewee in the wine industry pointed out that when GIs were being introduced into the Australian wine industry the regulatory scheme was not complicated by an FTA with the United States that favoured a first-in-time, first-in-right approach to the protection of trademarks.

Owners of trademarks are in a position to influence the workability of a GI system for food in Australia. One can imagine for example that in a hypothetical world where it was possible to register 'King Island' as a GI for cheese, the owner of the trade mark 'King Island Dairy' would have legitimate concerns about the confusion that this might cause amongst consumers and so object to the grant of the GI. As an aside, one can observe that in a world where consumer interest in provenance has grown, those who have secured trademark protection involving placenames have obtained a potentially significant competitive advantage.

5.3.1.3 Enforcement

As we have seen, the regulatory footprint of Australia's GI system, in terms of rules, is light. Nevertheless if a GI system is to win the trust of consumers, producers have to follow its rules. This is something well understood by the Australian wine industry. In the words of the industry regulator, the Australian Grape and Wine Authority:

> Australia is increasingly recognised as an abundant source of regionally distinctive wines made from an array of both traditional and recently introduced grape varieties. Maintaining the integrity of region and variety claims has, therefore, never been more important (AGWA n.d.).

This in turn raises questions about how compliance with the rules is best achieved, how breaches of the rules are to be detected, and what the strategies for enforcement should be. AGWA provides information and educational services to the wine industry about its labelling obligations. It also has responsibility for monitoring compliance with the Label Integrity Program, GI claims forming part of that program.[10] Whether or not Australia had a GI system it would still have a Label

---

[8]See Schedule 3 of the *US Free Trade Agreement Implementation Act 2004*.

[9]*Objection by Rothbury Wines Pty Ltd to determination of geographical indication filed in the names of Murray Tyrrell, Tyrrell's Vineyards Pty Ltd and Trevor Drayton* (2008) ATMO 13 June 2008.

[10]Part VIA of the AGWA Act contains provisions relating to the Label Integrity Program.

Integrity Program for wine because the industry understands the foundational importance of truth in labelling.

The Label Integrity Program imposes record keeping obligations on those in the wine supply chain. There is a range of sanctions available under the law including the suspension or cancellation of an export licence, as well as terms of imprisonment and/or fines.

Compliance with the Label Integrity Program is monitored by a small group of auditors (four at the time of interviews). This audit team aims to cover all Australian wine producers once every 3–4 years. Audits may be cold-calls or by appointment. Wine producers are obliged to keep detailed records that help meet the goal of full traceability. Traceability is one of the things that compliance inspectors comment upon in their audit reports. Auditors appear to follow what is known in the regulatory literature as an enforcement pyramid. The key idea behind the pyramid is that punishment and persuasion should be linked in a certain sequence that always begins with dialogue and persuasion at the base of the pyramid and ends with the most punitive sanction at the apex of the pyramid (see Ayres and Braithwaite 1992). At the time of interviews approximately 400 audits per year were being carried out. Resorting to coercive levels of the enforcement pyramid by the regulator appears rare, with few licence suspensions and only two prosecutions in the last 15 years. Only one of those prosecutions related to alleged breaches of provisions relating to geographical indications. A lot of the effort invested in obtaining compliance takes the form of presentations, advice and the provision of manuals and templates to wine producers and exporters. An important factor in the wine industry’s compliance culture is that many of its participants agree with the industry’s strategic objective of increasing Australia’s reputation for premium wines and the important role of GI regions within that objective.

5.3.1.4 Costs

A range of costs accompany the wine GI system; the costs of application, the costs of running the system and the ongoing costs of compliance. The costs begin with a group’s time in putting together an application for the determination of a GI. The costs of application include a fee and those arising from assembling the evidence/information required as part of the application process. The application fee is currently set at $27,500, a fee that approximates the average cost of the GIC hearing an application. The fees for wine GIs have changed over time. Initially they were met out of industry levies and then were kept at a low rate. The potentially deterrent effect of the present fee was seen as a desirable consequence by many interviewees.

The real costs of application, however, lie in the evidence that is required to convince the GIC how it should determine the boundaries of a GI. As we indicated earlier the applicant has to address a wide range of criteria (specified in the AGWA

Regulations)[11] as part of the application for a GI, and provide evidence in relation to those criteria. Providing evidence on matters such as the history of an area, its discreteness and homogeneity by reference to attributes such as climate and geology requires expert evidence. The estimates we were given suggest that these types of costs run into the order of tens of thousands of dollars. However, costs might rise to a six figure sum if, for example, they ran into opposition to the GI from a trade mark owner. This sum might be larger still if, as in the case of the dispute over the determination of the boundary for Coonawarra GI, the matter ends up before the Federal Court with many years of legal expenses having to be met.

The Coonawarra case is a spectacular example of a bitter and prolonged dispute over a GI boundary, but it is also something of an outlier. Settling GI boundaries has been a story of consensus rather than conflict. Most other Australian GI boundaries were determined without recourse to battalions of lawyers and appeals to the Federal Court.[12] One reason may be that before the introduction of the formal GI system, some winemakers in Australia were organising themselves on a regional basis and so perhaps there was, at least in some regions, a customary sense of boundaries. For example, there were about six regional wine associations in Victoria in the 1980s. Also in the 1980s some winemakers had organised an appellation program in Tasmania.

Obviously a GI system for food would require a GI register and a registrar to administer it. These are relatively modest salary costs. More significant are the costs of maintaining a compliance unit to monitor obligations under whatever GI system was designed for food. In the case of wine GIs there are only a small number of compliance inspectors and audit costs are passed on to producers. In the case of the wine industry the costs of an individual audit relate to compliance with the Label Integrity Program and various other regulations and do not relate just to the GI rules. If a food GI system was to be implemented the costs of audits to individual growers for compliance with the rules would very likely be incorporated into the cost of audits these growers already pay for their participation in various quality management and environmental systems. In short, existing audit systems could be adapted for use in checking records for compliance with the food GI rules.

5.3.1.5 Has the Wine GI Game Been Worth the Candle?

As the quote from AGWA in the previous section suggests, many people in the wine industry believe that GI regions are important. How important turns out to be difficult to say, and the reasons vary between regions and participants. Two

---

[11]The Australian Grape and Wine Authority (AGWA), trading as Wine Australia, is the single Australian Government statutory service body for the Australian grape and wine community. Regulatory activities available at https://wineaustralia.com/en/Production%20and%20Exporting.aspx as consulted September 2nd 2016.

[12]Another example is *Baxendale's Vineyard Pty Ltd v The Geographical Indication Committee* (2007) 160 FCR 542.

perspectives on wine GIs emerged from the interviews. One was a winemaker's perspective that, not surprisingly, looked to the role of GIs in helping to define and communicate the connection between the region and the wine. The other was a marketing perspective in which assessing the value of GIs comes down to answering one simple question. As one interviewee put it, 'does the consumer care?'.

If we look at consumer preferences as a guide for the importance of GIs, separating the influence of region from a range of other purchasing influences such as age, awards, brand, expert or other recommendation, price, previous experience and so on, it is not clear where region sits in a ranking exercise. The studies suggest that region is probably recurrently important, but how important is difficult to isolate (Goodman et al. 2006). The answer is likely to vary depending on which segment of the consumer market is being studied. Those we interviewed tended to think that GIs were needed above a certain price point: 'From $15.00 onwards you need a region'. Others nominated higher price levels. As prices for wines increase, winemakers can 'make them taste like where they are grown'. GIs were also seen as being important in export markets, especially as the rise in the Australian dollar began to affect the ability of wine producers to compete in the lower end of the market.

Our interviews with those in the wine industry who took a marketing perspective on GIs tended to confirm the difficulty of isolating the effects of regions on consumer purchasing decisions. Some interviewees made the point that industry surveys showed that the recall by Australian consumers of even high profile GIs, such as the Barossa, the Hunter Valley, and Coonawarra, is weak and essentially non-existent for many lesser Australian GIs. Amongst consumers in international markets there is also very little recall of Australian GIs. The survey results we were told about would have many winemakers reaching for antidepressants. There is 'massive ignorance and confusion out there' when it comes to GIs with consumers, according to one survey, thinking that Chardonnay and Jacobs were GIs. In Europe, consumer awareness of the European GI scheme was also reported as being very low with only 8% of consumers recognising the relevant symbols (European Court of Auditors 2011, p. 25). This is a telling example as it shows how difficult it is to familiarise consumers with new GI label categories (in this case the Protected designation of origin (PDO) and Protected geographical indication (PGI) symbols).

The interviews also suggested that the export success of Australia's wine industry in the 1990s was built around creating an image for Australian wine, providing consumers with information about the variety of grape they were drinking and the application of modern brand building techniques that allowed the winemakers of the New World to communicate with European consumers in ways that the Old World looked upon with disdain. The marketing of the Yellow Tail wine is a good example of the kind of creative strategy that was employed. Yellow Tail is reported to sell more in the United States market than all French producers combined.

Australian GIs appear not to have been central to this marketing strategy. The Australian wine export figures for the period 1989–1998 show remarkable growth;

an average increase of 26% per year (Hira and Aylward 2013, p. 402). They also show that Australia's export success was occurring well before the introduction of the GI system. GIs functioned as a Trojan horse allowing the Australian industry to gain greater access to European consumers. Once in the market, the combination of quality, price and marketing of Australian wine did the rest.

It is probably also fair to argue that GIs became progressively more important in the Australian wine industry's strategic thinking as it considered its future as an industry. In its 1996 publication *Strategy 2025*, AGWA identified 'wine regions as defined by geographic indications as a brand marketing opportunity'. The growing emphasis on GIs and regions can also be seen in its 2007 publication, *Directions to 2025*:

> Today, Australian wine is rightly best known for its Brand Champions. We will be able to consider the job near-done when we can say the same for our Regional Heroes - wines which reflect the remarkable number of successful combinations of classic grape varieties with Australian wine regions.

And even more recently AGWA, outlining its strategic plan for 2015–2020 in a discussion paper, indicated that 'increasing the demand and premium paid for Australian wine' was a key strategic priority to be underpinned by research into 'Australia's unique terroirs' and compliance with its labelling laws (AGWA 2014, p. 8).

There was a strongly held view amongst interviewees that having too many new GIs could create marketing problems. Australia's well-known GIs were reputable wine regions before they became GIs. The Barossa Valley, McLaren Vale, the Hunter Valley, the Yarra Valley and Rutherglen grew in prominence as wine regions throughout the 20th Century. These regions progressively changed from an exclusive reliance on producing cheap fortified wine and bulk wine, to making improved quality table wine, particularly as better grape varieties were developed and the industry placed more emphasis on technological innovation of all kinds (Hira and Aylward 2013, p. 401). When Australia adopted the GI system major regions had the benefit of sunk reputational costs. Taking a largely unknown region and turning it into a consumer brand is a much more difficult and expensive exercise irrespective of whether or not the region is defined by a GI. Some GIs were described as 'unwieldy and impossible'. Thinking about how to lift some of Australia's GIs out of anonymity remains a marketing challenge for the wine industry.

Our interviewees, whether speaking from a winemaker's perspective or a marketing perspective, were generally not in favour of a further sub-delineation of Australia's existing wine GIs. For example, Tasmania has been a registered GI since 1994 with no sub-regions appearing on the register. The Tasmanian winemakers we interviewed saw no gains in the creation of GI sub-regions. They saw Tasmania as a strong and recognisable brand. They characterised their industry as a small industry that had to work together if it was to grow and keep the focus on producing quality wines. Their worry was that sub-regional GIs might have a splintering effect on an industry that could not at this point in its evolution, lose the cooperation it had achieved: 'At this stage of our maturity we've got more to lose than to gain by going our own separate ways'.

The wariness about sub-regional GIs was not confined to Tasmania. The view that Australia had enough wine GIs was reported by our interviewees to be a general view within the industry. That said, some interviewees in Tasmania and elsewhere did hint at the desire of some smaller winemakers to have their own identity, suggesting that there is at least some debate within the industry about sub-regional GIs. An overall view from our interviewees was that while sub-regions could certainly be identified, something not surprising given the size of Australia's wine GIs and the diversity of soils within those GIs, at this stage of the industry's growth it was better not to formalise those boundaries.

The more general lesson here is that if an industry is to work successfully with a GI system it will have to arrive at a general consensus about its use, avoid its potentially destabilising effects and make it an ongoing part of its strategic planning objectives. The Australian wine industry has many levels and institutional actors from the public, private and research sectors that produce internal circles of deliberation and debate about its future (Hira and Aylward 2013). From those internal circles it has been able to forge common strategic objectives that include an increasing use of GIs to meet the goal of increasing the premium for Australian wine. If the potential benefits of GIs as regional assets are to be harnessed by other industries, then GIs will need to be part of the broader planning dialogues that take place within a region, a theme we return to in our conclusion.

The Australian wine industry is part of an industry that has continued to globalise. Here globalisation refers to the expansion by Australian companies into export markets and the incorporation of these companies into more complex corporate structures in which ultimate ownership may be vested in a foreign multinational. By way of example, Orlando Wyndham, which owns some of Australia's most recognisable wine brands such as Jacob's Creek, is owned by the French Pernod-Ricard Group (Australian Wine Online n.d.). Accolade Wines, the largest company in Australia by grape intake, has its corporate origins in Thomas Hardy and Sons and a lineage of mergers and acquisitions that has turned it into a global wine company with markets in some 80 countries. Treasury Wine Estates, the corporate product of the Fosters' demerger of 2011, has more than 50 brands as part of its portfolio, including brands located in Australia, New Zealand, the United States and South America.

The future of wine GIs in Australia and globally is more likely to be influenced by decisions in the designer boardrooms of multinationals than in the cramped offices of the many hundreds of small wine producers in Australia, 75% of whom crush less than 100 tonnes per year. Of Australia's 2573 wine companies the top four account for 48% of the national crush. It seems clear that the future of Australian wine GIs will be influenced by calculations about global demand, production and corporate managers' perceptions about the importance of region in consumer purchasing behaviour. One view might be that those wine GIs with a reputation for quality in which large corporate players have a stake will continue to attract investment as part of a strategy of maintaining a quality portfolio of wine brands, but that the support for new GIs will be modest at best.

However, our study is not focused on where GIs rank in terms of influence on consumer behaviour or on the role that they play in portfolio decision-making by multinationals, but rather on whether GIs might confer benefits for regions. On this issue the interviews suggest that some wine GIs have generated two effects that may bring regional benefits. One is what we call a reciprocal spillover effect and the other is an investment-in-quality effect. The next two sections expand on these effects.

5.3.1.6 Reciprocal Spillovers

As part of our investigations for this case study, we examined a wine GI in the Granite Belt, Queensland. Wine GIs do not commonly occur in Australia's north, but the Granite Belt, because of its elevation (600–1000 m) and soils, is quite suitable for viticulture. Centred on the town of Stanthorpe in Southern Queensland, the Granite Belt example suggests that a registered GI may create regional development benefits beyond that of other more generic region specific, provenance-branding options.

Queensland does not enjoy a great reputation for wine because its tropical and sub-tropical climate and soils are largely unsuitable for growing wine grapes. The granite soils around Stanthorpe area are an exception. The problem for the winemakers in the region was the reputation of Queensland wine; as long as they were under the banner of 'Queensland wine', success was limited. Queensland was simply not known as a traditional grape-growing area. Winemakers had to be skilled in creating interest in their product first and not mentioning Queensland in their marketing. One winemaker described how they explained the location of their winery in the Granite Belt to customers, commenting, 'notice how I don't use the word Queensland'. The registration of the Granite Belt GI in 2002 was seen as an important breakthrough: 'it's given us our own identity'.

In the interviews the wine producers were very clear that it was the registration of a GI that had been crucial to their success. It created the opportunity for them to escape the poor reputation of Queensland wines. The Granite Belt GI became the basis for a marketing strategy for wine that allowed winemakers to distinguish their wine from the rest of Queensland. It also became the basis for creating a broader identity for the region as a tourist destination. Importantly, there was an organisational vehicle that helped to back the promotion of the Granite Belt in the form of Granite Belt Wine and Tourism (GBWT), a peak local body representing wineries and other tourist operations.

According to interviewees the GBWT works with the GI boundary as a guide for its promotional activities. These activities are funded from member levies and fees, with businesses being able to buy different promotional packages. Critical to the stability of the GBWT has been finding a fee structure acceptable to the different types of businesses involved. Establishing a regional identity was described as a 'slog'. The challenges include settling on an identity since there are always options, alternatives and a diversity of views.

The Granite Belt wine GI seems to have played an important role in the region's growth by helping winemakers gain an independent identity and allowing them the opportunity to forge a reputation for the quality of their wines that was not distorted by an association with Queensland. As the region's wines began to achieve recognition for their quality, tourists became increasingly interested in visiting the region to sample the wines. This in turn generated demand for food and accommodation services; services that both drew on the success of the wine industry and contributed to its drawing power. In this particular case the GI was integral to a process of reciprocal spillovers in which the benefits of the wine GI spilled over into various other food and tourist ventures, which in turn created benefits for the local wine industry. This process of reciprocal spillovers has continued to intensify with the establishment of the Queensland College of Wine Tourism in the region, a specialist education and training provider relating to the wine tourism industry.

The Granite Belt case suggests that a GI may, under a set of conditions, help a region gain the benefits of reciprocal spillovers that might not otherwise occur. In the case of the Granite Belt two conditions emerged as crucial: distance to a large urban centre (Brisbane can be reached in three hours by road) and the creation of an owner-operated regional marketing organisation (the GBWT). In terms of the proximity to Brisbane, what emerged from our interviews was that the marketing focused upon the Granite Belt as a weekend destination, rather than a day-trip destination. This had important implications of course for the accommodation businesses in the region. Without a regional marketing body the Granite Belt region would have enjoyed much less success. As one non-wine producer who was paying $1600 to GBWT pointed out, 'where would $1600 go if we had to market ourselves?'.

Proximity to tourism was mentioned as a factor in other interviews in other regions. In the Yarra Valley one winemaker pointed out that Yarra Valley shiraz grapes commanded a higher price than Grampian shiraz grapes (the difference being in the range of $400–$600 per tonne), despite the premium excellence of the Grampian grapes. They attributed this to the fact that the Grampians did not get the tourist exposure of the Yarra Valley. Conversely, if we consider the situation where no Granite Belt GI was registered and the winemakers in the region continued to operate under a Queensland identity, then on their own account they would have continued to struggle and the region would have attracted fewer tourists.

The Granite Belt case example points to the possibility of a GI triggering or accelerating a process of reciprocal spillovers. It is not the only case study where we encountered the spillover process. In the Yarra Valley we heard that the success of Yarra Valley wine in China had also triggered an interest in other Yarra Valley products such as cheese.

One important issue is whether a product other than wine can trigger this spillover process. Wine is a complex cultural product. It can be a bulk commodity (beverage wine), as well as a Veblen good (fine wine), a term used by economists to describe the purchase of goods which are linked to social status. In the case of goods with Veblen attributes consumers are prepared to pay more because these goods have status effects. The number of food products like wine that can generate strong reciprocal spillover effects may be quite restricted. It seems implausible, for example, to think that a GI for Granite Belt apples would generate a process of reciprocal spillovers of the same intensity as one for wine. For Northern Australia it is clear that the role of a signature or iconic GI cannot be filled by grape wine. That said there is an emerging tropical fruit wine industry in North Queensland, the Northern Territory and Western Australia with good export potential into markets such as China and Japan. GIs would be of long-term importance to this industry. While the possibility exists that a GI based on a non-wine iconic regional or local product might generate similar spillover benefits for local tourism ventures and for other local products, finding plausible candidates for this iconic status remains a challenge.

5.3.2 Case Study 2: The King Island Story—Have the Cows Bolted?

Our interest in King Island off the coast of Tasmania as a case-focus site arose from two factors - its geographical significance in being a naturally defined and bounded space as an island, and its reputation for iconic products (in dairy and beef). Place importance and iconic produce are both classic attributes of GIs in terms of potential to place-brand uniqueness, and thus we were keen to examine the King Island 'story' in some detail.

5.3.2.1 King Island Agriculture

King Island has an economic history underpinned by agrifood production, primarily in dairy and beef (Coates 2014). In 2011, total gross value of agriculture on King Island was $31 million; 66% of which came from beef production and 28% from dairy (Coates 2014). King Island's population in 2011 was approximately 1500 people, with roughly 700 in employment; 200 of these were employed in agriculture.

Interestingly, agriculture on King Island has not grown necessarily from any comparative advantage of good soils and climate attributes such as high rainfall, but from political imperative. As Coates (2014) points out, King Island was decreed and settled as a farming community as a result of the distribution of soldier resettlement blocks in the early part of the 20th Century. There has also been a significant history of government support and intervention on King Island (Coates 2014). More detail and a different perspective on the KI story are provided in Chap. 11.

5.3.2.2 The Beef Story

King Island beef is a highly sought after and iconic product with a reputation for high quality and excellent taste, attributable to its longer growing season and the absence of extreme summer temperatures to degrade pasture:

> The longer growing season means we have more control over the quality. We are grading in the top 10% for pasture-fed beef and this means KI beef brings around a 40% price premium on the shelf (Respondent #105).

King Island has some obvious disadvantages in terms of agricultural production, the most significant of which lies in its isolation. Similarly, the cost of power and ageing port infrastructure (Coates 2014) are factors that impact on potential profitability of agricultural enterprises on King Island. Combined, these factors add significantly to production costs. Being able to attract a premium price for a recognised high quality product is essential to maintaining market share and the viability of production on King Island, as our interviewees recognised and pointed out. The logical response to these challenges therefore lies in ensuring that value-adding is maintained in situ as much as possible and to transport 'finished' products to market. This is essentially what was occurring for a number of years, where beef cattle were slaughtered through the abattoir located on King Island.

The abattoir had been run initially as a subsidised enterprise, with heavy investment from the Tasmanian Government. In the early years the abattoir processed the meat but did not have any role in marketing or purchasing of the product (Coates 2014). A relatively tumultuous history followed, with the abattoir firstly leased by private interests in the 1980s (King Island Exports), until its subsequent sale—initially to a Tasmanian enterprise, RJ Gilbertson Pty Ltd, which upgraded the facilities in the 1990s resulting in production growth. SBA Foods bought the business in 1996, focusing on marketing to Japan, and then sold on to the Tasman Group, who developed a relationship with Coles. In 2008 South American TNC JB Swift (JBS) acquired the Tasman Group's six Australian plants, further developing the relationship with Coles but ultimately closing down the abattoir in 2012 (Felix Domus 2013, p. 18). Reportedly, all operators continued to seek support from the Tasmanian Government in relation to the ongoing operations of the abattoir, in the form of either subsidies to support local jobs, or (as was the case for JBS) a no-interest loan to support a compliance upgrade in 2011 (Coates 2014). Despite this, having closed the on-island facility in 2012, JBS moved its processing operations to Tasmania. We now turn our attention to the perceptions of our interviewees, with regard to the impact of these events on their businesses.

Our interviewees were very aware of the challenges facing the beef industry on King Island, and were able to point to critical periods and decisions which have impacted upon the ongoing sustainability of their livelihoods. There was firstly recognition that the development of the dairy industry on King Island and the production of high quality cheeses had a major impact on the development of the beef industry: 'We were lucky there for a while, really. Beef tagged along behind KI Dairy and we benefited from the associations of quality. Dairy led the way,

I would say'. This is an example of the reciprocal spillover process that we referred to in Sect. 5.4.

A further example of the impact of critical decisions appears to be with the entry of a second processor of beef from King Island, for example:

> Yeah, we had a choice then, when [processor] showed an interest in our beef. They would subsidise our transport costs to get the cattle processed by them on Tasmania. The split between producers on the island using the local abattoir and [processor] was about fifty/fifty, I reckon.

This producer talked about the efforts of some to get local producers to 'stick together' and to maintain their 'loyalty' to JBS and the local abattoir, and their belief this was critical for the good of the industry on King Island:

> We did talk about it, yeah, but it's hard to do that when you're offered a better price elsewhere. I can't blame them. You're under pressure to provide for your family and you do what it takes. Some of us were concerned that it wasn't going to be good in the long-term though, and that's exactly what happened. JBS closed down in 2012. Of course, the subsidy from [processor] only lasted until JBS shut down.

Our interviewees reported on the subsequent challenges faced by the beef producers on King Island and the major impact the closure of the local abattoir has had on them. They related efforts to develop a local beef producer's cooperative and plans to either redevelop the local abattoir (JBS refused to sell) or to develop a new, smaller multi-species abattoir ($30 million estimated costs were prohibitive). Interviewees were very aware of the value of their product, and the value of the 'King Island' name; 'KI brand puts a floor in the price'.

They also relayed stories of 'passing off' and concerns about 'losing control' of the King Island brand:

> Even if we could get a local abattoir up and running, JBS won't let us use the 'King Island' on our beef. They own that name. Without the name, how do we get the premium price?

> Have you ever heard of 'King Island rabbits'? There was a story in the paper about them. There are no rabbits on King Island! We worry about things like that – people using the King Island name.

> We also heard about [major retailer] selling 'King Island beef stock.' Apparently the beef is coming from [processor], but how do we know it is all King Island beef that it is getting made from? They don't just buy beef from us. There's no way to keep track of it.

5.3.2.3 The Dairy Story

There is a long history of dairy production on King Island with the first operation being built in 1902. Coates (2014) reports on a patchwork of successes and failures and tells of an overall bleak economic outlook that included a prediction by the Bureau of Agricultural Economics (1977) that 'sustained and substantial' financial assistance would likely be required to support dairy production on King Island for the longer term.

In 1978, King Island Dairy was purchased by a local entrepreneur, Bill Kirk. Kirk, with the assistance of the Tasmanian State Government through a loan arrangement established the dairy as a viable business (Coates 2014). Improvements aimed at producing high value, high quality products appeared to be the focus of Kirk's business strategy and our interviewees described him as:

> An entrepreneur. In the first place, he chased up delis and specialty shops [and] didn't go down the usual road for distribution. He started to build the reputation and to establish the market that way.

However, as described in the introduction to this case study, the dairy became less viable over time, and was subject to many of the challenges discussed, not least of which was the deregulation of the dairy industry in 2000. Increasing production costs and global competition saw diminished production on King Island, and with it, a shifting of any power to negotiate that farmers may have previously had:

> South Cape [cheese] used to be made on King Island from second grade milk. Then it moved to Burnie, so it doesn't make sense to use the milk from King Island anymore. We're really price takers now. It used to be that we could negotiate our contracts [with King Island Dairy]. This year though, it pretty much just got slapped on the table. There was no room to negotiate. As dairy farmers, we don't get any more for our milk than any dairy farmer on the mainland selling to anyone, even though our milk goes into premium products.

There has been a need to seek external investment to support King Island Dairy, which has seen it change hands, a number of times. Currently, it is owned by Lion Dairy and Drinks, a subsidiary company of Lion, having been acquired by National Foods in 2001/2. Lion Nathan National Foods was formed in 2009, when Lion Nathan was purchased by Kirin Holdings Company Ltd and merged with National Foods. Lion was restructured in 2011, to become one company with three businesses: Lion Beer, Spirits and Wine Australia; Lion Beer, Spirits and Wine New Zealand; and Lion Dairy and Drinks (Lion n.d.). Lion has invested significantly in building an iconic brand in 'King Island Dairy.' Currently, the King Island Dairy website makes much of the island location, climate and environment and their collective influence on the quality of the products produced by the business:

> No ordinary island. No ordinary cheese. Guarding the stormy western entrance to Bass Strait is a lonely, rugged island. Situated at 144° longitude and 40° latitude, this isolated place is cut off from the rest of the world by jagged reefs, raging seas and howling Roaring Forties winds. With mineral rich soils, cool annual temperatures, abundant rainfall and a consistent salt spray thanks to the constant westerly winds, the environment contributes to quality milk production which helps us create some of the finest cheeses on earth. Welcome to King Island.

Our subsequent interviews with representatives of Lion reinforced the importance of provenance to their product, and the conscious decision made by the company to differentiate King Island products as high quality products aimed at discerning consumers from the premium end of the market: 'We would never ever use milk that wasn't from King Island, for King Island products. There's a lot of rigour behind our claim'; 'Maintaining the integrity of the brand is critical to maintaining the value'.

5.3.3 Discussion

There are a number of significant points that both these case studies emphasize. First, the certainty that GIs bring to the placement of boundaries appears to have played an important contributory role in attracting investment in quality in the case of wine. Wine GIs in Australia, subject as they are only to rules or origin and not additional quality standards, do not by themselves trigger an interest in the production of quality wines. Nevertheless once a GI precisely defines an area with a reputation for the production of quality wines it also defines who is a member of the resultant community of shared commercial fate and who is not. A winemaker who produces a poor quality wine and who is in a region known for its quality wine generates a negative externality for the region. A winemaker who wins prizes generates a positive externality for their region. Our interviews with winemakers from regions with a reputation for premium wines indicated that all were sensitive to the negative and positive effects that individuals could have on a region's reputation.

Second, informal mechanisms and practices have evolved amongst winemakers to try and ensure that the production of wine in a region does not fall below a certain quality. A common practice is the use of tasting groups or clusters in which winemakers bring along samples for evaluation by their peers. The goal is constructive criticism aimed at improving the wine prior to bottling. Wine competitions also seem to play a role in creating a culture of quality. Winemakers care about winning prizes. The esteem in which their peers hold them and the broader industry matters to them. There are also the more obvious benefits that accrue from winning prizes – the capacity to price accordingly. Wine competitions are places where they can find out from judges in informal conversation why they did not win or place and how they might improve. The leadership of more established wineries in a region is also important. These wineries keep a weather eye out for newcomers, finding ways to give them advice about what works best in the region. The response is generally positive: 'nobody wants to make bad wine—they jump on board immediately'.

In the second case study, one can see that for the dairy farmers on King Island, there are positive outcomes from Lion staking its reputation on King Island products. Investing in the brand and building reputation means greater security of the market for King Island products and thus the ongoing need for a consistent supply of milk from King Island cows. King Island Dairy (by its own admission) is entirely dependent on locally produced milk. However, because of the isolation of King Island, dairy farmers have no alternative but to sell their milk to King Island Dairy as the sole operator on the island. Thus they are also completely dependent on King Island Dairy and this dependency has had the impact of ensuring they remain price takers rather than becoming price makers, as one would normally expect with producers of high quality inputs to differentiated, premium products. None of the benefits associated with premium production are accruing to the farmers; 'we don't get a premium price on milk on King Island. We have our price aligned to a competitor's price'. The relationships between farmers and their buyer are purely transactional, and not embedded in the social relationships (Granovetter 2005) of the island. Granovetter (1973, 1983, 2005) argues that strong social ties create trust norms which can mitigate the impacts of power imbalances in business relationships. This is an important point, and one very well understood by our interviewees:

> We do have a pretty good relationship with the local guy, but he can only do what his masters tell him to do and they don't care. Actually makes it a bit hard for him here.

In terms of the regional economy, the dairy industry has become extractive, in that rents that may have previously contributed to the local economy under local ownership of the dairy are leaking from the region.

Continued market recognition of the 'King Island' brand provides a measure of safety for these regional producers. However, such safety is also limited by the fickleness of the market. A large player like Lion pursues efficiencies and market share on a scale measured in tens or hundreds of millions of dollars. Stopping the production of particular lines at the King Island Dairy for cost reasons or placing more of the emphasis on the cheese factory located at Burnie in order to exploit the Tasmanian brand are all part of restless, ever-changing market calculations. There is no doubt that should King Island Dairy products fail to make market share or lose their 'iconic' status, then the rational response by Lion would be to withdraw from King Island. Should this occur, then the 'King Island' name (in relation to dairy products) is arguably lost as well since a future investor would have to deal with the trade mark ownership of 'King Island Dairy'.

Arguably, it is the potential for local initiation and local control that are the most important features of a GI for regional SMEs and communities of which they are a part. Local control that can mitigate the impacts of power imbalances makes GIs important (in specific circumstances) to regional economies. Maintaining local control of livelihood assets including land, infrastructure and livestock in relation to both dairy and beef on King Island for example, through collective ownership of the brand 'King Island' would be of great benefit to the people of King Island, in a way that a monopoly-like ownership of the trade marked name cannot guarantee. GIs, as a public and locally embedded property right, therefore have the potential to mitigate the 'either/or' choice: *either* external investment *or* local control. With the option of having a GI registration available as a tool, there is potential to have both external investment (as has been essential in the case of King Island and for many other regional economies) while maintaining some measure of local control (which has been significantly eroded over time on King Island).

In the case of the beef producers there is a realisation that a more cooperative approach to marketing is going to be crucial to their farm businesses. As our interviewees pointed out, the widespread uptake of quality assurance programs by beef producers around Australia is having a 'levelling effect' on prices, a point that has resonance for many Australian agricultural sectors both nationally and globally.

Having a strong brand is one of the key ways in which to counter this levelling process. At the time of our interviews some producers were looking at registering a trade mark, a process that has required them to look at the legal impact of JBS

Australia's ownership of the King Island Beef trade mark (TM No. 1385474). Had the GI option been available to beef producers at some earlier point in time these kinds of complications could have been avoided. Moreover, through the use of a GI, beef producers could have settled issues such as how long cattle would have to remain on King Island in order to be classified as King Island beef, rules that would have also benefited Australian consumers.

Of course, if the GI option had been available there is nothing to say that beef producers would have chosen to register a Beef GI for King Island. Planning to create a long-term regional brand asset does require leadership, a willingness to find common ground amongst people who come to meetings with different mindsets. Nevertheless it would have been an option on the table to be argued for by those on King Island who were looking for new ways in which to retain the value from the King Island name. To some extent a GI does represent a test of a region's social ties. If 'people are not as tight as they think', a GI will not save them.

GIs would clearly have been a useful option for the dairy and beef industries on King Island to consider at an earlier point in their history. It may be that they still have a role to play in future value chain dynamics should they become available. A 2013 feasibility report, which discussed the difficult economics of establishing a new abattoir, concluded that ultimately the key for better financial conditions for beef producers lies in the development of supply chain options that 'maximises their share in the value of the King Island name' (Felix Domus 2013, p. 69), marketing rather than processing has to be the focus. With considerable experience of the role of trade marks in the hands of multinational processors, as well as how decisions in the headquarters of those processors directly impact on the viability of their businesses, beef producers on King Island might well throw their support behind a Beef GI, if it were available, aiming at a GI that has strong credibility with consumers and providing producers with leverage in negotiations with other players in the value chain. While the registration of such a GI might have to face objections from existing trade mark owners such as JBS, David might well triumph over Goliath, backed by good lawyering and effective social media campaigns.

As we have suggested GIs may help with King Island's existing industries. They would also be relevant to any new industry on the island. Our interviewees indicated that the future may well hold such an industry and that 'confidential' discussions were occurring. Hypothetically, if entrants into this new industry were able to work cooperatively to build the provenance branding backstory for their product —for example, 'high quality, uniqueness, clean, green' attached to King Island, then they may have some measure of control over the development of the subsequent business model. The GI affords a collective good where (potentially) higher returns accrue at the local scale, rather than the private good of the trade mark, where value accrues to the trade mark owner. Capturing value and the associated economic rents and retaining these in the local economy may well be enhanced by the implementation of a GI, should the legislative option become available.

5.4 Conclusion

We argue that it is the joint-asset nature of GIs that may encourage the emergence of collective cost-sharing approaches to marketing by farmers and regions, that may well increase options for greater branding freedom. However, there is an element of timing that must also be considered. It would seem that consumer awareness and the demand for provenanced food and beverages is growing (Hurst 2007). In the past 10 years or so, we have seen a proliferation of retail marketing practices aimed at making a direct connection between consumers and producers. One such example is the national 'Jamie Oliver' campaign implemented by Woolworths. Other examples include the growing 'slow food' movement and 'eat local' campaigns, together with an increasing number of farmers' markets. All are responding to a growing middle-class consumer preoccupation with understanding where, when and how their food is produced. In such an environment, the direct attribution to place and quality that can be captured in a GI relatively easily may well appeal to consumers, and provide benefits to producers of the goods so marketed.

Where a region has committed to a quality product, a GI boundary creates certainty for future investors. They know precisely where they must buy in order to gain the benefit of the GI. For example, winemakers who want to expand their business into premium wines benefit from having precisely defined regions that are associated with quality winemaking. A wine company such as De Bortoli, which has its headquarters in Griffith, has expanded into the Yarra Valley, as part of a strategy to increase its production of premium wines. The Yarra Valley GI is seen as generating an important halo effect. The more general point here is that once a GI region has achieved a reputation for quality, the embeddedness of the GI means that investors have to invest in the region if they wish to benefit from the GI.

We contend that GIs should be a part of the policy options for development for Australia's regions. Obviously we are not claiming that all regions or all industries will benefit or see benefits in GIs. Many variables affect whether or not a GI system will bring regional benefits. We saw for example, that some wine GIs provide regional benefits in the form of reciprocal spillovers. In other cases having a wine GI in place may help to drive further investments and improvement in the quality of the wine produced within the bounds of the GI. The case of wine GIs also shows that even if GIs are not important to an industry at one stage of its development they may become so. GIs with stricter standards may assist in combatting the free rider problem which otherwise deters quality-enhancing investment.

We conclude that there is not a uniform story to tell. Thus, Tasmanian whisky producers might benefit from the advantage of having uniform and mandatory regulation as to what can legitimately be referred to as Tasmanian whisky. Mareeba mango growers might benefit from delineating precisely the boundaries of Mareeba, and from taking advantage of the established reputation of their mangos in China more effectively. Beef producers on the Atherton Tablelands could benefit from specifying the exact boundaries of the natural grasslands that survive on high rainfall and volcanic soil.

GIs themselves do not trigger greater cooperation, but they do place a boundary around those who are potentially part of the conversations about regional identity and differentiation for given products. The effectiveness of such conversations and subsequent planning will in part depend on the strength of the social relationships that exist within the region and the capacity for such relationships in turn, to drive agency through competitive advantage. In consideration of such competitive advantage, can a GI branding strategy enhance and maximise those competitive advantages? Can the development networks and agencies in a region deliver the kind of persuasion and cooperation that is needed to implement such a strategy? The answer to these and other questions will vary from region to region. The asset-capital in regions, i.e. those stocks of human, built, social, environmental and financial capital that a region has at its disposal vary greatly, and the capacity to subsequently convert such capital into agency that can in turn maximise competitive advantage also varies.

However, what attracted some of those we interviewed to the idea of exploring a GI system for their region was the fact that a GI would not be privately owned but would remain embedded in its defined territory, serving as an asset for producers and (in the case of reciprocal spillovers) for others in the region. In that sense, it could become in itself part of the available regional asset-capital. For some regions there may be an advantage in thinking about ways in which to use GIs and quality specification standards to communicate in a credible way a commitment to 'clean and green' production. Michael Porter's influential book, *The Competitive Advantage of Nations* (1990), argues that firms will often do better by locating to jurisdictions where tough regulatory standards create the opportunity for the long-term improvement of products and processes. In a world where consumers are becoming interested in both product regionality and environmental sustainability, a GI system offers a region a credible way of signalling to consumers and to companies that it is a region that should be part of their purchasing and investment decisions.

A food GI system for Australia that maximises the possibility of benefits to Australia's regions will require creative design, that is to say, one that takes a 'New World' approach rather than simply following an 'Old World' model. It should be a model that draws on the insights of regulatory scholarship such as Porter's analysis of competitive advantage, as well as one that draws lessons from Australia's experience with the wine GIs. As QA, Food safety, EMS and other agricultural regulatory schemes proliferate globally, any competitive advantage gained through compliance will be diminished as more producers are instructed into such systems. Moreover as these systems are rolled out in developing countries by large investor companies, producers in those countries will reach compliance standards much more quickly than Australian producers might anticipate, thus further eroding any premium that may have accrued to Australian products in such markets.

In a world saturated by such systems, the importance of branding as a source of comparative advantage will increase. For farmers and regions the imperative to plan around branding has never been greater. However, there is no easy way out of such structures for many Australian farmers. Perhaps as one farmer said to us after we described the principles of GIs 'it's all a pipe dream'. Others were more optimistic seeing a chance to improve in some modest way their position in the value chain. The stakes are high for Australia's regions.

Finally, for Australian producers and the regions in which they are embedded, then, we argue that GIs can be contingently beneficial. They may suit some industries at a particular point in their evolution or they may bring benefits to some regions but not others. The important point is that like other policies for regional, rural and remote Australia, there is no 'one-size-fits-all' broad-brush application or design strategy that will fit all regions or industries. We suspect that there are sufficient analogies between Australia and some other countries with similar agricultural and regional conditions for our conclusion to have significant resonance in other countries as well.

Acknowledgements Much of the work presented in this chapter is drawn from research conducted between 2010–2015. This research was funded in part by the Rural Industries Research and Development Corporation (RIRDC) and first published in 2015 in two reports.[13,14]

We thank RIRDC for their kind permission to reproduce parts of those publications here. We also gratefully acknowledge the work of our colleagues Professor Peter Drahos for his contributions to the first work, and Ms. Madeline Taylor and Dr. Brenda Marshall, for their contributions to the second work.

Bibliography

ABS 2002, *Agriculture, Australia*, Cat. no. 7113.0, Australian Bureau of Statistics, Canberra. Ahn, T & Ostrom, E (eds) 2003, *Foundations of Social Capital*, Edward Elgar, Cheltenham.

Akerlof, GA & Kranton, RE 2005, 'Identity and the Economics of Organisations', *Journal of Economic Perspectives*, vol. 19, no. 1, pp. 9–32.

Alston, M 2004, 'Who is down on the farm? Social aspects of Australian agriculture in the 21st Century', *Agriculture and Human Values*, vol. 21, pp. 37–46.

Anderson, J 2004, 'Dairy Deregulation in Northern Queensland: The End of Traditional Farming?', *Anthropological Forum*, vol. 14, pp. 269–82.

Andrée, P, Dbiden, J, Higgins, V & Cocklin, C 2010, 'Competitive Productivism and Australia's Emerging "Alternative" Agri-food Networks: Producing for farmers' markets in Victoria and beyond', *Australian Geographer*, vol. 41, no. 3, pp. 307–22.

Augustin-Jean, L, Saavedra-Rivano, N & Ilbert, H 2012, *Geographical indications and international agricultural trade: the challenge for Asia*, Palgrave MacMillan, Great Britain.

Australian Bureau of Statistics 2013, *National Regional Profile: King Island (Statistical Area Level 2)*, http://www.abs.gov.au/AUSSTATS/abs@nrp.nsf/Previousproducts/604031093Industry12007-2011?opendocument&tabname=Summary&prodno=604031093&issue=2007-2011 viewed 3 November 2014.

[13]Van Caenegem, William, Drahos, Peter & Cleary, Jen (2014). *Provenance of Australian food products: is there a place for Geographical Indications?* Rural Industries Research and Development Corporation, Research Report No. 15/060. Canberra, ACT.

[14]Van Caenegem, William, Taylor, Madeline, Cleary, Jen, Marshall, Brenda (2015). *Collective Bargaining in the Agricultural Sector.* Rural Industries Research and Development Corporation, Research Report No. 15/055. Canberra, ACT.

Australian Government, The Australian Grape and Wine Authority (2014), Geographical indication Guidelines, A guide to labeling with geographical indications.
Australian Wine Online n.d. 'Australian Wine Industry Overview', http://www.winetitles.com.au/awol/overview/wineries_orlando.asp, viewed 31 October 2014.
Ayres, I & Braithwaite, J 1992, *Responsive Regulation*, Oxford University Press, Oxford.
Bain, J S 1956, *Barriers to new competition*, Harvard University Press, Cambridge.
Banerjee, R & Majumdar, M 2011, 'In the mood to compromise? Extended protection of geographical indications under TRIPS Article 23', *Journal of Intellectual Property Law & Practice*, vol. 6, pp. 657–63.
Barham, E 2003, 'Translating terroir: the global challenge of French AOC labelling', *Journal of Rural Studies*, vol. 19, pp. 127–38.
Bashaw, B 2008, 'Geographical indications in China: why protect GIs with both trademark law and AOC-type legislation?', *Pacific Rim Law & Policy Journal*, vol. 17, no. 1, pp. 73–102.
Benz, M & Frey, BS 2007, 'Corporate governance: What can we learn from public governance?', *Academy of Management Review*, vol. *32*, pp. 92–104.
Beverland, M 2006, 'Can co-operatives brand? Exploring the interplay between cooperative structure and sustained brand marketing success', *Food Policy*, vol. 32, pp. 480–95.
Blakeney, M 2013, 'Food Labelling and International Trade', *International Trade Law and Regulation*, vol. 19, no. 1, pp. 14–24.
Blakeney, M 2014, *The Protection of Geographical Indications, Law and Practice*, Edward Elgar Publishing Limited, United Kingdom.
Blakeney, M, Coulet, T, Mengistie, G & Mahop, M 2012, *Extending the Protection of Geographical Indications, Case Studies of Agricultural Products in Africa*, Routledge, United Kingdom.
Botterill, L 2005, 'Policy change and network termination: The role of farm groups in agricultural policy making in Australia', *Australian Journal of Political Science*, vol. 40, no. 2, pp. 207–19.
Bowen, S 2010, 'Development from within? The potential for geographical indications in the global south', *The Journal of World Intellectual Property*, vol. 13, pp. 231–52.
Braithwaite, J & Drahos, P 2000, *Global Business Regulation*, Cambridge University Press, Cambridge.
Brand Tasmania 2015, 'Tasmania, Australia – a beautiful place to grow', http://www.brandtasmania.com/, viewed 5 November 2014.
Braun, V & Clarke, V 2006, 'Using thematic analysis in psychology', *Qualitative Research in Psychology*, vol. 3, no. 2, pp. 77–101.
Bryman, A & Bell, E 2007, *Business Research Methods*, Oxford University Press, Oxford.
Bureau of Agricultural Economics 1977, *King Island: A viability study of the rural economy*, Commonwealth of Australia Canberra.
Bureau of Rural Sciences 2008, *Country Matters: Social Atlas of Rural and Regional Australia, 2008*, BRS, Canberra.
Burke, T 2008, 'European deal improves trade access for Australian wines', http://www.maff.gov.au/burke_media_releases/2008/december_2008/european_deal_improves_trade_access_for_australian_wines.html, viewed 1 September 2014.
Burton, R 2004, 'Seeing through the "good farmer's" eyes: towards developing an understanding of the social symbolic value of "productivist" behaviour', *Sociologia Ruralis*, vol. 44, no. 2, pp. 195–215.
Carson, D & Harwood, S 2007, 'Tourism Development in Rural and Remote Areas: Build it and they may not come!', *Queensland Planner*, vol. 47, no. 4, pp. 18–22.
Carsten, F & Maskus K, 2006, 'The Debate on Geographical Indications in the WTO', in R Newfarmer (ed), *Trade, Doha and Development: a window into the issues*, World Bank, Washington DC.
Cleary, J 2014, *Recycling dependency: The grant economy propping up rural, regional and remote Australia*, Proceedings of the SEGRA Conference, October 7–10, Alice Springs.
Coates, L 2014, 'Can contemporary Regional Development identify a future for islands? King Island: a case study', PhD thesis, University of Tasmania.

Cobia, DW (ed) 1989, *Cooperatives in Agriculture*, Prentice-Hall, New Jersey. Coldwell, I 2007, 'Young Farmers, Masculinities and the Embodiment Of Farming Practices In An Australian Setting', *Rural Society*, vol. 17, no. 1, pp. 19–33.

Collits, P 2015, 'Regional Policy in Post-war Australia: much ado about nothing?' in A Hogan & M Young (eds), *Rural and Regional Futures*, Routledge, London.

Congues, J 2014, 'Promoting collective well-being as a means of defying the odds: Drought in the Goulburn valley 2014', *Rural Society*, vol. 23, no. 3, pp. 229–242.

Corones, S 2014, 'Competition policy review: draft recommendations on competition law', *Australian Business Law Review*, vol. 42, no. 6, pp. 479-486.

Crotty, M 1998, *Foundations of Social Research: Meaning and Perspective in the Research Process*, Sage, London.

Curtis, A, Mendham, E & Mitchell, M 2015, 'Assessing the Efficacy of Transition Theory to Identify Industry Transformation: a case study examining the deregulation of Australia's dairy industry', *Australian Geographer*, vol. 46, no. 1, pp. 113–29.

Daniell, K A, Hogan, A & Cleary, J 2015 (in press), 'Breaking down the 'one-size-fits-all' approach to rural and regional policy: enhancing policy initiatives through multi-level governance' in K A Daniell & A Kay (eds), *Multi-Level Governance: conceptual challenges and case studies*, ANZSOG Series, ANU Press, Canberra.

de Chernatony, L, McDonald, M & Wallace, E 2011, *Creating Powerful Brands*, 4th edn, Butterworth-Heinemann, Oxford.

Department of Agriculture 2015, 'Senate Economics References Committee inquiry into milk pricing and competitiveness in the Australian Dairy Industry', http://www.agriculture.gov.au/agriculture-food/meat-wool-dairy/dairy/deregulation-of-the-Australian-dairy-industry, viewed 31 October 2014.

Dibden, J & Cocklin, C 2010, 'Re-mapping regulatory space: the new governance of Australian dairying', *Geoforum*, vol. 41, pp. 410–22.

Drahos, P 2014, *Intellectual Property, Indigenous People and their Knowledge*, Cambridge University.

Drahos, P & Braithwaite, J 2002, *Information Feudalism*, Earthscan, London.

Drahos, P & Frankel, S (eds) 2012, *Indigenous Peoples' Innovation: Intellectual Property Pathways to Economics*, vol. 62, pp. 383–7.

Drummond, F & Jouguelet, C 2009, 'First case on conflict between trade mark and geographical indication', *Australian IP Law Bulletin*, vol. 21, pp. 144–146.

EU-China Project on the Protection of Intellectual Property Rights 2009, *Comprehensive Feasibility Study for Possible Negotiations on a Geographical Indications Agreement between China and the EU*, IPR2.

European Commission 2012, *EU-China Geographical Indications – "10 plus 10" project is now complete*, Press release, http://europa.eu/rapid/press-release_IP-12-1297_en.htm, viewed 21 November 2014.

European Commission 2014, *EU-Canada Comprehensive Economic and Trade Agreement (CETA)*, http://ec.europa.eu/trade/policy/in-focus/ceta/, viewed 20 November 2014.

European Court of Auditors 2011, *Do the design and management of the geographical indications scheme allow it to be effective?*, Special report no 11/ 2011, European Union, Luxembourg.

Felix Domus Pty Ltd 2013, *King Island Abattoir Feasibility Study: Final Report*, http://www.kingisland.tas.gov.au/webdata/resources/files/KIAFSG%20final%20report%2017%20Jun.pdf, viewed 3 November 2014.

French, B 1987, 'Farm Price Estimation When There is Bargaining: The Case of Processed Fruit and Vegetables', *Western Journal of Agricultural Economics*, vol. 12, no. 1, pp. 17–26.

Frey, BS, Kucher, M & Stutzer, A 2001, 'Outcome, process and power in direct democracy: New econometric results', *Public Choice*, vol. 107, pp. 271–93.

FSANZ 2014, *Approval Report – Primary Production & Processing Requirements for Raw Milk Products*, FSANZ.

Gangjee, D 2007, 'Say Cheese! A sharper image of generic use through the lens of feta', *European Intellectual Property Review*, vol. 29, no. 5, pp. 172–79.

Geuze, M 2009, 'The provisions on geographical indications in the TRIPS agreement', *The Estey Centre Journal of International Law and Trade Policy*, vol. 10, no. 1, pp. 50–64.

Ghouri, N 2010, 'Chardonnayfication: how the New Zealand wine boom lost its sparkle - and what we're drinking today', *Mail*, 28 March, http://www.dailymail.co.uk/home/moslive/article-1260364/Chardonnayfication-New-Zealand-wine-boom-lost-sparkle–drinking-today.html.

Glaser, B & Strauss, A 1967, *The Discovery of Grounded Theory: Strategies for Qualitative Research*, Aldine Publishing, Chicago.

GlobalG.A.P n.d, 'GlobalG.A.P History', http://www.globalgap.org/uk_en/who-we-are/about-us/history/, viewed 30 November 2014.

Goebel, B & Groeschl, M 2014, 'The long road to resolving conflicts between trade marks and geographical indications', *The Trademark Reporter*, vol. 104, pp. 829–35.

Goodman, S, Lockshin, L, & Cohen, E 2006, 'What influences consumer selection in the retail store?' *The Australian & New Zealand Grapegrower & Winemaker*, December, pp. 61–3.

Graham, S 2014, 'A new perspective on the trust power nexus from rural Australia', *Journal of Rural Studies*, vol. 36, pp. 87–98.

Granovetter, M S 1973, 'The Strength of Weak Ties', *American Journal of Sociology*, vol, 78, pp. 1360–80.

Granovetter, M S 1983, 'The Strength of Weak Ties: A Network Theory Revisited', *Sociological Theory*, vol. 1, pp. 201–33.

Granovetter, M S 2005, 'The Impact of Social Structure on Economic Outcomes', *Journal of Economic Perspectives*, vol. 19, no. 1, pp. 33–50.

Gray, J 1998, 'Family farms in the Scottish borders: a practical definition by hill farmers', *Journal of Rural Studies*, vol. 14, no. 3, pp. 241–356.

Greenhill, J, King, D, Lane, A & MacDougall, C 2009, 'Understanding Resilience In South Australian Farm Families', *Rural Society*, vol. 19, no. 4, pp. 318–25.

Groth, TM, Curtis, A, Mendham, E & Toman, E 2014, 'Farmer Identity in Multifunctional Landscapes: using a collective identity construct to explore the nature and impact of occupational identity', *Australian Geographer*, vol. 45, no.1, pp. 71–86.

Guest, G, MacQueen, K & Namey, E 2012, *Applied Thematic Analysis*, Sage, USA.

Gunningham, N, Grabosky, P & Sinclair, D 1998. *Smart Regulation: Designing Environmental Policy*.

Gustafson, CR & Nganje, WE 2006, 'Value of social capital to mid-sized northern plains farms', *Canadian Journal of Agricultural Economics*, vol. 54, no. 3, pp. 421–38.

Hallett, P 2012, 'Certification Marks: Are They Really Worth the Hassle?', *Watermark Journal*, vol. 29, no. 2.

Halpin, D 2003, 'The Collective Political Actions of the Australian Farming and Rural Communities: Putting Farm Interest Groups in Context', *Rural Society*, vol. 13, no. 2, pp. 138–56.

Hawthorne, M, 2015, 'Milk wars likely to claim more victims yet', *Sydney Morning Herald*, 24 January, http://www.smh.com.au/business/milk-wars-likely-to-claim-more-victims-yet-20150123-12wirn.html.

Hernandez-Espallardo, M, Arcas-Lario, N & Marcos-Matas, G 2013, 'Farmers' satisfaction and intention to continue membership in agricultural marketing co-operatives: neoclassical versus transaction cost considerations', *European Review of Agricultural Economics*, vol. 40, no. 2, pp. 239–60.

Hira, A & Aylward, D 2013, 'Australia as a Triple Helix exemplar: built upon a foundation of resource and institutional coordination and strategic consensus', *Prometheus*, vol. 31, pp. 399–417.

Hird, MJ 2003, 'Case Study', in RL Miller & JD Brewer (eds), *A-Z of Social Research*, Sage, London.

Hogan, A, Cleary, J, Lockie, S, Young, M, & Daniell, K 2015, 'Localism and the policy goal of securing the socio-economic viability of rural and regional Australia', in A Hogan & M Young (eds), *Rural and Regional Futures*, Routledge, London.

Houghton, K 2012, 'Characteristics of economic sustainability in regional Australia', in *Scoping a Vision for the Future of Rural and Regional Australia*, Collection of Papers presented at the Sustaining Rural Communities Conference 2012, National Institute for Rural and Regional Australia, Canberra.

Hurst, B 2007, 'Accreditation: Is it Really Worth It?', *Just Food*, 22 March http://www.just-food.com/comment/is-it-really-worth-it_id97896.aspx.

IP Australia 2014, 'Trade mark fees', http://www.ipaustralia.gov.au/get-the-right-ip/trade-marks/time-and-costs/fees/, viewed 29 October 2014.

Isaac, J 2008, 'Collective Bargaining under trade practices law', *The Economic and Labour Relations Review*, vol. 19, pp. 39–56.

James, S 2015, forthcoming, 'Beyond "local" food: How supermarkets and consumer choice affect the economic viability of small-scale family farms in Sydney', *Arena*.

Kench, J 2003, 'Collective bargaining: The Dawson Review's assistance package for business cartels', *The University of New South Wales Law Journal*, vol. 26, no. 1, pp. 257–67.

Kontogeorgos, A 2012, 'Brands, quality badges and agricultural cooperatives: how can they co-exist?', *The TQM Journal*, vol. 24, no. 1, pp. 72–82.

Lawrence, G 1995, *Futures for Rural Australia: From Agricultural Productivism to Community Sustainability*, 1st edn, Rural Social and Economic Research Centre, Central Queensland University, Rockhampton.

Lawrence, G 1996, 'Rural Australia: Insights and Issues from Contemporary Political Economy', in G Lawrence, K Lyons & S Momtaz (eds), *Social Change in Rural Australia*, Rural Social and Economic Research Centre, Central Queensland University, Rockhampton.

Lawrence, G, Gray, I & Stehlik, D 1998, 'Changing Spaces: The Effects of Macro-Social Forces on Rural Australia', in B Kasmimas & A Papadopoulos (eds), *Local Responses to Global Integration: Toward a New Era of Rural Restructuring*, Ashgate, London.

Leonard-Barton, D 1992, 'Core capabilities and core rigidities: A paradox in managing new product development', *Strategic Management Journal*, vol. 13, pp. 111–25.

Mazzarol, T, Reboud, S, Limnios, EM & Clark, D 2014, *Research Handbook on Sustainable Co- Operative Enterprise: Case Studies of Organisational Resilience in the Co-Operative Business Model*, Edward Elgar, Cheltenham.

Millar, J & Curtis, A 1997, 'Moving farmer knowledge beyond the farm gate: An Australian study of farmer knowledge in group learning', *European Journal of Agricultural Education and Extension*, vol. 4, no. 2, pp. 133–42.

Moran, W, Blunden, G & Bradly, A 1996. 'Empowering Family Farms Through Co-Operatives and Producer Marketing Boards', *Economic Geography*, vol. 72, no. 2, pp. 161–78.

Muenstermonn, I 2009, 'Cross boundary farming: can this challenging farming method save the Australian family farm?', *Rural Society*, vol. 19, no. 3, pp. 262–74.

Mullainathan, S., & Thaler, R. 2001, 'Behavioral Economics', in N. Smelser & P. Baltes (Eds.), *International encyclopedia of the social and behavioral sciences*. Amsterdam: Elsevier.

Nilsson, J 2012, 'Are Large and Complex Agricultural Cooperatives Losing Their Social Capital?', *Agribusiness*, vol. 28, no. 2, pp. 187–204.

Nourse, E 1922b, *'The outlook for cooperative marketing'*, *Journal of Farm Economics*, vol. 4, no. 2, pp. 80–8.

Nourse, E 1945, 'The Place of the Cooperative in Our National Economy', *American Cooperation 1942 to 1945*, American Institute of Cooperation, pp. 33–39.

Oczkowski, E 2004, 'Agricultural bargaining co-operatives and associations: Australian issues and prospects', *Third Sector Review*, vol. 10, no.2, pp. 67–84.

Oxford University Press, Oxford. Gutman, P 2007, 'Ecosystems Services: Foundations for a new rural-urban compact', *Ecological*.

Pannell, DJ, Marshall, GR, Barr, N, Curtis, A, Vanclay, F & Wilkinson, R 2006, 'Understanding and promoting adoption of conservation practices by rural landholders', *Australian Journal of Experimental Agriculture*, vol. 46, no. 11, pp. 1407–24.

Pengilley, W 2006, 'The competition tribunal and chicken meat: Do exclusionary provisions have any future hope of authorisation?', *Competition and Consumer Law Journal*, vol. 14. pp. 196–213.

Plunkett, B & Kingwell, R 2001, 'New Generation Co-operatives for agricultural marketing and processing in Australia: Principles, practicalities and a case study', *Agribusiness Review*, vol. 9.

Porter, PK & Scully, GW 1987, 'Economic Efficiency in Co-operatives', *Journal of Law and Economics,* vol. 30, pp. 489–512.

Pritchard, B 1996, 'Restructuring in the Australian Dairy Industry: The Reconstruction and Reinvention of Co-operatives', in D Burch, R Rickson & G Lawrence (eds), *Globalisation and Agri- Food Restructuring Perspectives from the Australasia Region*, Avebury, Aldershot.

Pritchard, B 1999, 'Australia as the supermarket to Asia? Governments, territory and political economy in the Australian agri-food system', *Rural Sociology,* vol. 64, no. 2, pp. 284–301.

Putnam, RD 1993, 'The Prosperous Community: Social Capital and Public Life', *The American Prospect,* vol. 4, no. 13, pp. 35–42.

Putnam, RD 1995, 'Bowling Alone: America's Declining Social Capital', *Journal of Democracy,* vol. 6, no. 1, pp. 65–78.

Rabin, M 1998, 'Psychology and economics', *Journal of Economic Literature,* vol. 36, pp. 11–46.

Sarantakos, S 2005, *Social Research,* 3rd edn, Macmillan Education, Melbourne.

Sacconi L. and Degli Antoni G 2011 (eds), *Social Capital, Corporate Social Responsibility, Economic.*

Saitone TL & Sexton RJ 2009, 'Optimal cooperative pooling in a quality-differentiated market', American Journal of Agricultural Economics, vol. 91, no. 5, pp. 1224–1232.

Schaper, M 2010, 'Competition law, enforcement and the Australian small business sector', *Small Enterprise Research*, vol 17, no. 1, pp. 7–18.

Sharma, M 2013, 'Economics of Exemptions from Competition Law', 24 *National Law School of India Review*, vol. 62, pp. 62–74.

Sinclair, K, Curtis, A, Mendham, E & Mitchell, M 2015, 'Assessing the Efficacy of Transition Theory to Identify Industry Transformation: a case study examining the deregulation of Australia's dairy industry', *Australian Geographer*, vol. 46, no. 1, pp. 113–29.

Smith, R & Duke, A 2014, 'Agreements and competition law in Australia', *Competition and Consumer Law Journal,* vol. 22, pp. 54–79.

Thompson, P 2014, 'Inghams co-op plans continue', *Stock Journal,* vol. 10, pp. 1–2.

Tonts, M & Jones, R 1997, 'From state paternalism to neoliberalism in Australian rural policy: perspectives from the Western Australian wheatbelt', *Space and Polity,* vol. 1, no. 2, pp. 171–90.

Tonts, M, Argent, N & Plummer, P 2012, 'Evolutionary perspectives in rural Australia', *Geographical Research*, vol. 50, no. 3, pp. 291–303.

Van Caenegem, William. (2003). Registered Geographical Indications: Between Intellectual Property and Rural Policy—Part I. William Van Caenegem, William Van Caenegem.

Van Caenegem, William, Cleary, Jen A, & Drahos, Peter. (2014). Pride and profit: Geographical indications as regional development tools in Australia. *Journal of Economic & Social Policy, 16*(1), 90–114.

Author Biographies

Jen Cleary is a human geographer with research interests in regional development and agricultural value chain analysis. Jen holds a number of leadership positions in national and international organisations concerned with rural, regional and remote development and is recognised for her expertise, experience and extensive networks in this field.

William van Caenegem is a Professor at the Faculty of Law, Bond University, the author of a number of books on intellectual property law and comparative law. He has published on many areas of intellectual property law, including IP and innovation and also IP and the protection of commercial reputation. The second edition of his textbook on Australian Intellectual Property law was published by LexisNexis in 2014. He is a Barrister-at-Law of the Supreme Court of Queensland.

Chapter 6
Protecting Appellations of Origin: One Hundred Years of Efforts and Debates

Genevieve Teil

Abstract The phylloxera crisis, which started in the 1870s, coincided with a wine shortage period as well as winemaking innovations and wine diversification. This increased variability of wine products was perceived as harmful because it impeded the acknowledgement of wine quality. The French authorities struggled to contain the variability of wine products and their many subcategories. Numerous legislative initiatives took place: the idea of *origin* was first interpreted as a traceability link to the vineyard and later, with the 1935 law on Controlled Denominations of Origin (*fr.* AOC) was perceived as heritage, mirroring superior taste-quality, which required constraining the vintners' wine-growing and wine-making practices. Different interpretations of the notion of heritage led to a new heterogeneity in the quality of wines originating from the AOCs, and disputes about *terroir* quality persist in modern times. This chapter explores the notion of *terroir* beyond its interpretation as a simple resource and notes that these perceived incompatible interpretations should coexist and be reconciled.

6.1 Introduction

Today, denominations of origin are widely seen as rural development instruments to support alternative commercial regimes.[1] However, this has not always been the case; in fact it has been far from it. For centuries, commodities used places of origin to indicate remarkable product quality: for example, in antiquity, wines were often labelled according to the region of the vineyard. Sometimes other non-food

[1]Bowen (2010), Feagan (2007), Jenkins and Parrott (2002), Kan and Gulcubuk (2008), Lamine (2005), Paxson (2010), Selfa and Qazi (2005).

G. Teil (✉)
Institut National de la Recherche Agronomique SADAPT Agroparistech, Paris, France
e-mail: genevieve.teil@agroparistech.fr

G. Teil
Institut Français de Pondichéry, Pondicherry, India

W. van Caenegem and J. Cleary (eds.), *The Importance of Place: Geographical Indications as a Tool for Local and Regional Development*, Ius Gentium: Comparative Perspectives on Law and Justice 58, DOI 10.1007/978-3-319-53073-4_6

commodities have also been labelled this way, such as the Alep soap, Le Thiers knives, Mysore silk and many others. For that reason, many studies have sought to validate or invalidate the foundation of the "link to the place", also sometimes referred to as the "link to the terroir".[2] This process consisted in clarifying what this link is or should be.[3] Moreover, its natural determinism or, on the contrary, its existence as a social construct has been debated.[4] When the latter (social construct) is considered to take precedence over the former (natural determinism), the link to the place is said to be constructed and artificial.[5] This challenges its groundedness, particularly in view of the barriers that it may create for competition. However, as analyses from the sociology of science have shown, this analytical approach overlooks the fact that all determinism is itself the outcome of a very human formatting of experiencing world.[6] This has the particular effect of making it virtually impossible to distinguish between a natural or artificial link.[7]

Other authors have judged the groundedness of appellations of origin based on what they consider to be the specificity of the products labelled this way. Such assessments always involve debatable arbitration on the singularity of all the goods carrying an appellation of origin. This difficulty has also affected the actors involved; it is even one of the reasons why they have tried to regulate appellations and appellations of origin in particular and, in so doing, have tried to interpret the reference to the place involved in the denomination of products. The present article discusses the link to place, not to validate (or invalidate) its pertinence, but on the contrary, to highlight the diversity of contrasting interpretations surrounding this notion, which have given rise to a variety of regulatory frameworks.

In the 19th century, the concept and nomenclature of wines was challenged: the word wine in general, but especially the names of the different wines that referred to the place of production in order to differentiate between varieties. Significant disagreements arose as to what should constitute wine fuelled controversies, both between producers themselves, and between producers and their customers. This resulted in frequent accusations of fraud and unfair competition that went hand in hand with the existence of market exchanges: buyers had quite precise ideas about the quality they would have liked but could not find or doubted they could find in the products they available for purchase. They communicated this by expressing doubts as to the quality of products or the sellers' reliability.

Those concerned about fraud or unfair competition followed a similar line of reasoning to that of G. Akerlof, who explained in his seminal article (Akerlof 1970)

---

[2]Berard and Marchenay (2006), Ilbery et al. (2005).

[3]Brunori and Rossi (2007).

[4]Bérard and Marchenay (1995), Dessaux (2006), Fonte (2008), Letablier (1997), Selfa and Qazi (2005), Tellstrom et al. (2005), Wilk (1999).

[5]Ilbery et al. (2005).

[6]*See* Callon and Latour (1981) for fundamental pioneering work on Science and Technology Studies (STS).

[7]Latour (1983).

that variability in the qualities of a product sold under a same name led to the market's collapse, because potential buyers, lacking information on the quality of the good they wanted to buy, aligned the prices they were willing to pay with the minimal quality. Products of a higher quality that could not expect a price corresponding to *their quality* therefore withdrew from the market. All actors saw this as allowing for discrimination between the different qualities, in other words for creating new categories of products to regulate their variations more closely.

This economic reasoning, often found under the name of unfair competition, analysed the question of the variety of interpretations of quality as an issue of information on the 'real' quality of the products. According to this reasoning, situations of fraud arose when information on products was inaccessible or was hidden from buyers.[8] The effect was to prevent thus economic transactions.[9] For markets to function better, the emergence of 'real' quality therefore needed support, particularly by way of reduction the variety of ways in which producers interpreted quality.

Commercial brands existed in the 19th century already, but they seemed incapable of putting a stop to fraud and unfair competition.[10] After framing the notion of wine, a large number of regulations followed. Some of these were very quickly replaced, whereas it took decades for others to be challenged. They all contributed to regulating the qualities of wines and their variations. Over the years, they also gave shape to an innovative category of commercial designations, based on the geographical source and no longer just on the human source of production. The historical detour in this article sheds light on current conflicts surrounding designations, which oppose widely diverging interpretations of the notion of denomination of origin.

This article analyses the last 150 years of wine quality regulations and the efforts made to frame a variety of products, bottles, and the use of terminologies to designate them. To do so it examines the hypothesis of information on the 'real' quality of products articulated in the debates. Sociology posits the true quality of wine as a pure social construct, a simple convention or a fruit of the imagination, denying it any kind of reality that could be put to the test. The difficulties that actors encounter in their economic transactions are then interpreted as social or class struggles. One may therefore reproach the sociological analysis with sidestepping these difficulties instead of accounting for them. Whether one posits that information can be real or purely socially constructed, the heterogeneity of qualities is always flagged up, because the actors involved consider this variability to be ultimately harmful. This cross-cutting issue of acceptable diversity of wines sheds light on these same actors' difficulties and efforts, without explaining their actions

---

[8]*See* the economic theory of information asymmetry and the seminal article by Akerlof (1970), for whom this real quality stems from a "real" knowledge of goods. But this "real" quality can also be seen as "conventional", in other words as the result of a social agreement on objects Eymard-Duvernay (1989), Eymard-Duvernay (2006a, b), Young (1996).

[9]Garella and Petrakis (2008), Grossman and Stiglitz (1980), Marette (2007), Shapiro (1982).

[10]As commercial brands have been studied extensively Di Palma (2014).

in terms of economic or social behaviours. This question will therefore guide us throughout this examination of regulatory activity and the reactions it sparked.

6.1.1 Merchants' Regulations of Qualitative Diversity

Few vineyards sold their wine themselves in 19th century France, unlike present day practice. A few famous domains did, like some châteaux in the Médoc, but the overwhelming majority produced for personal consumption or sold their production to merchants who took care not only of selling the wine, but also of promoting its quality and its improvement through ageing. However, not all wines produced by wine growers were suitable for ageing, nor for immediate sale; some lacked the appropriate quality for this purpose, and therefore had to be improved. Merchants were thus also experts at blending and mixing wines with different characteristics to make them correspond to specific quality criteria. Wines bought from producers were therefore considered as the raw material needed to develop a finished product.

> The nature of wines needs to be preserved as much as possible [...]. But when a wine is altered or naturally bad, it is often impossible to make it drinkable without mixing it with other wines. [...] When the grapes have not ripened, wines, even good vintages, lack quality and long retain an unpleasant pungency, such as those from the 1805 and 1809 harvests, and especially those from 1816 and 1817. (Jullien 1999 [1822]: 142)

Wines were then sold as brand wines from a winegrower, or as ordinary wines with no particular distinction, under the merchant's brand, or else as wines with a name of origin: Bordeaux, Bourgogne, Coulée de Serrant, etc. Interpretations regarding the significance of these names of origin varied. Wines made in countries other than France readily carried names that enjoyed a flattering reputation: Champagne, Bordeaux, Bourgogne, etc. These denominations were denounced by French vintners vineyards, which saw them as counterfeits of a genuine product made in the region of origin. For many merchants, however, these appellations corresponded to a particular type of wine, a recipe, a type of taste, etc. associated with a geographical region. Most of them contributed to regulating the diversity of wines by selecting the best representatives, and then proceeding to complex blends and improvements of the wines' quality so that the resulting wines would conform as closely as possible to what their denomination suggested.

> The aim of blending is generally to offset flaws or contrary qualities. Black wine is mixed with wines lacking colour, or with white wines; light wines with little ageing potential are mixed with full bodied wines to ensure their preservation; wines with a high alcohol content but heavy and pasty, with lively and light wines, etc. (Lenoir 1828: 534)

For most merchants, the denomination of origin attested to a product's association with a certain style of wines. The denominations of wines were arranged within a table of equivalences of appellations or vineyards. Merchants used this table to perform wine blends and substitutions, depending on availability. Broad geographical regional denominations referred to styles of quality, while the more restricted appellation of vineyards indicated a hierarchical level of quality within this style of wine. Thus, within the Beaujolais family, the distinction was made between the vineyards of Brouilly, Moulin à Vent, Saint-Amour, etc., informed by the table of equivalences of vineyards, which guided possibilities of substitution. In a book in defence of merchants' work, F. Laneyrie wrote:

> Excellent Brouilly was served under the name Moulin-à-Vent, because it was the best Beaujolais wine that could be found to honour the appellation under which it was served. (Laneyrie 1926: 19)

The Moulin à Vent was usually the best Beaujolais. The vineyard table of equivalences then stated that a Brouilly, with the right qualitative level, could become a Moulin-à-Vent. Merchants thus played a decisive role in the regulation of wine quality. H. Sempé, a lawyer who published a thesis on the wine economy in 1898, detailed and highlighted the importance of their work and know-how.

> We know how valuable the combination of capital, premises and the science of these skilful and precious auxiliaries [merchants] is for wine growers. These auxiliaries are not simple, more or less passive intermediaries: they are responsible for giving a harvested raw product the finish expected by consumers; in their cellars the wine receives the constant care it requires. Its limpidity, its perfect preservation and the development of its bouquet are largely the work of the practitioner. The slightest negligence can noticeably reduce the quality: wine is a delicate product; it requires the consummate skill of our merchants to reach its maximum value. (Sempé 1898: 105–106)

6.1.2 *Challenged Regulations*

The regulation of certain varieties in quality certainly contributed to conferring a certain unity to the style of wines with an appellation of origin. However, it did have a flipside. At the time, wines were widely accused of sophistication. This word denoted all the operations and treatments through which wines acquired their market quality as well as the practice of blending, which allowed merchants to guarantee certain conformity in the style of wines from a region of appellation.[11] These accusations of wine sophistication targeted merchants in particular, along with their efforts to improve wines:

> As for the wine trade, I posit that actually, in Paris, we drink few bottles of vintage wines, and there are more than one of which I would not want to put half a glass in my stomach. You are lucky not to have witnessed, as I have, the filth of all this jiggery-pokery at night. F.-V. Raspail, Annuaire de la santé, 1860, cited in Argenson (1871: 3)

---

[11]This commercial quality involved an absence of flaws and especially a good stability of wines, which should not evolve too fast in the bottle or the tanks. For a detailed study see Stanziani (2003a, b, c).

Customers and wine growers sometimes saw the practices that merchants defended as crucial to the qualitative polishing of wines, as alterations of their quality. They suspected merchants of hiding behind quality improvement to adjust the poor quality of poor products through debatable practices, in order to give them a better commercial appearance. Merchants like L. Maurial confirmed the existence of dubious practices, but proved incapable of imposing best practices in this respect:

> This question is the most sensitive part of the task I have undertaken: it raises so many objections in the world of consumers, such wariness, and one has to admit that it is so conducive to deception, that it is difficult to set the limits where good faith stops and those where fraud starts. (Maurial 1865: 134)

Drinkers found a way around this by seeking vintage wines from grapes exclusively sourced from a specific estate, which had to be 'pure', neither blended nor improved. They preferred a poorer taste quality to improvement practices deemed dubious. V. Argenson, a wine merchant, provided a small information blurb on his vineyard and wine for his customers, in which he recounted one of his suppliers confiding in him:

> 'I would not sell you my harvest', a good old peasant told me in 1858, accustomed as he was since childhood to winegrowing. 'I would not sell you my harvest if I thought that you blended my wine.' He seemed reassured when I cited the following proverb, so mundane and so true: However bad be a natural wine, It is always better than blended wine. (Argenson 1871: 5)

Merchants tightly regulated the quality of wines, but they resorted to controversial practices. Challenging these practices intensified with the phylloxera crisis.

6.1.3 The Challenge Heightened by the Phylloxera Crisis

In the early 1870s an aphid began to gradually ravage French vineyards. Without suitable treatment, the vines massively died, first in the south, and then further up north as the aphid's invasion spread. The lack of grapes led French merchants and winemakers to purchase grapes, particularly raisins, or wines from countries that were not yet hit by the aphid, especially Algeria. Winemaking gained new processes like raisin fermentation, and the reconstitution of wines using the main components of musts: sugar, tartaric acid and tannins. Accusations of sophistication intensified along with the defence of 'natural', i.e. unsophisticated wines. Winemakers who were challenged responded that they were just adapting to the shortage, doing their best to carry on supplying wines at reasonable prices, and that customers appreciated them for what they bought.

With the gradual development in the late 1880s and the 1980s of means to control the aphid, some of the French vineyards were replanted which led to a new situation: raisin wines and grape wines cohabited. The phylloxera crisis also brought on great change, not only in winegrowing and winemaking but also in consumption habits and France's balance of trade. Exports plummeted as raisin and wine imports rose, particularly from Algeria. This became a growing concern within the French administration: should it uphold or oppose this trend in French wine production?

6.2 The Fight Against Unfair Competition

French wines enjoyed a certain prestige and greatly contributed to the French balance of trade; but this prestige was under threat. Worse still, the fact that raisins were imported further reduced that balance. Although the two products had a very different taste, many consumers seemed to have become accustomed to the shortage and to have changed their consumption habits, opting for raisin wines instead of grape wines. Winemakers wished to return to old production techniques, especially for grape winemaking. They complained about the lower production cost of raisin wines preventing them from selling their grape wines. The expression *unfair competition* became widely used not only by producers but also by merchants.[12] These challenged the diversity of goods and production costs within the wine category.[13]

Likewise, the questions of unfair competition targeted the denominations of wines according to their origin.

> Finally, as a means of spreading the use of certain wines, we can cite the habit of selling them under names indicative of their origin, names which constitute real brands requested by consumers. Thus wines from Champagne, Bourgogne and Bordeaux enjoy a marked preference; they are found on all tables, though with very diverse qualities.The pitfall of these generic appellations is the tendency to use a brand name for wines of another origin, of a lower quality, at a depreciated price. This danger has presented itself with Gironde wine, in particular. It is difficult to appreciate the enormous quantity of so-called 'Bordeaux' made available for consumption. (Sempé 1898: 179)

The author called these appellations generic brands for, like common nouns, they belonged to no one. Just like language, they were subject to each person's free interpretation, as long as they were comprehended. H. Sempé was concerned about buyers only looking at the price of wines and paying little attention to the specific quality of the goods, preventing authentic, more expensive Bordeaux from finding a place on the market. The French government decided to act supportively in restoring the former prestige of French wines. Bringing back this prestige proved to

---

[12]Many consumers were likely also concerned about this situation, but they left virtually no direct traces in the archives.

[13]I prefer to talk of variability and not of "uncertainty regarding quality", like Lamine (2005) for example, as the idea of uncertainty implies that a more or less hidden or concealed "real" quality exists, when the object of disputes surrounding regulation is to determine the modalities of existence of commodities with their acceptable variability, regardless of any "truth" about what they are.

be a difficult task, as winegrowers did not share all relevant information about the quality of the products they commercialized. Moreover the government could not rely on commercial interactions as they took place, as they evidently left no room for recognizing and developing old products, and even seemed detrimental to restoring the success and prestige of French wine production.

The problem identified by the authorities, in the case of wines both with and without appellations of origin, was a problem of wine production and heterogeneity, as it hindered the recognition of the quality of good wines. This heterogeneity therefore needed to be reduced, which involved facilitating the identification of good wines to allow drinkers to make choices that better supported the recognition and therefore the development of quality wines. At the end of the 19th century, the French public authorities thus worked on compiling a set of rules to regulate the diversity of production processed. A long list of articles specified acceptable wine improvement practices that enhanced its quality without denaturing it. The first regulation, distinguished wines from raisin wines by creating a specific category that was duly labelled. Raisin wines were then excluded from the wine denomination, along with pomace wines and other plonks. The 1889 Griffe law stated:

> Art. 1. It is prohibited to dispatch, sell and put on sale a product other than that of grape fermentation under the denomination of wine (Griffe 1889)

6.2.1 *The Field as a Wine's Unit of Production of Origin*

Deciding on acceptable wine production practices was a long and difficult process, but it gained clarity throughout the late 19th century.[14] The regulation of the products' origin raised more complex issues. There were already laws in place to associate and link manufactured products with their own manufacturer or merchants.[15] But were agricultural products manufactured products?

Could a field's productive land be seen as a business's production factory? In his *Traité des marques de fabrique et de la concurrence déloyale en tous genres* (treaty on trademarks and unfair competition of all sorts), E. Pouillet presented excerpts of discussions between legal experts regarding the branding of agricultural products:

> [...] natural products of the soil that are sold without prior preparation or handling are excluded from the provisions of the 1824 law. [...] their generic designation cannot privately and exclusively belong to anyone. (Pouillet 1883: 413–414).

According to E. Pouillet, grapes were a product of nature and humans could not claim any right or ownership over them. Others thought that grapes, and especially wine, were elaborated products and not a pure product of nature.[16] As such, they

---

[14]See Stanziani (2003b).

[15]1824 Law completed by the 1857 Law.

[16]Legal experts say "manufactured", but this term could cause confusion if used in a different way to winemakers and merchants. That is why we prefer the term "elaborate".

could be trademark-protected. Winegrowers thus gave their wine its "origin" through the magnitude of their contribution to agricultural production.[17] Moreover, as winemakers produced wines using grapes from clearly identified production sites, they had the right to choose the name of the vineyard from which the grapes originated and use it as a brand name.

On 8 June 1847, it was ruled that:[18]

> Wines must be placed in the category of manufactured products, and that owners and winegrowers can enjoy the protection, for wines from their own harvest, granted to the producers of manufactured objects by the 1824 law; it follows that only and all vineyard owners have the right to brand the vessels containing their wine with a stamp signalling this vineyard. (Pouillet 1883) Article 424, p. 414

Vineyard plots that were all differentiated by name allowed for a non-ambiguous identification of production and served to identify commercial brands. In order to ensure the smooth functioning of markets, particularly in terms of reputation building, wines could therefore be associated with a geographical origin.

> It is important [for farmers and growers] to be able to secure the exclusive use of a brand to distinguish their products and gain the public's trust. (23 June 1857 law on brands, Art. 20)

Contrary to certain interpretations, which have now become widespread, this association was not linked to natural particularities of different vineyards or

[17]The term "origin" is understood not in geographical terms, but as the production unit from which the wine comes.

[18]The Champagne appellation played a driving and decisive role in the reinforcement of the legal interpretation. E. Clémentel described the stages of this journey towards turning appellations of origin into wine "law":

"On 12 July 1845, the Criminal Chamber of the Court of Cassation had to rule on a trial against wine merchants from Touraine who had affixed the name of the widow Clicquot and those of Ay and Verzy on bottle corks. The court considered that this constituted personation under Article 1 of the 28 July 1824 law.

In 1847, the civil appeal court of the Court of Cassation ruled that owners and winegrowers were entitled to the protection of this law for wines coming from their harvests.

On 4 March 1870, the Court of Angers recognized the merit of legal proceedings instituted by owners of the Ay, Bouzy and Sillery winegrowers and by merchants who purchased their products, against merchants from another region who sold champanized wines in the Anjou under these names. On 19 July 1887, the same Court ruled that the appellation "Champagne" had to be reserved for wines from and produced in the Champagne region.

On 11 April 1889, it declared that the fraudulent indication of the word Champagne for wine that had not been produced and harvested in the Champagne region constituted an offence provided for and punished by Article 1 of the 28 July 1824 law. It considered that Champagne or Champagne wine should be considered as wine both harvested and produced in Champagne, an old province of France […] on 26 July 1889, the Court of Cassation rejected the appeal lodged against this decision.

Finally, on 18 November 1892, through the adoption of motives, the Court of Paris confirmed a judgement by the Reims commercial court on 17 July 1891 […]: 'it follows that since the designation of Champagne or Champagne wine does not fall under the public domain, it cannot apply to sparkling wine that is not from Champagne'…" Clémentel (1914).

vineyard plots, but rather with each winegrower's *know-how*, thus endowing their products with particular qualities by processing 'fruit'.

6.2.2 A Variety of Meanings for the Name of Origin

Debates surrounding the notion of appellations of origin highlighted the existence of a range of possible interpretations. It referred sometimes to styles of wines that merchants endeavoured to reproduce, and at other times to simple categories identifying different qualities of wines for consumers. In commercial regulation, an appellation of origin was a brand, a link to the producers and their production units which added a quality control or guarantee tool for the consumer: an appellation of origin indicated the winemaker, to allow consumers to choose between different wines if they so wished. Nevertheless, in the particular case of wines, this also established a geographical link with the place of production, since the vineyard was considered as a unit where the production of grapes, the crucial raw material of wine, originated. These different interpretations of origin were often incompatible. In particular, merchants' practices, regulating wines' quality of origin through blending, were incompatible with the interpretation of the appellation of origin as a geographical link with the production vineyard.

6.2.3 Regulating the Name of Commodities

Wines, like other commodities, traded goods or objects, were designated with names. The name *wine* designated a range of products, an umbrella term that allowed to differentiate and restrict to a greater or lesser degree other products: fruit wine, liqueur wine, Alsace wine, etc. These variations in nomenclature sparked accusations of fraud and unfair competition: some actors deemed that some wine qualities should not fit within a certain category of commodities, be it "wine" or a specific wine with a name of origin. Those challenged responded by upholding their products' belonging to the category in question based on factors such as demand, ethics, history, the common good, health, economics, etc.

It could be left up to those involved to resolve these debates by developing communication strategies or purchasing strategies allowing them to sell or acquire products of the desired quality. However, as in the case of wine in the late 19th century, public authorities deemed it necessary to intervene. This shift in arbitration from the actors to public authorities resulted in a few interesting particularities. The significance of names was no longer the result of customs and interactions between concerned parties; through regulation, the public authorities sought to steer all the actors, and thus to orient the significance of the goods' names. Commercial denominations, recorded by the government, were associated with the fraud repression services, with a view to constraining actors to conform or to adjust their production or purchasing practices to these denominations, as much as the effectiveness of the checks and repression exercised allowed, of course.

Note that this regulation was still partial; quality was never totally regulated, as it was impossible to describe the quality of wine exhaustively. The rules stipulated that it was up to the producers or buyers to reach agreements regarding the qualities, which the general regulation did not cover. More importantly, they imposed standards for the verification or investigation of specified qualities, which had significant consequences on the products at stake. All these arbitration techniques eventually provided different forms of regulation of the variability of the goods and therefore of their capacity to have an identity.

6.3 The Public Authorities' Regulatory Attempts

6.3.1 The 1905 General Fraud Repression Law

In 1905, a general fraud repression law was passed to regulate the commercialization of commodities.[19] In regards to wine, it reiterated the Griffe law, which defined the nature of wine as being the exclusive result of grape fermentation. Moreover, it reasserted the obligation to respect the appellations of origin of products, already found in the 23 June 1857 law on brands.

> Art 1. – Anyone who shall have deceived or attempted to deceive the contracting party: Either about the nature, the substantial qualities, the composition and the useful principles of any good;
>
> Or about their variety, their origin when, according to convention or customs, the false designation of the goods' variety or origin will have to be considered as the main cause for the sale; […] will be liable to three months to a year of imprisonment, and with a fine of at least F100 and F5000 at most, or with one of these two sentences only. […]
>
> Art. 11. – This will be governed by public administration rules on the measures to take to ensure the execution of the present law, particularly regarding:
>
> […] 2° inscriptions and marks indicating either the composition, or the origin of the goods, or regional appellations and appellations of particular vineyards which buyers might request on invoices, on packaging or on the products themselves as a guarantee from sellers, as well as necessary exterior or apparent indications to ensure the honesty of the sale and offer. (Loubet and Ruau 1905)

The 1905 law, like the various regulations on acceptable oenological practices, restricted wine quality improvement practices for merchants, in particular,

[19]For a detailed account of the elaboration and then implementation of this law, which P.-A. Dessaux said was "probably largely adopted because it offered a general framework for the protection of wine appellations", see Dessaux (2006). Note however that the analysis is more focused on the "substantial qualities" of commodities than on the question of appellations of origin.

prohibiting them from blending wines from different regions of appellation. The latter voiced strong discontent, claiming the right to blend wines, but supporters of the law gave priority to a duty of responsibility over the freedoms they demanded.

> Asserting the principle of responsibility over the principle of freedom, consecrating prior dispositions against fraud by specifying and completing them, such was the aim of this important law. (Clémentel 1914: 78)

E. Clémentel, Minister of Agriculture from 1913 to 1914, challenged the excessive freedom in merchants' practices, which raised suspicion among customers. In his eyes, there was a tendency to reduce appellations of origin to an interpretation of wine-tastes left to individual appreciation and a trade-off between cost and quality.

6.3.2 Debates Surrounding the Significance of Appellations of Origin

Applying the law required consensual delimitations of the wines' regions of origin and was eventually a process that faced major difficulties. A second law, passed on 5 August 1908 (Fallières and Ruau 1908), specified how origins should be delimited, stating:

> Art 1 - [...] This delimitation will be based on local and constant customs. (Fallières and Ruau 1908)

On the basis of this specification, the State Council adopted the first delimitation Decree, regarding Champagne, on 17 December 1908. The Decree triggered violent riots in 1911, due to some producers' exclusion from the grape supply areas of Champagne making.[20]

The law was not just problematic for the producers excluded, but also for merchants who saw their possibilities for wine improvement tightly restricted. Deputy J. Capus argued that the delimitation of the main grape-sourcing regions (Bordeaux, Bourgogne, etc.) was too vague, and did not allow for reducing unfair competition and the qualitative drift it was allegedly encouraging.

> The law is dangerous, for by administratively attributing a fictional value to everything that is harvested within the perimeter, it encourages those who might be tempted to plant dubious varieties in bad soil, resulting in mediocre quality wines which, because of the administrative delimitation, would nevertheless be legally fully entitled to the renowned names their producers want to give them. (Capus 1947: 15–16).

The 1905 law interpreted appellations of origin as the place of origin of the grapes used to produce the wine. Yet customarily, J. Capus noted, appellations of origin were especially a symbol of wines' superior quality, differentiating them from ordinary wines without a particular appellation. The idea that wines with an

[20] Wolikow (2009).

appellation were of a higher quality did not imply that *all* wines made within a specific area of appellation were so, but that *exceptions aside* they were indeed of a superior quality, considering the history, professionalism of the winegrowers, the competence of the drinkers to appreciate the wines, etc. However there needed to be a minimum assurance that the land up for delimitation had the potential required to produce virtuous wines.

J. Capus' interpretation conflicted with another one, supported by a new bill: the Pams Dariac Bill. Specific to wine, it advocated an interpretation of the denomination of origin simply as a way to trace a producer, as envisaged in the debate prior to passing the 1905 law. Supporters of this interpretation highlighted the difficulties raised by the delimitations of appellations and the definition of honest and constant customs. If it was so difficult to trace geographical boundaries or to define traditional practices, what hope was there for defining the qualitative superiority that delimitates good and bad wines?

In 1919 a subsequent law on appellations of origin was passed. This law, which provided producers with a tool for commercial differentiation, was a great success: the number of geographical wine appellations registered skyrocketed. Producers registered the name of their vine plots to endow their production with a commercial brand, alongside collective appellations enjoying a local, national or international reputation. The volumes of appellation-wines rapidly increased, as did yields. The law also resulted in a very wide diversification of wine qualities, particularly with the use of certain innovations created during the phylloxera crisis, like growing new varieties of vines, hybrids, with an often very high yield, which gave the wines new taste features and qualities.

6.3.3 Innovating to Avoid Competition?

The 1905 law moved away from defining the terms "Bordeaux wine" or "Champagne wine" based on the style of the wines, as merchants did, and turned to the wine production and especially grape production process. Later, the 1919 law allowed all winegrowers to use the geographical names of grapes' places of origin to designate their products. Many producers thus acquired a commercial brand name that made them identifiable for consumers.

Difficulties arose, however, when a region encompassed several producers. There again, producers proved to be highly creative, perhaps to escape the competition they faced from the same name of "wine from the X region" imposed by the denomination. The law on appellations of origin protected producers from what was seen as unfair competition from merchants who supplied themselves at a low cost in vineyards outside of the region of appellation. But at the same time, it introduced strong competition between producers of a same wine, the wine from their region of production. To survive, these producers sought to differentiate themselves and used all the means that the 1919 law provided or did not regulate: yields, grape varieties, etc.

6.4 Safeguarding a Heritage

J. Capus considered the success of this law to be calamitous, arguing that it increased the qualitative heterogeneity of wines and could only bring their quality down. Meanwhile, in the 1920s, there were other laws adopted that attempted to regulate the production of ordinary wines, without appellations of origin. The 19 August 1921 Implementing Decree (République Française 1921) forced winemakers to mention the alcoholic content of their wines, a mention considered pointless in the case of wines with a designation of origin assumed to be of superior quality. J. Capus and his supporters maintained that with these measures to systematically regulate the quality of wines without appellations, wines of appellation of origin paradoxically risked becoming a haven for bad wines, due to the lack of qualitative constraints. In 1935, he proposed another bill defending a new conception of appellations of origin. According to him, the simple traceability of origin introduced by the 1919 law could not contribute to restoring the quality of French wines. To counter this risk, J. Capus called for regulating the quality of production and, to better inform buyers, stimulating an upward quality trend. This required producers to understand that appellations' reputations and that the fruit of their work was now a collective good that needed to be preserved.

> Winegrowers saw legislation on appellations of origin only as a means to protect themselves from encroachment by their neighbours or usurpation of the trade, and nothing could make them happier than a theory equating the appellation of origin to a particular property. It did not cross their minds that producers could join forces, that the appellation of origin was a common good and that it just took one of them putting a defective product on the market to taint the appellation's reputation.
>
> By showing that the appellation is a collective right, Jenouvrier turned it into a moral issue and highlighted every producer's duty more distinctly. But he overlooked the question of quality. By turning the appellation into the collective property of the producers who use it, he did not see that these producers, under certain influences, could jeopardize it. (Capus 1947: 61)

In his writing, Capus turned appellations into a common good; if the practices of a few led to the destruction of this common good at the expense of all the others, a regulation had to prevent it.[21] He asked that appellations be considered a national heritage that the State also had a responsibility to protect:

> According to the theory developed in the 'Jenouvrier' report, which I discussed earlier, appellations of origin are the collective property of winemakers from the region of appellation. It is fair to say that they are part of the national heritage and that the State has the same rights and responsibilities regarding them as it does regarding certain sites and historical monuments, for example, protected by particular law. (Capus 1947: 37)

---

[21] Torre (2002) developed a utilitarian economic analysis of AOC's reputation as a "club" common good. For J. Capus, the common good was not reputation but the name of origin, which included all interpretations and meanings of the term, whether material or not. He did not seek to regulate the use of words, just one of the procedures supporting reputation building: the quality check based on the product and the gustative experience of wine.

The 1935 government Decree that he pushed through parliament established compulsory specifications for vitivinicultural practices in order to obtain the right to use the name of appellation of origin. The regulation of practices was designed to protect the quality of wines of appellation of origin by limiting acceptable innovations. It was the responsibility of the State to control its application.

These specifications involved winegrowing practices, the choice of varieties, their plantation density, their pruning and maximum yield in particular, key winemaking elements, and controlling the results, particularly regarding the alcoholic strength and the volatile acidity. The 1935 government Decree let go of gustative control of wines' quality, although taste did remain a major indicator of the quality of appellations of origin:

> A known wine is expected to provide certain pleasant gustative sensations already experienced. This legitimate requirement just needs to exist among a certain number of connoisseurs to be entitled to legal protection. Wine tasting, unfortunately, is not a science; it is helpless when it comes to expressing its appreciations in objective and concrete terms. (Capus 1906: 21), reproduced in (Capus 1947: 12)

J. Capus argued that wine owed its quality to an identifiable taste of choice, what today we could call the *typicity* of wine. But it seemed impossible for him to define its elements and thus build a wine quality recognition test. He supported the idea that in appellation vineyards, known for their exceptional growing conditions, the best practices made compulsory by the specifications ensured a minimum quality level.

With its controlled denominations, the 1935 government Decree was also a great success. In the 1980s, production under *appellation d'origine contrôlée* (AOC, controlled denomination of origin) exceeded that of ordinary wines. A significant price difference distinguished ordinary wines from a large number of appellation wines. The AOC became a sign of wines' superior quality. The government Decree spread to many countries, in a more or less amended, revised, tailored form. This success soon raised new concerns, and new controversies. AOCs differentiated superior quality wines for consumers but what real guarantees did it provide? The quality regulation by the 1935 government Decree was only concerned with limited elements of the production process and a few result indicators. Were these sufficient to distinguish superior quality products?

6.4.1 The Quality Guarantee Issue

Winegrowing practices were a subject of interest throughout the late 19th century. Regulations and the 1905 law managed to appease concerns about fraud and the subject of winegrowing practices vanished. AOCs and their control became quality guarantees that have been gradually challenged since the 1970s.

J. Capus argued that the authorized practices and poor choices of winegrowing land were responsible elements for the qualitative deterioration of wines. Stronger constraints on practices, coupled with more precise delimitations, would allow for improvement in the quality of wines with AOCs, the superiority of which would prevail once again. However, drinkers could get impatient; the improvement expected showed weaknesses in some aspects, which made the appellation of origin a prospective more than an effective quality guarantee. A wine critique emerged from this debate, with wine tasters evaluating the quality of wines and the general effective quality of AOCs. This question of the potential guarantee of the quality of AOCs led to a debate on the opportunity to introduce complimentary wine tasting.

6.4.2 Is Wine Tasting a Guarantee of Quality?

The success of AOCs led to the creation of intermediary categories of wine quality, between *table wine* and AOC wines: *Vins Délimités de Qualité Supérieure* (VDQS, delimited wine of superior quality), particularly the *Vins de Pays* (country wines). The purpose of these categories was to spur the qualitative improvement, within the table of wine variations, that had been fuelled among AOC wines. These *Vins de Pays* and VDQS were subject to less stringent specifications than those of the AOC, but also to compulsory wine tasting.

Some appellations of origin had already long integrated complimentary wine tasting into their specifications, and their numbers were rising. Public authorities were considering a reform to generalize this practice. This trend raised concerns among some producers, who set up a defence Committee, the *Comité Français pour la Défense et la Promotion qualitative des vins et eaux-de-vie à AOC* (COFRADEP, French Committee for the defence and qualitative promotion of AOC wines and spirits).

> Controlling the quality of the wine produced in AOC regions through compulsory tasting would complement the appellation.[...] thus the very substance and foundation of the legal doctrine of the AOC are being challenged by prioritizing an eminently subjective, fickle and imprecise criterion.[22] (COFRADEP 1974b: 370)

The authors also challenged the bias of the assessment procedure, which made winegrowers the judges of their own produce and furthermore paved the way for dishonest compromise.

> It leaves too much room for error, for the illusion of misconstrued camaraderie, for economic interest, for various influences, for organizational difficulties [...], for possible concealment in cellars, for AOCs to be willing to adapt. (COFRADEP 1974a: 473)

Getting delisted for lack of taste quality was a heavy and stigmatizing sanction. Tasting could not serve for more than simply detecting major consensual flaws. Who would implacably eliminate their colleague, neighbour or friend through an evaluation of the taste of their wine?

---

[22]I would like to thank P. Baudouin for sharing these documents with me.

The COFRADEP saw wine tasting as a faulty solution to a real problem: the challenging of the AOC's ability to guarantee the quality of their wines. Ultimately, the Committee produced a charter reasserting the "spirit" of AOCs with regard to the 1967 law: quality wines were the result of a remarkable adjustment between humans and their work tool, the vineyard. The practices resulting from this adjustment were therefore the ones that needed to be controlled by preventing improvements that denatured wines, unknown to any taster. It rejected tasting as a test control and demanded that regulation keep to the initial principles of AOC control.

The COFRADEP lost the battle: EU regulation introduced the compulsory tasting of all AOC wines. The EU administration was concerned with obtaining proof of the 'objective' quality of AOC wines, so these appellations could not be seen as barriers to competition. However, this compulsory tasting did not prevent the perceived worrying qualitative drifts. They explained that the purpose of appellations of origin related to the particular quality of the wines from a specific region of appellation was due to the remarkable dispositions of the vineyard that was exploited through vitivinicultural practices invented, tailored and selected through time. However, since the creation of AOCs, vitivinicultural practices had experienced many changes with the arrival of fertilizers, herbicides, and the systematic fight against all diseases and pests in vineyards. The situation was even worse in cellars: a long list of products to assist winemaking allowed oenologists to steer the final characteristics of wine as they pleased. All these innovations contributed to artificializing the style of wine at the expense of the expression of *terroir*, in other words the expression of the vineyard itself, representative of what is grown in a particular pedoclimatic environment.

6.4.3 *The Dispute Surrounding the Identity of Appellations and the Expression of* Terroir

As illustrated, at the end of the 19th century, winegrowers and merchants defended pure wine, free of sophistication. A century later, winegrowers are driven by a similar quest, "returning to the *terroir*", seeking the authentic quality of origin of their wines. Both in vineyards and in cellars, they are gradually banning all interventions likely to mask or denature the expression of their *terroir* in their wines. While this renewal of practices is often based on a return to old methods form before the last agricultural revolution, it reopens the idea of inventing a vitiviniculture that is more respectful of the land and protective of the environment. These changes in practices also induce a change in the tastes of wines. This new variability of wines is of concern to other winegrowers, who fear a loss of identity for appellation wines. There is a growing distance between advocates of the need for clear identification, which involves styles of wine having a stabilized identity, and winegrowers in search of *terroir* who justify changes in wine tastes in the name of rediscovering the taste of a *terroir*.

However, concerns regarding the excessive variability of AOC wines are no longer a matter of unfair competition against wines in search of *terroir*, which tend to be more expensive than the rest. On the contrary, their elitism is denounced, with its niche strategies which confine the market to the rich few when, still according to supporters of the stabilization of wines' identities, the focus should be on expanding demand for the consumption of appellation wines. One could think that the terms of the problem have been reversed, that the problem today relates to qualitative excess. But this would be a dangerous claim. In particular, it would imply that a natural scale of quality exists, independently of instruments or protocols to measure it. While the price of wines is very widely—and roughly—used to this end, the opposition does not boil down to a dispute about prices. Advocates of the return to *terroir* products are calling for the resulting quality to not be constrained, and rather for acceptable practices to be better regulated. They wish to increase the variability of the result, to allow for rediscovering the real, untouched *terroir*. To do so, they need to reinforce the regulation of practices in order to eliminate those that fail to sufficiently respect the land. Their challengers call for the exact opposite: in the name of market imperatives, they want a better framed result, recognizable by consumers, and greater flexibility in practices in order to achieve this targeted result each year, irrespective of vintage variations.

At the turn of the 20th century, the French administration was concerned with reconquering the market. It supported producers who defended qualitative excellence as a commercial differentiation. Today, 100 years later, it is once again concerned with foreign production encroaching on the market shares of French vitiviniculture. Nonetheless, the appellations of origin administration, the *Institute National des signes d'origine et de la qualité* (INAO, National Institute for Signs of Origin and Quality) and appellation unions elected by winegrowers rather tend to support advocates of the stabilization of wines' taste identity. Defenders of *terroir* production have the advantage however of often being recognized professionals, sometimes owners of prestigious domains, able to intercept winegrowing critique so that very gradually, a few journalists have been echoing their claims since 2004.

Throughout the 20th century, the notion of acceptable variability of products gathered within a same commercial category evolved. In 1919, the law on so-called simple appellations of origin stated that any variation was acceptable provided that the link to the vineyard of origin, which was in a way responsible for production, could be tracked. Diversity increased in terms of both the number of appellations of origin and variability within each category. In 1935 the government Decree on AOCs restricted the number of appellations and diversity within each category by generalizing stricter specifications and analytical tests. Regulation of the resulting wine quality then constantly grew throughout the 20th century, in the name of the effective quality guarantee that signs of appellation of origin were to provide. In the 1970s, the specification and analytical tests were further complemented with compulsory complimentary wine tasting, followed by the critical tasting organized by the winegrowing critique. It reinforced control of the result and not of practices; critics thus judged the taste of wines primarily on its own merits and, until very recently, had little interest in vitivinicultural practices.

6.5 Two Diverging Interpretations of Heritage

In the early 2000s, repeated incidents affected AOCs and complimentary tasting in particular. Wines produced by defenders of *terroir* were excluded from appellations for lack of typicality. The winegrowers in question responded and accused complimentary tastings. In particular, they reproached tasting panels for being victims of commercial or technical excesses and for having lost all notion of wines' *terroir* quality. The debates on AOCs' capacity to provide a quality guarantee intensified. In 2008, a EU reform of AOCs required all producers to rewrite their link to *terroir*, in other words everything that, in areas of appellation of origin, contributes to the singular quality of wines. The purpose of this rewriting was to build a more reliable system to verify wines' *terroir* and its quality.

6.5.1 The Question of the Standardization of Production

Implementing the reform raised issues that the 1919 and 1935 laws had avoided, concerning the difficulty of defining the taste of productions carrying an appellation of origin as well as the question of the standardization of production. Was it desirable, winegrowers asked, for all wines of an appellation to be identical? Would this cause a price war and therefore a drop in quality? Moreover, defining quality carried the risk that the elements of the conformity test would allow for reproduction of the taste quality. Insofar as oenological technology allowed for it to be reproducible elsewhere, the definition of quality of origin would make it a generic taste quality, a standard or a style of taste, and would encourage the emergence of especially fierce competition due to a style enjoying a strong reputation. They argued that by trying too hard to reduce the resulting quality to the consequence of a winegrower's establishment in a particular region, the 2008 reform risked turning quality into a standard achievable everywhere else.

Advocates of terroir quality feared that this standardization would prevent the necessary variability of appellation wines' quality. They argued that quality could not be reduced to the predetermined effect of a set of *terroir* factors. Defenders of *terroir* products proposed that there must be a link to the author, the vineyard itself. As conveyed by the term expression of *terroir*, which they constantly use, the production of quality *terroir* wine is more like the musical interpretation of a musical author: a winegrower's wines are interpretations of his or her *terroir* which must be expressed therein. For them, the link to the *terroir* is not a causal determination link, but a link to the one responsible for quality. To ensure that this would be the case for AOCs, they ask for controls to guarantee that winegrowers, through their respective practices and winegrowing projects, give importance to the *terroir* and facilitate its expression.

The 2008 reform did not manage to subdue the advocates of *terroir* wine. On the contrary, the difficulties surrounding its application have stirred up opposition between two competing interpretations of heritage, which AOCs are responsible for protecting. Should the advocates of *terroir* wine be vindicated by loosening control of the quality of the result and tightening control of *terroir*-friendly practices? Alternatively, should the 2008 reform be maintained, particularly the control of the conformity to *terroir* quality, which it seeks to establish.

6.5.2 Two Interpretations of AOC Heritage and of the Link to Origin: The Determined Link to Be Reproduced, and the Author Producer Link

Current conflicts within appellations do not just relate to the question of control. They revisit the notion of heritage introduced by the appellation of origin sign. When winegrowers or the administration conceive a link to origin as a result of a predetermined cause, the product owes its characteristics to the causes that determine it. But should these causes be treated in a particular way? For those who see the *terroir* as an asset, these causes determine a necessary quality. In this case, the AOC should not be concerned with guaranteeing a particular quality. It simply should inform buyers of the particular local determination, which is the source of the specific quality of wines with a particular appellation. In this first case, signalling the quality of origin amounts to providing information on the final quality of the product. Products with similar characteristics can be produced without constituting reprehensible copies of the quality of origin, but the indication of origin signals original, *authentic* wines, as opposed to *copies*. For others, on the contrary, this determination is valid only if the winegrower is competent and does not denature the wine. *Terroir* quality must therefore be subjected to a conformity test to verify if the appellation of origin is of genuine quality. For the interpretation of *terroir as reproduction* to make sense, according to this approach, the label must provide *guarantees* regarding the quality of the reproduction and its conformity to the original quality of the wine appellation, which the 2008 reform addresses.

The advocates of quality expressing *terroir* propose an entirely different approach to the link to origin. Heritage refers not to the predetermined production of a set of factors or particular causes called terroir, the reproduction of which needs to be monitored, but to a production created by a set of authors, human and natural, the terrior. Their concern, given the excesses they have set out to fight, is to ensure the expression of their co-author, the vineyard. Therefore, they want the wines' quality of origin to be guaranteed by accrediting the winegrower and their practices.

The guarantee of results is in this case pointless. Each winegrower's commercial brand and their reputation are what ensures consumers of the products' quality and protects producers of poor productions from their colleagues.

6.6 Conclusion

Wine production appellations have always sparked different interpretations among interested actors: drinkers, tasters, producers, merchants, public authorities, and so on. These interpretations have evolved over the course of years. Merchants used the appellation of origin to refer to a particular taste quality. Diverging interpretations appeared and the quality of their wines of origin was challenged from all sides. During these debates, merchants defended their technical freedom and especially wine blending, which was necessary for them to produce this particular quality identified by a name of origin. They were not heard and the following laws, of 1905 and particularly 1919, associated the appellation of origin to a place of production of the grapes grown by the winegrower.

This attachment partly resembles that of a brand, associating commodities to a company producing them, with the difference that the company is a geographical place univocally identified by a land registry, which can be owned by a large number of winegrowers, nevertheless all independent in the way they conduct their trade. Moreover, a vineyard is not a being *responsible* for its production. In the debates surrounding appellations of origins, the winegrower is the one growing the vineyard and thus endowing the production with its own quality, who answers for it.

In the 1905 and 1919 laws, the appellation of origin introduces a twofold traceability link to the field and to the winegrower; neither relates to each one's respective contributions. J. Capus' reflection on appellation wines' superior quality is a step in that direction. With the notion of heritage, he introduces the idea of protection of a common good coupled with duties controlled by public authorities. Thus the AOC no longer designates a simple link to a geographical place, but a common good, a name that associates a production to a reputation. This fragile object is put in the sole hands of the winegrowers cultivating the vineyards in question and whose practices, regulated by specifications are monitored by public authorities. This notion of heritage gives a status to the vineyard and specifies the relations between the vineyard and the winegrower: the winegrower has to commit to properly grow the vineyard in order to not damage the appellation and its reputation. The vineyard is no longer a simple source of raw material associated with a place of production; the winegrower has duties towards it.

But once again the question of the process and the result has led to two different interpretations of this ephemeral heritage, which subsists through the continued production of wines of appellation of origin. According to one interpretation, this production is a reproduction, which targets a predetermined or pre-regulated quality of result, whereas another inrepretation puts forward that production resembles a work with regulated production techniques. Are we back to square one, with the management of quality based on the style of wines required by merchants, who were opposed to the regulation of production techniques, including grape supply? The situation has its similarities, however over the last hundred years the link to the place of cultivation has developed. Debates on the good quality of wines have led winegrowers to consider the respective contributions of the vineyard and of the winegrower, leading AOCs towards an embryonic status as producer for the vineyard, as a determining element in the acquisition of reputation. The two interpretations, based on the production or reproduction of heritage to be protected by the appellation of origin, diverge on this question of status. When the quality of origin is that of a good to reproduce, the vineyard is a set of production factors, a source of raw materials with predetermined characteristics, which the winegrower knows how to produce and process in order to develop a product with explicit characteristics. The link to origin is a link to original raw material exploited following a recipe or practice over the course of history. When, on the contrary, quality is to be produced, the vineyard becomes a source of creative expression. The link to place becomes a link to an author who is a captive of his or her roots in the land. This examination of the history of interpretations of the link to place shows how the regulation of meanings of the name shifted them away from genericity, giving particular weight to certain people in the emergence of what wines with appellations of origin are or can be. But it also shows that this attachment has been formatted and interpreted in different ways, closely linked to the management of the acceptable diversity of products designated by a commodity denomination, and to the hierarchy of threats that everyone thinks they have to guard against. Some have prioritized the diversity of result and strong regulation of acceptable means; others have made the opposite choice.[23] Do we need to distinguish between these conceptions or consider that both are serious threats—excessive rigidity at the cost of adaptability, and excessive diversity eroding identity? Alternatively, those in charge of regulation should focus on the coexistence of interpretations of the link to origin, which are not always particularly compatible.

Bibliography

Akerlof, G.A. (1970) The market for "lemons": Quality uncertainty and the market mechanism. *Quaterly Journal of Economics*, 84(3) pp. 488–500

Argenson, V. (1871) *Le parfait vigneron: Ouvrage inédit à l'usage des vrais gourmets, contenant une notice sur les soins à apporter aux vins depuis la maturité du fruit jusqu'à la mise en bouteille*, (Lyon: impr. de A. Vingtrinier)

Berard, L. and P. Marchenay (2006) Local products and geographical indications: Taking account of local knowledge and biodiversity. *International Social Science Journal*, 58(187) pp. 109–116

[23]Yet others reject any constraint, such as those who argue that the determination of terroir quality is a natural determination, who received little attention in this article. But they do not do so in the name of a threat weighing on the long-term survival of appellations of origin.

Bérard, L. and P. Marchenay (1995) Lieux, temps et preuves – la construction sociale des produits de terroir. *Terrain,* numero-24 - La fabrication des saints(mars 1995, [En ligne], mis en ligne le 08 juin 2007. URL: http://terrain.revues.org/3128. (Consulted 22 March 2013.)

Bowen, S. (2010) Embedding local places in global spaces: Geographical indications as a territorial development strategy. *Rural Sociology,* 9999(9999)

Brunori, G. and A. Rossi (2007) Differentiating countryside: Social representations and governance patterns in rural areas with high social density: The case of chianti, italy. *Journal of Rural Studies,* 23(2) pp. 183–205

Callon, M. and B. Latour (1981) Unscrewing the big leviathan; or how actors macrostructure reality and how sociologists help them to do so? Pp. 277–303 in K. K. a. A. C. (eds) ed., *Advances in social theory and methodology: Toward an integration of micro and macro-sociologies* (London: Routledge an Kegan Paul)

Capus, J. (1906) La défense des appellations régionales du département de la gironde. Pp. 31 in, Présenté à la session générale de la Société des Viticulteurs de Ftance et d'Ampélographie)

Capus, J. (1947) *L'evolution de la législation sur les appellations d'origine; genèse des appellations contrôlées,* (Paris: INAO– Editions Louis LARMAT)

Clémentel, E. (1914) *Un drame économique: Les délimitations. Le passé, l'avenir,* (Paris: P. Lafite et Cie)

COFRADEP (1974a) Appellations d'origine contrôlées. *Le progrès Agricole et Viticole,* 15 juin 1974) pp. 440–443

COFRADEP (1974b) Appellations d'origine contrôlées – trois mots pour un programme (suite). *Le progrès Agricole et Viticole,* 15 mai 1974) pp. 364–373

Dessaux, P.-A. (2006) Comment définir les produits alimentaires? L'élaboration des références pour l'application de la loi du 1er août 1905 entre expertise et consensus professionnel. *Histoire, économie et société,* 25(1) pp. 83–108

Di Palma, S. (2014) *L'histoire des marques depuis l'antiquité jusqu'au moyen âge,*Société des Ecrivains)

Eymard-Duvernay, F. (1989) Conventions de qualité et formes de coordination. *Revue Economique,* 40(2) pp. 329–359

Eymard-Duvernay, F. (ed.) (2006a) *L'économie des conventions. Méthodes et résultats tome 1 débats,* (Paris, La Découverte)

Eymard-Duvernay, F. (ed.) (2006b) *L'économie des conventions. Méthodes et résultats. Tome 2 développements,* (Paris, La Découverte)

Fallières, A. and J. Ruau (1908) Loi du 5° août 1908 modifiant l'article 11 de la loi du 1° août 1905 sur la répression des fraudes dans la vente de marchandisis et des flasifications des denrées alimentaires et des produits agricoles et complétant cette loi par un article additionnel. Pp. JO du 11 août 1908 p. 5637–5638 in, Journal officiel de la république française, N° 217, du 7? août 1905)

Feagan, R. (2007) The place of food: Mapping out the 'local' in local food systems. *Progress in Human Geography,* 31(1) pp. 23–42

Fonte, M. (2008) Knowledge, food and place. A way of producing, a way of knowing. *Sociologia Ruralis,* 48(3) pp. 200–222

Garella, P.G. and E. Petrakis (2008) Minimum quality standards and consumers' information. *Economic Theory,* 36(2) pp. 283–302

Griffe, J. (1889) Loi du 14 août 1889. in,

Grossman, S.J. and J.E. Stiglitz (1980) On the impossibility of informationally efficient markets. *Am. Econ. Rev,* 70(3) pp. 393–408

Ilbery, B., C. Morris, H. Buller, D. Maye and M. Kneafsey (2005) Product, process and place: An examination of food marketing and labelling schemes in europe and north america. *European Urban and Regional Studies,* 12(2) pp. 116–132

Jenkins, T. and N. Parrott (2002) The commodification of heritage and rural development in peripheral regions: Artisanal cheesemaking in rural wales. *Ecolabels and the greening of the food market. Proceedings of a Conference Boston, Massachusetts, USA, 7–9 November 2002,* pp. 51-61

Jullien, A. (1999 [1822]) *Manuel du sommelier ou instruction pratique sur la manière de soigner les vins – avec trois planches. Première édition 1813. Edition revue et augmentée.*, (Paris: Bibliothèque des Introuvables)

Kan, M. and B. Gulcubuk (2008) Geographical indications for recovery and local owned of rural economy. *Ziraat Fakultesi Dergisi, Uludag Universitesi,* 22(2) pp. 57–66

Lamine, C. (2005) Settling shared uncertainties: Local partnerships between producers and consumers. *Sociologia Ruralis,* 45(4) pp. 324–345

Laneyrie, F. (1926) *Résultat en bourgogne des premières années d'application de la loi dite de protection des appellations d'origine. Étude documentaire (première étape sur la voie de sa suppression),* (Macon: author's printing)

Latour, B. (1983) Comment redistribuer le grand partage? *Revue de Synthèse,* III(N°110) pp. 203–236

Lenoir, B.A. (1828) *Traité de la culture de la vigne et de la vinification,*Rousselon)

Letablier, M.-T. (1997) *L'art et la matière - savoirs et ressources locales dans les productions spécifiques,*Centre d'Etudes de l'emploi)

Loubet, E. and J. Ruau (1905) Loi du 1° août 1905 sur la répression des fraudes dans la vente des marcahndises et des falsifications des denrées alimentaires et des produits agricoles. in, Journal officiel de la république française, N° 210, du 5 août 1905)

Marette, S. (2007) Minimum safety standard, consumers' information and competition. *Journal of Regulatory Economics,* 32(3) pp. 259–285

Maurial, L. (1865) *L'art de boire, connaître et acheter le vin et toutes les boissons: Guide pratique du producteur, du marchand et du consommateur,* (Paris: Librairie du "Petit journal")

Paxson, H. (2010) Locating value in artisan cheese: Reverse engineering terroir for new-world landscapes. *American Anthropologist,* 112(3) pp. 444–457

Pouillet, E.-. (1883) *Traité des marques de fabrique et de la concurrence déloyale en tous genres,* (Paris: Marchal et Billard)

République Française (1921) Décret du 19 août 1921 pris pour l'application de la loi du 1er août 1905 sur la répression des fraudes. in République Française ed., (JORF du 19 août 1921 page 9753

Selfa, T. and J. Qazi (2005) Place, taste, or face-to-face? Understanding producer–consumer networks in "local" food systems in washington state. *Agriculture and Human Values,* 22(4) pp. 451–464

Sempé, H. (1898) *Régime économique du vin.* Faculté de droit de l'Université de Bordeaux

Shapiro, C. (1982) Consumer information, product quality, and seller reputation. *Bell journal of economics & management science,* 13(1) pp. 20–35

Stanziani, A. (2003a) Action économique et contentieux judiciaires. Le cas du plâtrage du vin en france, 1851–1905. *Genèses,* 50(pp. 71-90

Stanziani, A. (2003b) La construction de la qualité du vin, 1880-1914. Pp. 123-150 in A. Stanziani ed., *La qualité des produits en france, xviiie-xxe siècles* (Paris: Belin)

Stanziani, A. (2003c) La falsification du vin en france, 1880-1905: Un cas de fraude agro-alimentaire. *Revue d'histoire moderne et contemporaine,* 50-2(2) pp. 154-186

Tellstrom, R., I.-B. Gustafsson and L. Mossberg (2005) Local food cultures in the swedish rural economy. *Sociologia Ruralis,* 45(4) pp. 346–359

Torre, A. (2002) Les aoc sont-elles des clubs? Réflexions sur les conditions de l'action collective localisée, entre coopération et règles formelles. *Revue d'économie industrielle,* 100(pp. 39–62

Wilk, R. (1999) 'Real belizean food': Building local identity in the transnational caribbean. *Am. Anthropol.,* 101(2)(p. 244

Wolikow, C. (2009) La champagne viticole, banc d'essai de la délimitation (1903-1927). in, *Territoires du vin [en ligne], Pour une redéfinition des terroirs*

Young, H.P. (1996) The economics of convention. *Journal of Economic Perspectives,* 10(2, spring 1996) pp. 105–122

Author Biography

Genevieve Teil works, as a social scientist at the French National Institute for Agronomic Research in Paris. Her research focuses on quality, taste, perception and more generally on the different qualification and valuation devices or processes with a longstanding field experience on wine and the many controversies it has raised about its quality, taste or environmental namely, but also terror. This particular field expertise has fuelled more theoretical work about the experience of material objects and the variety of their presences to the experience in line with the pragmatist notion of experience developed by W. James. More recently her interest regarding the status of non-human like terroir has enlarged to milk microbes, pasteurisation and raw milk cheese. She has written and coauthored two books on wine (2004) De la coupe aux lèvres–pratiques de la perception et mise en marché de vins de qualité. Octarès, Toulouse and Teil, G., Barrey, S., Floux, P., Hennion, A., 2011. *Le vin et l'environnement: faire compter la différence.* Presses de l'Ecole des Mines, Paris.

Chapter 7
From Geographical Indications to Collective Place Branding in France and Morocco

Mechthild Donner, Fatiha Fort and Sietze Vellema

Abstract Geographical Indications (GIs) protecting the origin of specific food products are expanding worldwide and are promoted as a first order tool for agricultural and local development in developing countries. At the same time, collective place brands are adopted by the public and private sectors in Europe as a strategy to jointly valorize and promote a package of place-specific products and services, which supports a distinctive territorial identity. We hereby present two cases of collective place-branding from France and Morocco, with the objective of comparing them with Geographical Indications and to uncover their potential to contribute to territorial development. Results indicate that collective place brands can be used as an alternative (or complementary) strategy for supporting agricultural and territorial development. Contrary to GIs, these brands represent an integrated approach to territorial development, involve various local actors, are more flexible in use and less subjected to legal and administrative rules. They can create synergy effects and increase the overall visibility of places and their endogenous assets. On the other hand, collective place brands need leadership and cooperation in order to be successful. Moreover, collective place brands require a long-term commitment from the public and private sector for inducing novel forms of territorial governance.

7.1 Introduction

Communicating the quality of products on the markets is a process as old as commerce itself. It becomes particularly important if quality is not visible for the buyer as in the case of geographical origin of products. The indication of

S. Vellema
Wageningen University, Wageninge, The Netherlands

M. Donner (✉) · F. Fort
INRA (French National Institute for Agricultural Research),
Montpellier SupAgro-Inra, UMR Moisa 2, Place Pierre Viala,
Montpellier 34 060, France
e-mail: mechthild.donner@inra.fr

W. van Caenegem and J. Cleary (eds.), *The Importance of Place: Geographical Indications as a Tool for Local and Regional Development*, Ius Gentium: Comparative Perspectives on Law and Justice 58, DOI 10.1007/978-3-319-53073-4_7

geographical origin of food products can be a collective intellectual property right and then underlies strict regulations in Europe, in order to protect producers against dishonest competition and to guarantee the product's origin to consumers.[1] While the objectives of the first quality signs in Europe were primarily to protect producers from misuse of the product name and increase their incomes via price premiums, the multiplicity of food crises has led to an explosion of different signs and labels that aim to inform and return confidence to consumers. Interventions of public authorities as guarantors of the faithfulness of the transactions and information towards consumers focus on quality signs whose implementation procedures and product specifications are thoroughly controlled.

Among the official signs in Europe, Geographical Indications (GIs) make an explicit reference to the geographical origin of products. The importance of certifying local food products as GIs has largely been documented. Several EU-projects (SINER-GI, Dolphins) have been executed and official guides have been developed (International Trade Centre: Giovannucci et al. 2009, FAO: Vandecandelaere et al. 2009), that show the potential of GIs to add value to typical local products by means of differentiation. However, GIs are not exclusively commercial or legal instruments, but can also increase revenues of small-scale actors, foster local values and preserve biodiversity and heritage (culture and tradition). Therefore, they are also used as policy tools to stimulate the sustainable development of territories and to create dynamics in rural areas.

In Europe, the majority of GIs concern food products of the five Mediterranean countries, i.e. Italy, France, Spain, Portugal and Greece. Certainly, the negotiations regarding the regulations of quality signs have led to a division between northern and southern European countries; the first group of countries being rather reluctant and considering these norms as a form of protectionism, the latter arguing in favor of defending their culinary heritage.

In the context of developing countries, territory-based approaches protecting local products are proposed as first order tools for agricultural and local development. Strongly promoted by international organizations, such as the World Bank and FAO, and national authorities, a growing number of initiatives try to protect and valorize traditional and territorial food products. These mainly use GIs as an instrument to contribute to a local, integrated and sustainable development of places.

At the same time in Europe, the success of origin-products on the markets, the political decentralization and regionalization processes and the globalization incite regions to launch collective place brands. Contrary to GIs, which deal with mono-product chains, these collective place brands form a diversity of local quality products and services (often food and tourism) in order to jointly promote them, based on a distinctive territorial identity. Therefore, we hereby question the nature,

[1] In the framework of the WTO, the 1994 TRIPS (Trade-Related aspects of Intellectual Property Rights) agreement classifies Geographical Indications (GIs) as new category of intellectual property right.

processes and potential outcomes of these collective place brands and their opportunities to be used as alternative development strategy to GIs.

The aim of this chapter is to compare collective place brands with GIs and to uncover their potential to contribute to territorial development, via two case studies. We present two cases from Mediterranean countries: the first one is the French region Languedoc-Roussillon and its place brand *Sud de France*, to show the process, dynamics, governance form and development outcomes of a collective place branding initiative; the second one is the rural province of Chefchaouen in Morocco with its various initiatives for valorizing local assets including food products linked to eco-tourism.

The chapter is structured as follows: first, we bring forward the current status and enabling conditions of GIs in Europe and their challenges regarding developing countries; then, we introduce collective place brands, highlighting their conceptual differences with GIs and explaining the interest of implementing them as the basis for a territorial development strategy; thereafter, we present the results of the two place branding cases and discuss these results contrasting the differences of collective place branding compared to GIs; this allows to introduce general conclusions and policy recommendations.

7.2 Geographical Indications in Europe and Developing Countries

In the present section, we introduce the current status of GIs in Europe and the conditions significant for their success. We also explain why GIs are particularly challenging in regards with developing countries.

7.2.1 GIs in Europe: Between Commercial Success and Territorial Divergence

The European Commission defines a GI as "a distinctive sign used to identify a product as originating in the territory of a particular country, region or locality where its quality, reputation or other characteristic is linked to its geographical origin".[2]

In the TRIPS agreements of the WTO (World Trade Organisation), GIs have been recognized as intellectual property rights on the same basis as commercial brands; this has strengthened their status as quality signs based on food origin and has contributed to the international expansion of product certification procedures.

[2]http://ec.europa.eu/trade/policy/accessing-markets/intellectual-property/geographical-indications/ as consulted March 31st 2016.

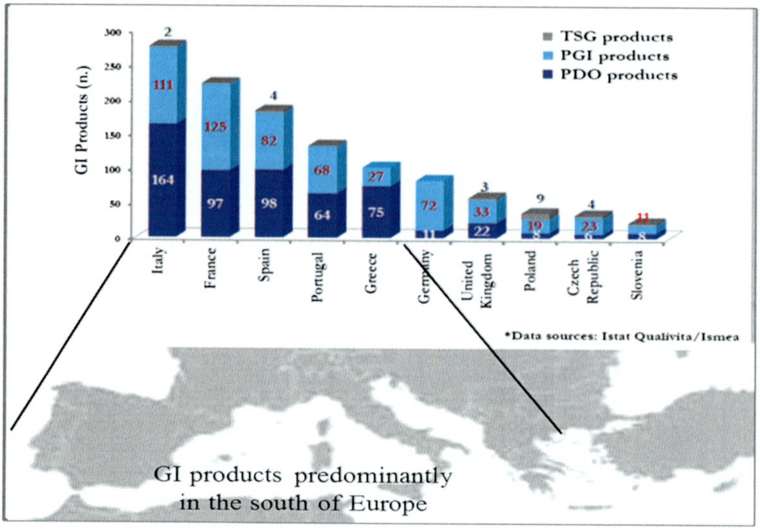

Fig. 7.1 The number of PDO, PGI and TSG products in different EU countries. *Source* Figure based on data resources from Istat Qualivita/Ismea, 2014

The EU, which has the largest number of agricultural and food products protected via GIs, disposes of a well-developed system of GI (sui generis) regulation, based on three types[3]:

- Protected Designation of Origin—PDO: covers agricultural products and foodstuffs, which are produced, processed and prepared in a given geographical area using recognized know-how.
- Protected Geographical Indication—PGI: covers agricultural products and foodstuffs closely linked to the geographical area. At least one of the stages of production, processing or preparation takes place in the area.
- Traditional Specialty Guaranteed—TSG: highlights traditional character, either in the composition or means of production

However, as illustrated in Fig. 7.1, the distribution of GIs among EU countries is fairly unequal, with a clear dominance of the Mediterranean countries, notably Italy and France.

The EU system of GI protection for agricultural products is in general considered a success story. Economic statistical data (Table 7.1) reveals that GIs have the potential to create value and contribute to long-term economic development. In 2014, their sales volume was estimated at €55 billion (estimated at wholesale stage in the region of production), with food contributing €16 billion, wine €31 billion and spirits €8 billion. This represents 6% of the total EU food and drink sector.

[3]http://ec.europa.eu/agriculture/quality/schemes/index_en.htm as consulted April 2, 2016.

Table 7.1 The economic importance of GIs for food and beverages in the EU

	Food	Wine	Spirits
Sales value (2014)	16 billion	31 billion	8 billion
Total	55 billion		
Export	11.5 billion (16% of EU food and drink industry exports)		
Premium price	2.25 times the price of a comparable non-GI product		

Data sources Istat Qualivita/Ismea, 2014

Regarding the premium expected for a GI from the market, compared to non-GI product, the price of a GI product is on average 2.25 higher than the price of a comparable non-GI product.

7.2.2 *Enabling Conditions for GIs in Europe*

Certainly, products with origin and quality signs complement standard production on national markets and are frequently consumed at a regional level. They even allow for a differentiation in export markets. This success is mainly due to the effectiveness of collective action, to the capacity of national control and to the sensibility of consumers.

First, the dimension of collective action for the development of GIs represents the core of the EU legal system, GIs being considered as a value creation process where numerous actors are involved (Pacciani et al. 2001). Producers, processors and distributors must be able to cooperate for developing, managing and assuring the protection of the GI specifications (Hayes et al. 2003). The diversity of actors participating in the management of GIs has led to different governance forms of GIs in Europe. Allaire and Sylvander (1996) show that depending on the amount and nature of the actors involved in GI development, the governance is either territorial or industrial, which is reflected in product specifications being more or less strict. In the case of the French Pelardon cheese e.g., the governance can be qualified as territorial because the GI is mainly managed by small producers and the requirements are very restrictive. In opposition, the governance of the Comté cheese is industrial due to the importance of refiners in the valorization of the GI; here, the specifications leave some flexibility to industrials in order to differentiate themselves via brands or other characteristics as e.g. the length of the maturing periods. Hence, the success of a GI depends on the capacity of local actors to find a balance between a liberal strategy allowing the production of a large product portfolio under the same GI (case of Comté) and a strategy with a unique and strong image that reduces differences between products with the same GI on the market (Bramley et al. 2009).

Secondly, the capacity to control the quality of a GIs is crucial in order to avoid the risk of free-riders (Ménard 2000) and to enhance the credibility of specifications

(Charlier 2007). In economic theory, the reputation of products is the best way to reduce information asymmetries between producer and consumer (Belletti 2000). However, for upholding the reputation of products, it is necessary to protect and maintain their quality; and this can only be done by means of efficient control mechanisms. In Europe, the registration of a GI therefore requires that specifications be referred to an officially recognized control body. The development and recognition of a GI by the EU allows fighting against misuse and imitations and thus to reserve benefits of GI usage only for authorized users. This protection is based on the TRIPS Agreements of 1994 and the European Council Regulation (No 510/2006). However, according to the European Court of Auditors (Special Report No 11/2011), the protection is insufficient to date, because:

- The provisions for checks of compliance of PDO and PGI products with the product specification do not set minimum requirements.
- There is a lack of a clear legal definition for the checks aimed at the detection and suppression of disallowed practices, with the consequence that member states' reporting concerning their checks of the GIs scheme is incomplete.
- Weaknesses exist in the commission's supervision of member states checks related to the Geographical Indications scheme.

Thirdly, consumers are a key factor for a GI's success. The price difference between a GI and a standard product in Europe is on average 2.25. The effect of the country of origin of products on consumers (COO-effect) and their willingness to pay (WTP) for a specific product quality and origin has largely been researched in academic literature. Table 7.2 lists several studies on the WTP across different

Table 7.2 Empirical studies on the willingness-to-pay of consumers

Authors	Type of data	Products	Main results
Loureiro and McCluskey (2000)	Consumer survey on consumption patterns and attitudes	Galician veal	PGI as a powerful marketing tool if combined with quality indicators. Marginal diminishing returns with respect to quality
Teuber (2010)	Internet auction data for single-origin coffee	Single-origin coffees	Single-origin coffees receive price premiums due to their reputation
Bonnet and Simioni (2001)	Scanner data on purchases of Camembert brands	French Camembert cheese	The PDO label is not significant. The brand appears to be more relevant in the consumer's evaluation of alternative products
Monteiro and Lucas (2001)	Consumer survey	Traditional Portuguese cheese	PDO is the most important attribute for consumers
Caputo et al. (2011)	Consumer survey	Olive oil	Consumers are willing to pay the highest price for extra-virgin olive oil, followed by PGI, PDO and Organic farming labels

Source Building on Bramley et al. (2009)

countries and products. Although these studies suggest that consumers attach value to the origin of food products and are in general willing to pay a premium price for them, the robustness of GIs compared to well-known brands and to other indicators as organic labels or fair trade may be questioned.

7.2.3 Some Challenges to Creating GIs in Developing Countries

As GIs have in general a strong socio-economic impact in the European Union, they are increasingly promoted in developing countries. However, little is known about the extent of the benefits that GIs can create in developing countries (Bramley 2011). Even more, the creation of GIs poses serious challenges for developing countries, as their implementation needs a wide range of activities and conditions, at a legal and political, financial and organizational level (Vandecandelaere et al. 2009).

First of all, a number of legal and policy issues arise. The WTO requires that national legal systems for GI protection be brought in accordance with the TRIPS agreement; GIs must be protected in their home country before they can be registered at an international level. In addition, developing countries have to claim rights for new GIs, i.e. proof that the quality of the product is fully attributable to its origin. As the majority of policy makers in developing countries have none or limited experience with the use of GIs, they are often given technical and regulatory assistance by developed countries (Anders and Caswell 2008). On the other hand, the EU and the USA seek to promote their version of a GI protection system and to conclude bilateral protection agreements with developing countries in order to assure best economic advantages. Hence, policy makers in developing countries need to consider regulatory issues as well as commercial priorities before putting GIs in place (Anders and Caswell 2008).

Furthermore, according to the CIRAD (the French Agricultural Research and International Cooperation Organization, Workshop 2009, in Bramley 2011), cost burdens for GI enforcement is one of the biggest challenges for developing countries. Apart from the financial efforts by national institutions for adopting laws and applying for GI registration, product specifications and quality controls entail costs for local producers. In addition, marketing and promotional campaigns are needed for origin-based products in order to make consumers aware of their specific value. This is even more urgent for products that do not yet enjoy a well-known reputation, as it is often the case in developing countries (Das 2009). Expensive niche marketing strategies are risky for any enterprise, but even more so for producers with scarce resources (Bramley 2011). Thus, cost-benefit ratios must be carefully considered in the context of GI development.

Finally, there are serious organizational challenges related to GIs. The efficiency of the collective organization of producers and of the cooperation between different supply chain actors is crucial for the success of GIs (Rangnekar 2004); these elements impact the fair distribution of the added value (Reviron et al. 2009). However, empirical examples from developing countries show that collective action among producers is often difficult to put in place (Fournier 2015; Vitry et al. 2015). The collective, voluntary and exclusive nature of GIs has led to a so called club theory among scholars (Bramley and Kirsten 2007). Members of a GI club are not only collectively producing a common good, but also facing internal competition. Therefore, the use of GIs that increase local competition may not be evident for local farmers. In addition, for farmers in developing countries often favoring direct sales or short distribution channels, it may seem as a market risk to create GIs that are mostly designated for export markets from which they derive the added value. Closely related to this, the lack of sophisticated supply chain structures that are significant for the success of GIs may reduce the possibilities for efficiently marketing GIs in developing countries.

7.3 Collective Place Brands

Below we introduce collective place brands, highlight their conceptual differences with the concept of GIs and present collective place brands as the basis for territorial development strategy.

7.3.1 What Are Collective Place Brands?

Collective place brands are strategies to jointly valorize and promote baskets of local (food) products and services, based on a distinctive territorial identity. Hence, they link regional identity, endogenous assets and various local actors (often from the food and tourism sectors). Place branding can be defined as "the practice of applying brand strategy and other marketing techniques and disciplines to the economic, social, political and cultural development of cities, regions and countries".[4] While in academic literature, a lot of attention has been given to nation and city branding; the regional scale has been rather neglected. Moreover, research has focused either on institutional place brands or on GIs, but little is known about the collective type of place branding which is based on public-private interaction and cross-sector synergies; and even less about its potential for territorial development (Donner 2016).

Place brands are increasingly developed in Europe, often as part of regional public agendas. In France, nearly each region has a collective place brand, sometimes even several ones. Place branding strategies face the following challenges: to

[4]The Journal of Place Branding and Public Diplomacy: http://www.palgrave-journals.com/pb/index.html as consulted April 2, 2016.

create a strong common brand identity, which incorporates the core values of the labeled territory and fits to different sectors (Kotler and Gertner 2002), to keep a coherence of quality within the portfolio of products and services in order to avoid damaging the brand image (Iversen and Hem 2008), to develop a strong governance structure, as e.g. a public-private partnership (Rainisto 2003); and to orchestrate the cooperation between different actors (Therkelsen and Halkier 2008).

7.3.2 Conceptual Differences Between GIs and Collective Place Brands

A collective brand is, unlike an individual brand of a company, simultaneously used by different enterprises (producers or processors) that are not necessarily owners of the brand. It allows bringing together enterprises while providing them with a common image and brand awareness among consumers.

The WIPO (World Intellectual Property Organization) and the INTA (International Trade Mark Association) differentiate between two types of collective trademarks: ordinary collective marks where members of a collective (a cooperative, association, or organization) share certain characteristics, and certification marks which are based on defined standards, controlled by external certifying agencies and open for anybody who meets the set requirements.

Collective and certification marks are registered with national trademark offices and protect the owners from fraudulent use of the mark. GIs are also in the vicinity of marks, but do not indicate the source of a particular manufacturer as trademarks do. They are neither indicative of the membership of a particular association as collective marks nor indicative of minimum standards of quality laid down by a certifying agency as certification marks (WIPO 2007). GIs guarantee the origin and the quality of a product based on where it originates. Unlike trademarks, GIs are inalienable (cannot be sold or given a using license) and based on public law. Collective place brands have in common with GIs that they both rely on the concept of *terroir*, which is associated by consumers with natural, authentic, traditional or tasty food. They aim to create value locally by valorizing and promoting typical and qualitative territorial products. Furthermore, like GIs, collective place brands need interaction and collaborative processes between private and public entities in order to be successful.

However, collective place brands are different from GIs with regard to their scope and liberty of choice: while GIs protect single products; collective brands can cover a wide range of different local food products as well as tourism or other endogenous assets. Thus, collective place brands concern not just a group of actors in a single value chain, but multi-clusters with cross-sector synergies. Moreover, while GIs are a formal, institutionalized tool, and therefore underlie strict administrative procedures, collective regional brands can be used in a flexible way,

Table 7.3 Conceptual differences between GIs and collective place brands

	Geographical Indications	Collective place brands
Ownership	Owned by the state (as public good), on behalf of the producers who govern and use the GI in conformity with the established specifications	The owner of the brand is a government or an association, which gives the right to use the brand to those who respect the specifications
Governance	Mainly the producers (bottom-up)	Various forms possible
Objectives	Legal protection of true origin of single products to avoid a fraudulent use of the product name	Valorization and promotion of several or all place-specific assets, often quality (food) products/tourism/crafts/natural landscape
Legal framework	WTO TRIPS agreement 1994 (Intellectual property rights) EU legislation for protection of GIs and National legal frameworks	Registration and protection for collective brands at the national trademark offices
Target	Mainly export, but GIs can also target local markets	Residents, tourists, export
Geographical scale	Established territory	Region or department
Examples from France	*Bordeaux wines* *Champagne* *Roquefort cheese*	*Produit en Bretagne* *Bravo l'Auvergne* *Le Pays Cathare* *Sud de France*
Anchorage in literature	– Country-of-origin – Origin and quality labelling – Food sociology & anthropology – Food law and regulations	– Strategic marketing – Place branding – Destination branding – Country-of-origin – Origin and quality labelling

Source Authors

enabling various strategies, types of governance and functioning. Table 7.3 shows the conceptual differences between GIs and collective place brands.

7.3.3 *Collective Place Brands: A Tool for Territorial Development?*

The process of collective place branding is similar to a territorial development project and it is based on a public-private interaction, specific territorial assets (or competences) and a common brand identity that unifies material and symbolic values and is distinct from those of competing territories (Rochette 2012).

Due to regionalization and decentralization processes, many European regions are conscientious of local development. They need to create suitable environments to stimulate economic growth, to support local enterprises and to ensure the welfare of the population. The importance of regions as drivers for economic development and growth has been widely recognized in theory and practice (Storper 1997; Pike et al. 2006; Porter 2003; Scott and Garofoli 2011, EU Cohesion Policy). Nevertheless, regions are facing strong national and international competition at economic, spatial and identity levels. This is reinforced by globalization, "exposing even the most remote spaces to competition and forcing firms, localities and regions to react and adjust to the new economic conditions" (Pike et al. 2006: p. 4). Competitiveness has become a new paradigm for territorial development (Camagni and Capello 2013). Therefore, many European regions have started to apply place marketing and branding strategies with the aim of increasing the attractiveness and competitiveness of places, to create dynamics and to stimulate local development, by promoting distinctive (material and immaterial) assets based on unique territorial identities.

Since the middle of the 1990s, the territorial development paradigm has appeared in France and Morocco, taking into account not only a national, but worldwide competitive environment. The territorial approach aims at making places more competitive and attractive. It introduces new forms of governance via a coordination and cooperation of multiple actors (also with the participation of the civil society) and includes top-down as well as bottom-up approaches (Baudelle et al. 2011; Campagne and Pecqueur 2014). Governance can be understood as a production of rules by actors to regulate the economic, productive, social etc. system (Angeon et al. 2006). Local actors then socially construct their territory, in contrast to a given, historically-administrative delimited territory (Leloup 2005).

While earlier approaches were seeking for an optimal allocation of resources, now the identification, creation and valorization of specific territorial resources, being different from those of competitors, became important (Côté 2003; Pecqueur 2001). Territories are considered as places with own particular histories and identities. Heritage and local assets are used as elements for cooperation and as drivers for the construction and sustainable development of territories (Francois et al. 2006, Landel and Senil 2009). Increased mobility, particularly in forms of tourism and leisure, can offer opportunities for the development of rural areas, often combined with agricultural models (Skuras et al. 2006). Rural development has changed into a "multi-level, multi-actor and multi-facetted process" (Van der Ploeg et al. 2000). The OECD New Rural Paradigm 2006 has also shifted from sectors to *places* and from subsidies to investments in *local specificities*, thus highlighting the heterogeneity of rural areas (Ward and Brown 2009).

In Mediterranean rural regions, agriculture and tourism often constitute the most important and reliable sectors (Vandecandelaere and Abis 2012). As GIs deal with mono-product chains, creating collective umbrella place brands as integrated territorial development strategies may offer new options for building cross-sector synergies and stimulating territorial dynamics. In this area, the use of place branding strategies may be advantageous for farmers and small economic (food and tourism) enterprises within a highly competitive economy, to collectively valorize and market diverse products and services based on a distinctive territorial identity. Even more, collective place branding may also serve to protect the natural and cultural heritage of rural areas, to reinforce the embeddedness of local actors, to create and exploit local networks and finally to stimulate the sustainable development of territories.

Therefore, it is relevant to assess options for developing collective place brands for jointly valorizing and promoting various endogenous assets and for supporting small-scale local actors, leading to an integrated development of rural areas. These ideas are considered in the next section on the basis of two case studies from France and Morocco.

7.4 Two Case Studies of Collective Place Branding in the Mediterranean

7.4.1 Methodology

Two in-depth case studies of the *Sud de France* brand in the Languedoc-Roussillon and the various emerging and interlinked territorial initiatives in the province of Chefchaouen in Morocco have been performed by the authors (Donner et al. 2014, 2015). These studies represent different local initiatives that work towards territorial development by valorising and promoting various place-specific assets; they have been completed in order to understand the complex processes of collective place branding as a tool for adding value to local food products and territorial development.

In the Languedoc-Roussillon, an extensive documentary research and 13 semi-structured interviews with key stakeholders of the place brand *Sud de France* (regional council, development agencies, enterprises, farmers, tourism providers etc.) have been realized, in regards to the evolution, governance and various development outcomes of the brand.

In Chefchaouen, we have performed 21 in-depth interviews with different stakeholders, such as the mayor of the city Chefchaouen, the director of the agricultural ministry of the province, non-governmental organizations (NGOs), restaurant owners and cooperatives. These interviews were conducted to investigate options for creating a collective place brand. Leadership, cooperation and governance issues have been paid particular attention, being considered as major determinants for the success or failure of collective place branding. Therefore, various cooperating initiatives have been studied in Chefchaouen.

We have focused on two rural Mediterranean areas, the French region Languedoc-Roussillon and the province Chefchaouen situated in the Rif Mountains, North of Morocco (Fig. 7.2). A view on both sides of the Mediterranean Sea seemed challenging as Mediterranean countries possess similar territorial (gastronomic, cultural and natural) assets and climatic conditions, their nutrition habits are similar and they trade agricultural products among them; however, they are embedded in different political and economic contexts.

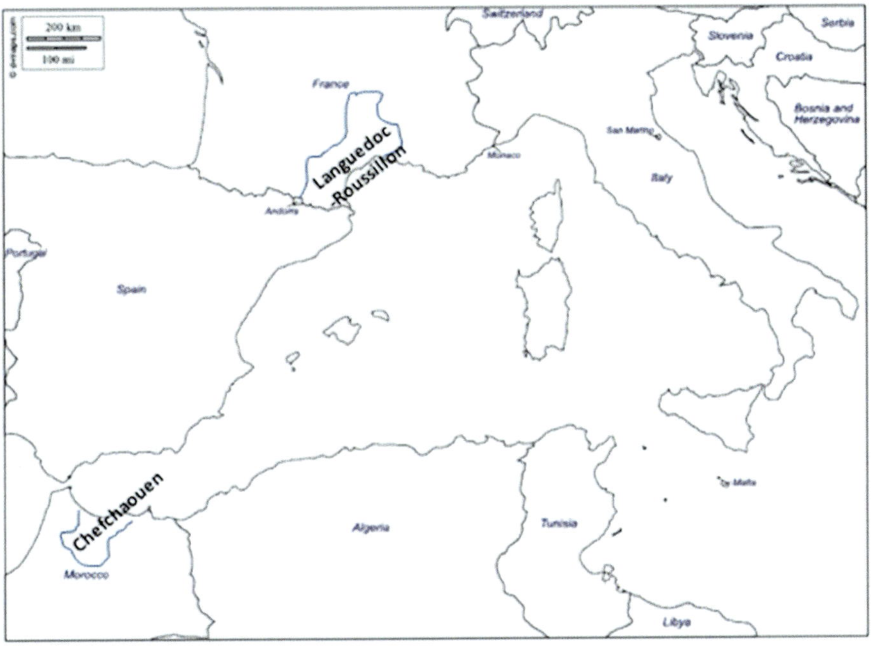

Fig. 7.2 Locations of the research areas. France: the region Languedoc-Roussillon and its Sud de France brand and Morocco: the province of Chefchaouen

In France, after a period of strong centralization, regions have progressively been given more administrative and financial autonomy since the 80's. Nowadays, competences between the central State administration and regions are partly shared and place-specific development projects (*projets de territoire*) co-financed. In the framework of decentralization, the number of French regions has currently been reduced from 22 to 14 (effective from 1 January 2016), to assimilate them with other European regions and enforce their economic power.

France has a long tradition and rich experience in the valorization and commercialization of local food products (*produits de terroir*). The INAO (Institut National de l'Origine et de la Qualité- National Institute of Origin and Quality) is the oldest institute for legal food protection in Europe. France also has the highest sales value of products with geographical indications and French retailers as Carrefour and Leclerc have since more than a decade their own trademarks for local food as *Reflets de France* and *Nos régions ont du talent*. At the same time, French regions increasingly develop place brands to highlight their place-specific assets.

The Languedoc-Roussillon region is bordered by the Mediterranean SEA, its capital is Montpellier. Compared to other regions in the South of France as the Provence or Côte d'Azur, Languedoc-Roussillon has been for a long time relatively unknown, despite its beautiful landscapes, cultural heritage (e.g. the Pont du Gard or the city of Carcassonne) and hundreds of kilometers of sandy coast. Agriculture and tourism are major economic sectors; Languedoc-Roussillon is the world's largest wine producing area and many small farmers and food processing enterprises are settled in the region.

Since 2006, the region has an umbrella brand called *Sud de France*, which is mainly used to valorise and promote local wine, food products and tourism services at a local as well as at an international level. The brand was launched by the former president of the region, Georges Frêche (2004–2010), in a time where especially small firms had difficulties to expand and the important wine sector was in crisis due to low awareness among consumers and strong competition of the new world's market producers. G. Frêche wanted to bring the region out of its seclusion, by activating endogenous resources for internationalization. Together with an organization for wine professionals, Inter Sud de France, he grouped for the first time all the different wines of the area (with and without appellations of origin).[5] Although *Sud de France* was initially intended as collective export label for wineries, the brand was very soon extended to local food products and tourism services in 2008. It also serves now to label other sectors, sport and cultural events as well as regional infrastructure and institutions.

The brand identity is constructed upon the values of the Mediterranean art of living, as an expression of a convivial lifestyle, with authentic, tasty and healthy food and wine. *Sud de France* is a tool to facilitate the consumers' choice among a multitude of products, which is particularly difficult as in the Languedoc-Roussillon 2800 wineries and 52 different wine appellations exist (26 AOC-AOP, Appellations d'Origine Contrôlée and 26 IGP, Indications Géographiques Protégés).[6]

> The average consumer is able to cite only five names of wines and doesn't make the difference between a brand and a PDO (protected designation of origin) … We think that Sud de France is the only name susceptible to be kept in mind by the average consumer. (Director, Wine Trade Organisation, cited from Galbrun 2012)

While in the beginning, a simple online demand was sufficient to become a brand member; criteria for using the brand have in 2014 been determined in a book of specifications *per* each chain. Several problems concerning food provenance—e.g. products from Spain were sold under the *Sud de France* label—, have resulted in a redefinition of the requirements and an introduction of quality controls. Since then, the number of brand members and labeled products from the agrifood sector has considerably decreased (approximately by one half). In 2015, *Sud de France* counted 4163 registered agricultural and food products from 1332 member enterprises and 1100 labeled quality tourism providers.[7]

The brand is financed and its strategy determined by the regional government as a brand owner; brand activities are delegated to the development agency *Sud de*

[5]Appellation of origin is a more restrictive category of GIs and one of the earliest forms of GI recognition and protection (Vandecandelaere et al. 2009).

[6]http://www.vin-vigne.com/vignoble/vin-languedoc-roussillon.html as consulted March 31st 2016.

[7]Press release 2015: http://www.sud-de-france.com/wp-content/uploads/2015/06/DP-sud-de-france_DOSSIER.pdf as consulted April 4, 2016.

France Développement. This organization defines actions for distribution, export and business development, ensures the worldwide promotion of the brand e.g. via trade fairs and festivals and assists enterprises with exports. It also plays a strategic role between producers and buyers, e.g. the big retailers. Using the brand doesn't offer a direct price premium for producers—as French foodstuffs are already highly priced and therefore less competitive at international level—, but the brand emphasizes a good price-quality ratio. Apart from some economic data proving the success of the brand for wine exports—Languedoc-Roussillon now being the first wine export region of France ahead of Bordeaux (cf. press release 2015)—, little is known about the various development effects of the brand. Therefore, we investigated the perceived brand benefits and drawbacks among different stakeholders via a number of interviews (Donner et al. 2014).

Results have indicated that the brand provides a number of economic as well as non-economic benefits. First of all, *Sud de France* contributes to the overall territorial development of the region as it increases its international awareness and creates a common territorial identity. It also helps to resist economic crises and to secure the existence of small firms.

> The brand has two values: an economic one for enterprises, as communication and commercialisation tool, and an institutional one that gives value to the region as a whole, by increasing the awareness and creating a common strong identity. This includes an extension of the brand to other sectors. (Member of the regional council, own interview, Montpellier 07/02/2013)

The benefits for enterprises mainly consist in cost savings for own marketing and access to various distribution channels, at local, national and international levels. The brand facilitates the commercialization of products and services by enhancing their awareness and image. The interviewed persons perceived whether or not the brand has created a local network differently. An important point mentioned was that common actions of the brand foster information flow and exchanges, within and across different chain actors. Something that has been critically regarded by brand members was the lack of strict quality requirements and origin controls for food products (before 2014). This was not only causing dissatisfaction among enterprises, but it was also perceived as a risk of damaging the brand image.

Thus, local development effects of the *Sud de France* branding initiative are various. They are principally economic but also entail social and cultural benefits such as the creation of a common territorial identity, a support of small-scale firms preserving a diversity of typical local products or the stimulation of local cooperation. A critical issue of the brand was to develop a clear definition of quality and product origin standards. Another important discussion point during interviews was the extent to which the brand can be applied to other domains. According to a brand manager, there are two opposite ways of thinking about the brand: as a restriction to the original key sectors agriculture and tourism vs. as an application to the whole region (as overall territorial marketing tool, holistic branding approach). The region now follows the second strategy, but it remains unsure how this may impact the further development of the brand. The future of the brand is also questioned by the recent integration of the Languedoc-Roussillon with the region Midi-Pyrénées. Altogether, the *Sud de France* case exemplifies that collective place brands are highly dynamic and can be used in a flexible way (e.g. the right to become a brand member or quality requirements). It shows that a place brand can create benefits for various stakeholders although a public institution retains it.

7.4.2 Morocco: Various Initiatives Towards Collective Place Branding in Chefchaouen

Morocco has been chosen as a research area because of its increasing international cooperation and opening of the economy (e.g. EU-Moroccan Free Trade Agreement), its decentralization and regionalization process (Plan de Régionalisation Avancée) and its initiatives for the development of the agricultural and tourism sectors (Plan Maroc Vert and Vision 2020), which represent strategic sectors for the socio-economic development of the country.[8] Notably, within the Pillar II of the Plan Maroc Vert, projects are performed concerning marginal zones (i.e. non-irrigated or mountainous zones and oases) that essentially depend on direct aid from the State. These projects consist of collective actions for valorizing territorial food products by means of GIs, following the European model. Some prominent examples are Argan Oil, Clémentine de Berkane or Safran de Taliouine. In Morocco a legal framework for quality signs exists since 2008, enabling recognition of the geographical origin of food products. In 2015, the EU and Morocco concluded a GI protection agreement.

The province Chefchaouen belongs to the region Tanger-Tetouan that has been defined as one of the strategic regions for the development of the agricultural sector. The province is mainly rural as 90% of its habitants are living in non-urban neighborhoods, countryside and remote villages. The capital of Chefchaouen is the city of the same name and unique urban commune with around 36,000 habitants. Living standards and infrastructure are highly contrasting between the urban area (the municipality of Chefchaouen) and the rural communes—which are characterized by the culture and people living in the mountains, called Jbala. Chefchaouen has a rich gastronomic heritage with typical terroir products such as goat cheese, olive oil, honey and dried fruits; it has a UNESCO label for immaterial heritage as one of the emblematic communities for the healthy and traditional Mediterranean diet. Furthermore, the province disposes of a rich biodiversity, with two national parks in the surroundings (Park Talassemtane and Park Bouhachem), and the city Chefchaouen possesses several cultural assets, e.g. the ancient medina and the Kasbah.

---

[8]Agriculture in Morocco constitutes 19% of the national GDP (15% agriculture and 4% agri-food) and has an employment rate of about 43% of the total workforce; the tourism sector contributes with 12% to the national GDP and has 5% of the workforce. Cf. http://data.un.org/CountryProfile.aspx?crName=MOROCCO as consulted April 4, 2016.

Due to these unique characteristics, it has already become a popular tourist destination. Many initiatives are taken to valorise and promote the various local assets. Therefore, Chefchaouen offers good options for developing a collective umbrella place brand based on the food and tourism sectors. However, Chefchaouen struggles with the predominance of the cannabis culture in the Rif and its negative externalities, amongst others on its image.

We identified different initiatives for valorizing and promoting various local assets that could lead to a collective and coherent place branding project. Activities for place branding at a micro-level in Chefchaouen included, for example, using the UNESCO label of the Mediterranean diet to increase the international awareness of the place and its typical food products, efforts to label Chefchaouen as eco-city and as child-friendly city, and elaborating a business plan for place branding by an external consultancy. Hence, a lot of ideas exist in Chefchaouen, but their implementation often seems to be difficult.

Several individuals were asked about who could lead a branding initiative. Interviews have shown that there is neither consensus about this issue nor confidence in a specific person or institution to play a vital role in the area. A major idea was that this should be financed by public institutions and elaborated in cooperation with the private sector. But the problem of the Chefchaouen area is a lack of political unity, because public actors such as the municipality, the provincial council and the ministries have heterogeneous competences and a very limited financial scope. The administrative units (city and province) e.g. do not correspond to the predominant mountainous culture of the Rif, the Jbala; and at a level of sectors, each representative (of the agricultural, tourism, craft ministry) is restricted to its own domain which makes a public financing of local transversal projects impossible.

Furthermore, various cooperating initiatives (meso-level) have been studied in Chefchaouen in order to investigate the potential for building partnerships and (cross-sector) synergies. One example is a project called *Les Restaurateurs Engagés,* which has been initiated in 2009 by the local association AFHTA Chefchaouen with the financial support of the French non-governmental organization GERES. The main aim of this initiative is to reinforce the agro-ecological system. The approach consists of valorizing and certifying agricultural products and selling them with added-value in local restaurants and guest houses. Short distribution channels are preferred for ecological reasons. In 2014, eleven farmers and six restaurants took part. Although some of the participants are highly dedicated and also involved in other initiatives or networks, the initiative as a whole suffers from an insufficient participation and also from a lack of financial means for marketing. In addition, because the financial support of GERES was ending in 2014, the continuation of the project was endangered.

Another cooperating initiative is the eco-tourism plan for Chefchaouen. Here, the local association ATED, in partnership with the Spanish foundation ETEA for cooperation and development has elaborated a tourism strategy for the province. ATED has brought together various public and private stakeholders, such as members from other local associations, from the municipality, and the private sector.

The complete strategy has been presented at the Government Tourism Advisory Service in Rabat. It is considered as an exemplary strategy, a pilot project in Morocco, including amongst others the creation of thematic circuits with rural houses in the province, and trainings for people active in the tourism sector. However, the tourism strategy for Chefchaouen has until now not been implemented. As the project coordinator explained, this is due to cooperation and federation problems among tourism actors, and above all, to the question of governance.

Public-private partnerships take also place within the Plan Maroc Vert, which is performed both at a regional and provincial level. Here, farmers' cooperatives and unions are created in order to stimulate cooperation, to label territorial products and increase production outcomes. Public support is given in form of various trainings with regard to organisational and commercial, plantation and quality issues and by financial means for infrastructure and material. This year, results of the PMV will be assessed by the number of products per region that have entered in a qualification process for Geographical Indications. Interviews in the Chefchaouen province indicate that the national top-down strategy of the PMV is not without barriers. The farmers' motivations to cooperate seem to be limited as potential benefits are not obvious to them, and distribution continues to take place via conventionalised channels.

The analysis of the Chefchaouen case has demonstrated that macro-environmental conditions for collective place branding in Morocco are favorable, as national policies increasingly give importance to global exchanges via free-trade agreements, to the agricultural or tourism sector via large development plans and to territorial development by means of regionalization and decentralization. At a micro-level in Chefchaouen, many efforts are undertaken to valorize endogenous resources (agricultural products, natural parks etc.) and to build a distinctive identity. However, a deeper analysis concerning leadership, political unity and cooperation has revealed that they are hindering the development of a collective place branding project. The initiatives taken at this level are often short-term, not coherent with other local projects and have difficulties to reach a full-scale. In addition, they are mostly stimulated and financed by external, international development organizations. This proves the importance of the meso-level, cooperating factors for place branding. Another insight gained from the case study was that although a collective place brand could be a useful tool for stimulating cross-sector cooperation and territorial development, Moroccan regions currently lack sufficient financial autonomy to fully develop their own territorial projects.

7.4.3 Discussion

The objective of this chapter was to compare collective place brands with GIs and to uncover the potential of collective place brands to contribute to an integrated territorial development, by presenting two case studies of the Mediterranean area. We argue that collective place brands can offer elements that are favorable for territorial development, which differ from GIs in four main points: collective place brands are an integrated territorial development strategy, implying public-private interaction and cross-sector cooperation; they create synergy effects; they can address both local and export markets; and they are flexible in use.

First of all, the major opportunity of collective place brands compared to GIs is that they represent a development strategy for a territory as a whole, as kind of a common territorial project. They include not only one product, but also several or even all endogenous assets of a place and different local stakeholders. Herein, cooperation can be fostered between chains and sectors and value co-created by different public and private actors. The Chefchaouen case has suggested that such a territorial approach could be a real chance to create coherence between many dispersed (external) development initiatives that are short-term and have difficulties to reach scale. In the Languedoc-Roussillon a collective place brand, bringing together all the different regional wines with local food products and tourism services under one common brand, has not only resulted to be an effective promotional and commercial tool, but also had other positive development effects such as a common territorial identity, the preservation of diverse local food products or the stimulation of a network.

The holistic territorial focus of collective place brands also makes it possible to integrate different forms of governance. Although *Sud de France* was a public initiative, the brand has successfully created regional dynamics and involvement of various public and private actors. Especially in developing countries, a brand can thus be a chance to stimulate public-private interactions and to develop a common idea or project adapted to specific territorial resources and conditions. In addition, collective place brands, fostering a common territorial identity, can create conditions for bringing together different mind-sets in urban centers and rural areas; thus connecting rural and urban environments.

Second, a joint sales and marketing strategy of different products and services is leading to economies of scale, while GIs often consist of small quantities. This also means that collective place brands can achieve a greater visibility of the place and its products than GIs through the promotion of one common, recognizable brand (vs. many different appellations in the Languedoc-Roussillon); this also allows an easier identification with the brand by the consumers. Furthermore, using a common brand implies that collective place brands have the potential of brand portfolio extensions, with possible spillover effects or multi-functional agriculture. In addition, place brands can contribute to change and readjust the image of a place and its products in the consumers' minds (e.g. the change of the initially bad image of the Languedoc wines and potentially the cannabis image of Chefchaouen).

Third, it can be advantageous for developing countries to sell products with added-value on local markets to tourists instead of targeting export markets and thereby facing complicated administrative procedures, as it is the case of GIs. Selling high-quality food products on home markets may even be easier as local consumers are in general more sensitive to their own local product quality than external ones and therefore willing to pay a superior price.

Fourth, a positive aspect of collective place brands is that they are flexible, e.g. the right to use the brand is defined by the brand owner, which may facilitate the access for small producers, while GIs have often the burden of strict standardized procedures. Here, the *Sud de France* case has revealed that product specifications depend on governance. They were flexible and changed over time: while in the beginning, the region as brand owner allowed any wine and food producer from the region to become a brand adherent (just by making a simple online demand) in order to achieve large and common territorial dynamics, in 2014 quality requirements have been redefined in accordance with the enterprises and must be demonstrated before obtaining membership. For GIs, differences in the governance have also been observed, e.g. the industrial Comté cheese having more liberal specifications is more successful than the very restrictive Pelardon cheese governed by local producers.

However, there are several challenges related to collective place brands, which are: product specifications, price premium, stakeholder involvement or leadership and cooperation.

First, as exemplified by the *Sud de France* brand, finding the right level of product specifications is necessary to avoid dissatisfaction among members and damaging the brand image; this also means deciding about inclusion and exclusion of members. Second, while GIs have the advantage that they guarantee a price premium for producers, this is not necessarily the case for collective place brands, whose economic benefits rather lie in increased sales due to brand awareness. Third, another challenge for collective place brands is to motivate and involve various local stakeholders. An appropriation by the private sector may particularly be difficult and take a long time if a brand is publicly governed as *Sud de France*; on the other hand in Chefchaouen, (institutional) leadership would have facilitated to start a collective place branding initiative. Fourth, cooperation between various actors from different chains and sectors is a critical point for collective place brands; this may be easier for GIs being mono-product initiatives.

Although the two approaches are different and implemented separately, it is essential for the debate to keep in mind that collective place brands can be (and in practice often are) combined with GIs; they are complementary. Thus, a joint use of GIs and collective place brands may exploit their respective advantages of having international legal protection, a guaranteed price premium and cooperating structures of single product chains on the one hand, and to profit from an integrated territorial approach, public-private interactions, economies of scale, increased visibility and cross-sector collaborations on the other hand.

7.4.4 Conclusion

As demonstrated through two case studies in France and Morocco, collective place brands can be used as an alternative (or complementary) strategy for supporting agricultural and local development. Contrary to GIs, collective place brands represent a holistic territorial development approach—including cooperation between various (public and private) local actors—, are more flexible in use and less subjected to legal and administrative rules. In addition, collective place brands have the potential to achieve synergy effects and to increase the overall visibility of places and their assets. These features make them particularly valuable for territorial development. On the other hand, collective place brands need leadership and cooperation in order to be successfully implemented.

This implies for national and local policy makers, especially in developing countries, that before deciding on a strategy for agricultural and local development, a detailed analysis of the territories is recommended. Thereby, not just a single (food) product chain, but several or all the available material and immaterial assets and their local market potential (including tourists) should be considered. This means taking into account the kind and diversity of local actors, as collaboration between public and private actors as well as across chains and sectors is important for territorial development. Such an approach may allow developing more tailor-made territorial development strategies instead of adapting existing standardized (Western) models. However, as we have only considered two cases of collective place branding, further research will be needed in developed and developing countries to assess the different strategies for valorizing and promoting food origin and stimulating territorial development in detail. Hereby, investigating the contextual conditions is crucial as they are decisive for the kind of governance forms, regarding GIs and collective place bands.

Bibliography

Allaire, Gilles and Bertil Sylvander. 1996. Qualité spécifique et innovation territoriale. *Cahiers d'Economie et Sociologie Rurales* 44: 30–59.

Anders, Sven and Julie Caswell. 2008. The benefits and costs of proliferation of geographical labelling for developing countries. University of Massachusetts, Amherst Department of Resource Economics Working Paper No. 2008-7. http://www.umass.edu/resec/workingpapers. Accessed 22 January 2016.

Angeon, Valérie, Patrick Caron and Sylvie Lardon. 2006. Des liens sociaux à la construction d'un développement territorial durable: Quel rôle de la proximité dans ce processus? *Développement durable et territoires* 7. https://developpementdurable.revues.org/2851. Accessed 22 January 2016.

Baudelle, Guy, Catherine Guy and Bernardette Mérenne-Schoumaker. 2011. *Le développement territorial en Europe. Concepts, enjeux et débats.* Rennes: PUR.

Belletti, Giovanni. 2000. Origin labelled products, reputation and heterogeneity of firms. *Actes et communications - Institut national de la recherche agronomique. Economie et sociologie rurales* 17: 239–259.

Bonnet, Céline and Michel Simioni. 2001. Assessing consumer response to Protected Designation of Origin labelling: a mixed multinomial logit approach. *European Review of Agricultural Economics* 28: 433–449.

Bramley, Cerkia. 2011. *A review of the socio-economic impact of geographical indications: Considerations for the developing world.* WIPO Worldwide Symposium on Geographical Indications. Lima, Peru, 22–24 June 2011. http://www.wipo.int/edocs/mdocs/geoind/en/wipo_geo_lim_11/wipo_geo_lim_11_9.pdf. Accessed 22 January 2016.

Bramley, Cerkia and Johann Kirsten. 2007. Exploring the economic rationale for protecting geographical indicators in agriculture. *Agrekon* 46: 47–71.

Bramley, Cerkia, Estelle Biénabe and Johann Kirsten. 2009. The economics of geographical indications: Towards a conceptual framework for geographical indication research in developing countries. In *The Economics of Intellectual Property*, ed. WIPO, 109–149. http://www.wipo.int/edocs/pubdocs/en/economics/1012/wipo_pub_1012.pdf. Accessed 22 January 2016.

Camagni, Roberto and Roberta Capello. 2013. Regional competitiveness and territorial capital: A conceptual approach and empirical evidence from the European Union. *Regional Studies* 47: 1383–1402.

Campagne, Pierre and Bernard Pecqueur. 2014. *Le développement territorial: Une réponse émergente à la mondialisation*. Paris: ECLM.

Caputo, Vincenzina, Maria Aprile and Rodolfo Nayga Jr. 2011. Consumers' valuation for European food quality labels: importance of label information provision. Proceedings of the EAAE International Congress, Zurich, Switzerland, 30 August to 2 September 2011. http://ageconsearch.umn.edu/bitstream/114324/2/Caputo_Vincenzina_434.pdf. Accessed 22 January 2016.

Charlier Christophe. 2007. La protection européenne des indications géographiques face au principe du traitement national de l'OMC. *Économie rurale*, 299: 70–83.

Cirad. 2009. *The challenges relating to geographical indications (GIs) for ACP countries*. A Joint CTA, AFD and CIRAD Workshop Report, Montpellier.

Côté, Serge. 2003. Développement local: Sur fond de proximité relationnelle, l'apport indispensable de l'extra-local. *Interventions économiques* 30. https://interventionseconomiques.revues.org/1002. Accessed 22 January 2016.

Das, Kasturi. 2009. *Socio-economic implications of protecting geographical indications in India*. http://wtocentre.iift.ac.in/Papers/GI_Paper_CWS_August%2009_Revised.pdf. Accessed 22 January 2016.

Donner, Mechthild, Fatiha Fort and Sietze Vellema. 2014. How to capture place brand equity? The case of Sud de France. *Place Branding and Public Diplomacy* 10; 145–157.

Donner, Mechthild, Fatiha Fort and Sietze Vellema. 2015. *Potential and limits for developing a place brand as tool for territorial development in Morocco: Case study insights from Chefchaouen*. Annual conference of the Regional Studies Association, Piacenza, Italy, 24–27 May 2015. http://www.regionalstudies.org/uploads/Donner_et_al_-_place_branding_Chefchaouen.pdf. Accessed 22 January 2016.

Donner, Mechthild. 2016. Understanding place brands as collective and territorial development processes. Phd thesis, Wageningen, Wageningen University, 178 p.

Fournier, Stéphane. 2015. Évaluer ex-ante l'impact d'une indication géographique? De la difficulté d'appréhender des processus d'action collective et d'innovation au sein de systèmes complexes. *Économies et sociétés, série "Systèmes agroalimentaires"* 37: 1167–1184.

Francois, Hugues, Maud Hirczak and Nicolas Senil. 2006. Territoire et patrimoine: La co-construction d'une dynamique et de ses ressources. *Revue d'Économie Régionale & Urbaine* 5: 683–700.

Galbrun, Claudine. 2012. Sud de France, entre identité individuelle et identité collective pour les vins du Languedoc-Roussillon. http://www.pleinchamp.com/viticulture/actualites/sud-de-france-entre-identite-individuelle-et-identite-collective-pour-les-vins-du-languedoc-roussilon. Accessed 22 January 2016.

Giovannucci, Daniele, Timothy Josling, William Kerr, Bernard O'Connor and May T. Yeung. 2009. *Guide to geographical indications: Linking products and their origins*. Geneva: International trade centre.

Hayes, Dermot J., Sergio H. Lence and Andrea Stoppa. 2003. 'Farmer-Owned Brands?' Briefing Paper 02- BP 39. *Center for Agricultural and Rural Development, Iowa State University*. http://www.card.iastate.edu/publications/dbs/pdffiles/02bp39.pdf. Accessed 22 January 2016.

Iversen, Nina M. and Leif E. Hem. 2008. Provenance associations as core values of place umbrella brands: A framework of characteristics. *European Journal of Marketing* 42: 603–626.

Kotler, Philip and David Gertner. 2002. Country as brand, product, and beyond: A place marketing and brand management perspective. *The Journal of Brand Management* 9: 249–261.

Landel, Pierre-Antoine and Nicolas Senil. 2009. Patrimoine et territoire, les nouvelles ressources du développement. *Développement durable et territoires* 12. https://developpementdurable.revues.org/7563. Accessed 22 January 2016.

Leloup, Fabienne. 2005. La gouvernance territoriale comme nouveau mode de coordination territoriale? *Géographie, économie, société* 7: 321–332.

Loureiro, Maria Luz and Jill J. McCluskey. 2000. Assessing consumer response to protected geographical identification labeling. *Agribusiness* 16: 309–320.

Ménard, Claude. 2000. Enforcement Procedures and Governance Structures: What Relationship? In *Institutions, Contracts and Organizations*, ed. Claude Ménard, 234–254. Cheltenham, UK: Edward Elgar.

Monteiro Monjardino de Souza and Maria Lucas. 2001. Conjoint measurement of preferences for traditional cheeses in Lisbon. *British Food Journal* 103: 414–424.

Pacciani, Alessandro, Giovanni Belletti, Andrea Marescotti and Silvia Scaramuzzi. 2001. The *Role of Typical Products in Fostering Rural Development and the Effects of Regulation (EEC) 2081/92*. 73rd Seminar of the European Association of Agricultural Economists, Ancona, Italy, June 28–30.

Pecqueur, Bernard. 2001. Qualité et développement territorial: L'hypothèse du panier de biens et de services territorialisés. *Économie rurale* 261: 37–49.

Pike, Andy, Andrés Rodríguez Pose and John Tomaney. 2006. *Local and regional development*. London: Routledge.

Porter, Michael. 2003. The economic performance of regions. *Regional studies* 37: 545–546.

Rainisto, Seppo K. 2003. *Success factors of place marketing: a study of place marketing practices in Northern Europe and the United States*. Phd Thesis, Helsinki University of Technology.

Rangnekar, Dwijen. 2004. *The socio-economics of geographical indications. A review of empirical evidence from Europe*. UNCTAD-ICTSD Project on IPRs and Sustainable Development, Issue Paper No. 8. http://www.ictsd.org/sites/default/files/research/2008/07/a.pdf. Accessed 22 January 2016.

Reviron, Sophie, Erik Thevenod-Mottet and Nadja El Benni. 2009. *Geographical indications: Creation and distribution of economic value in developing countries*. Swiss National Centre of Competence in Research (NCCR), Working Paper No.14-2009. http://phase1.nccr_trade.org/images/stories/publications/IP5/report_IP5_GI_Value_2009.pdf. Accessed 22 January 2016.

Rochette, Corinne. 2012. L'approche ressources et compétences comme clé de lecture du processus d'élaboration d'une ressource originale: La marque territoire. *Gestion et management public* 1: 4–20.

Scott, Allen and Gioacchino Garofoli. 2011. *Development on the ground: Clusters, networks and regions in emerging economies*. London: Routledge.

Skuras, Dimitris, Efthalia Dimara and Anastasia Petrou. 2006. Rural tourism and visitors' expenditures for local food products. *Regional Studies* 40: 769–779.

Storper, Michael. 1997. *The regional world: Territorial development in a global economy*. London: Guilford Press.

Teuber, Ramona. 2010. Geographical indications of origin as a tool of product differentiation: The case of coffee. *Journal of International Food & Agribusiness Marketing* 22: 277–298.

Therkelsen, Anette, and Henrik Halkier. 2008. Contemplating place branding umbrellas.The case of coordinated national tourism and business promotion in Denmark. *Scandinavian Journal of Hospitality and Tourism* 8: 159–175.

Van der Ploeg, Jan Douwe, Henk Renting, Gianluca Brunori et al. 2000. Rural development: From practices and policies towards theory. *Sociologia ruralis* 40: 391–408.

Vandecandelaere, Emilie and Sébastien Abis. 2012. *Alimentation, tourisme et collectivités territoriales*. Paris: Presses de Sciences Po.

Vandecandelaere, Emilie, Filippo Arfini, Giovanni Belletti and Andrea Marescotti. 2009. *Linking people, places and products, a guide for promoting quality linked to geographical origin and sustainable geographical indications*. Rome: Food and Agriculture Organization (FAO) and SINER GI.

Vitry, Chloé, Abdellaoui El Hassan, Patrick Dugué and Eduardo Chia. 2015. Apprendre a cooperer: Un defi pour l'adhesion des agriculteurs au Plan Maroc Vert. *New medit: Mediterranean journal of economics, agriculture and environment* 2: 13–21.

Ward, Neil and David L. Brown. 2009. Placing the rural in regional development. *Regional studies* 43: 1237–1244.

WIPO. 2007. WIPO Asia Sub-Regional Workshop on the Use of Intellectual Property (IP) by SME Support Institutions for the Promotion of Competitiveness of SMEs in the Food Processing Sector. Topic 4: The Role of Trademarks, Collective Marks, Certification Marks and Geographical Indications in Marketing Food Products. http://www.wipo.int/meetings/fr/doc_details.jsp?doc_id=81475. Last accessed 7 January 2016.

Author Biographies

Mechthild Donner holds an MBA in International Marketing from Reutlingen University (Germany). She has practical sales and marketing experience from working in an international agrifood company located in the Netherlands. In 2016, she has completed her PhD at Montpellier SupAgro and Wageningen University within the European Commission Erasmus Mundus Joint Doctorates program AGTRAIN (www.agtrain.eu), with a focus on territorial marketing for agri-food products in Europe and Morocco. Since then, she is working as researcher at INRA, the French National Institute for Agricultural Research (mechthild.donner@inra.fr)

Fatiha Fort is professor at Montpellier SupAgro (France) and belongs to the MOISA research unit. in the marketing field, she focuses on consumer behavior towards sustainable food and new product development. She is currently working on an information processing theory to understand the role of food product information (technological, nutritional, sensory), but also the effect of the interaction between brand and other quality labels in consumer choice. (fatiha.fort@supagro.inra.fr)

Sietze Vellema is Associate Professor at the Knowledge, Technology and Innovation group at Wageningen University (Netherlands), which combines technology and development studies with a scholarly interest in institutional theory. His research and teaching focuses on social analysis of the use of techniques and skills in food provision. He leads action research at the Partnerships Resource Centre, Erasmus University Rotterdam, which links business and management literature to development studies for understanding the role of partnering for the embedding of global sustainability standards. (sietze.vellema@wur.nl)

Chapter 8
GI Blues: Geographical Indications and Wine in New Zealand

John Overton and Warwick E. Murray

Abstract The introduction of geographical indications in New Zealand has been slow and problematic. There has been some enthusiasm for tighter definition of regions in the way New Zealand agro-commodities are produced and marketed, yet also some apparent resistance. In this chapter we examine the reasons for this dichotomy with reference to the wine industry in particular. Despite some attempts at using and codifying geographical indications in the wine industry, the broader agricultural political economy is marked by a basic distinction between large-scale commodity production in a neoliberal trading environment—characteristic of the dominant dairy, wool and meat export industries—and the emergent moves towards small-scale niche high value-added production. The wine industry straddles this divide with both large-scale industrial production of Sauvignon Blanc by often global beverage companies—which really only appeals to broad spatial descriptors such as New Zealand and Marlborough—and a small-scale sector which seeks much tighter areal descriptors and trades off notions of place-based production, artisanal techniques and higher price points in the market. Given these tensions, it is yet to be seen what type of GI landscape will emerge in New Zealand following the imminent introduction and regulation of legislation.

8.1 Introduction

New Zealand is poised to enact legislation that will gazette and protect Geographical Indications (GIs). This would seem like an obvious step for a country, which depends heavily on agricultural production and exports, and on a global image for integrity, provenance and sustainability in production. Indeed, it could be suggested that if New Zealand was not able to secure niche markets for its agricultural exports based on these putative qualities associated with its national image,

J. Overton (✉) · W.E. Murray
School of Geography, Environment and Earth Sciences,
University of Wellington, Wellington, New Zealand
e-mail: John.Overton@vuw.ac.nz

W. van Caenegem and J. Cleary (eds.), *The Importance of Place: Geographical Indications as a Tool for Local and Regional Development*, Ius Gentium: Comparative Perspectives on Law and Justice 58, DOI 10.1007/978-3-319-53073-4_8

it could not compete effectively against larger global producers. Yet it is a country that has been tardy in introducing such legislation: it is "currently the only major wine exporting country that does not have some sort of GI registration system or equivalent" (Barker 2016: 4). A closer look at the issue of GIs reveals some underlying tensions and differences regarding not only the structure and marketing of its wine industry but also of a wider debate about the very nature of its export economy. This chapter begins by outlining the history of both the New Zealand wine industry and the broader agro-commodity sector. It then analyses the evolution and current nature of GI policies and practices in the country's wine sector. Several case studies of wine regions and attempts there to define and protect their regional identities are then presented. The chapter concludes addressing a number of key issues behind the implementation of GIs in this case.

8.2 Wine and the Agricultural Political Economy of New Zealand

Although winemaking in New Zealand dates back to the nineteenth century, before the 1980s it was largely characterized by production for the highly regulated local market. Wine quality, in retrospect, was low and products often attempted to produce poor—and sweet—imitations of European styles, though they tended to use high yielding hybrid varieties. A revolution in the industry began in the 1980s with the gradual replacement of sweeter and higher yielding varieties such as Muller Thürgau with more classical European varieties: Chardonnay, Riesling, Cabernet Sauvignon, and Merlot and, increasingly Sauvignon Blanc and Pinot Noir. Production turned more to export and new winemaking and viticultural techniques helped raise quality significantly (Stewart 2010). With export success and an expanding local market, wine production increased exponentially, though such growth began to level off in 2010. The total area planted in vines expanded from 5800 ha in 1992 to 35,859 ha in 2015 and the value of exports grew from $NZ 34.7 million in 1992 to $NZ 1.4 billion in 2015 (NZ Winegrowers 2015: 30). Within this rapid growth, the dominance of the Sauvignon Blanc variety has become marked. The Winegrowers Annual Report in 2015 reported that 86.5% of the volume of wine exports was from this variety alone (NZ Winegrowers 2015: 29).

The spatial manifestation of this industry transformation was a move away from the former established North Island centers of production (Auckland, Gisborne and Hawkes Bay—though the latter expanded production significantly) to the cooler climate regions further south (Martinborough, Canterbury, Central Otago and, particularly, Marlborough) (see Figs. 8.1 and 8.2).[1] Increasingly, Marlborough

[1]Elsewhere we have mapped this southward shift in the 'center of gravity' of New Zealand wine production (Overton and Murray 2014a, pp. 29–33). Figures 8.1 and 8.2 available in the appendix of this chapter.

came to dominate the country's wine industry (Barker et al. 2001, Gwynne 2006a, b). In the twenty years between 1992 and 2012, the province increased its share of the national vineyard from one third to two thirds (see Overton and Murray 2014a: 32). Hawke's Bay, a region associated with red variety production tripled its area under vines over the same time but growth in Marlborough meant that its share of the national area fell from 27% to just 14%. Apart from Marlborough, only Central Otago managed to increase its share of the national vineyard area over this period. Thus, there has been a remarkable growth but also a significant concentration in the New Zealand wine industry in the last 30 years. Although much has been written with regard to different regions and varieties in the country—examples such as Pinot Noir from Central Otago or Martinborough or Syrah from Hawkes Bay and Waiheke Island have emerged as notable regional styles with outstanding reputations—the reality of the New Zealand wine industry is of a 'behemoth' (Overton and Murray 2014a), a single variety and single region (Marlborough Sauvignon Blanc) that has almost become synonymous with 'New Zealand wine' in the minds and glasses of global consumers.

Matching the growth and regionalization of New Zealand has been the emergence of a particular industrial structure. Wine production is often portrayed in marketing material and popular journalism as artisanal: skilled winemakers and viticulturalists carefully combine high quality fruit with both time-honoured and innovative wine making techniques to create distinctive wines that reflect the *terroir* of their origin and even the personalities of their winemakers (Howland 2014a). There are many small winemaking operations in the country that would seem to conform to, or at least aspire to, this ideal. In 2014, there were 2046 vineyards enumerated in the country but over half of these (1268) had less than 10 hectares planted in grapes (New Zealand Winegrowers 2014, p. 6). Similarly, there were some 673 wineries in the country in 2015 but 587 of these produced less than 200,000 liters of wine (NZ Winegrowers 2015: 31). By contrast, only 17 wineries produced over 4 million liters per year (and several of these were owned by the same company). Some prominent New Zealand wine brands appeal to this artisanal image: perhaps the country's best known wine, Cloudy Bay, is owned by the global luxury brand company LVMH but its operations in New Zealand have remained relatively small and select (in the medium-sized category). Similarly well-regarded brands, Te Mata, Dry River, Stoneyridge and the 'Family of Twelve' group of wineries attempt to promote smaller-scale high quality wines strongly associated with their region of origin and the (albeit short) family histories.

Yet only a handful of companies dominate the industry in terms of total output (Overton and Murray 2014a; Overton and Murray 2016a). There are three major foreign-owned conglomerates: Pernod Ricard (with the Brancott Estate, Stoneleigh, Church Road brands amongst others), Constellation Brands (Nobilo, Selaks, Kim Crawford) and the Fosters Group/Treasury Wine Estates (Matua). These are global companies with portfolios of wine brands from around the world and very well developed marketing networks. Only two New Zealand-owned companies (the Villa Maria Group and Delegat's) figure amongst this group of the largest producers—and all are heavily involved in the production and export of Marlborough Sauvignon Blanc.

Thus, whilst the artisanal image is well established in the New Zealand wine industry, and it is reinforced through wine tourism encouraging visitors to taste wine at cellar-doors throughout the country, the reality is of large-scale 'industrial' production by an oligopolistic set of companies (Overton et al. 2012). Although visitors may taste wine in small well-appointed 'wineries' replete with rooms of stained oak barrels, much of the wine they consume and purchase is produced in large factories in Blenheim, Hastings, Auckland or Christchurch where row upon row of large stainless steel tanks are located on industrial estates. In a similar vein, promotion of New Zealand wines tries to stress the high per bottle value, and New Zealand wine is claimed to rank only behind French wine in price per bottle on the UK market for example. There have also been calls to keep away from bulk exports of wine (either retailed in cardboard casks or exported in bulk and bottled overseas, for example in supermarket-branded labels) in order to preserve the high reputation (and price) of New Zealand wine. Yet, as Lewis (2014) argues, this is a flawed and singular view of the industry. Many exporters are now engaged in bulk wine exporting and New Zealand wine covers a very wide range of labels and prices in world markets. Stewart (2010) has noted some major differences and occasional tensions in the industry, often submerged within the discussions inside the Wine Institute and New Zealand Winegrowers.

One such probable tension has concerned the introduction of GIs. Whilst some sectors of the industry, as we see below, have been active in promoting local schemes for regional denominations in wine and it could be argued that GIs would do much to promote the small-scale distinctive and regionally specific image of New Zealand wines, perhaps oddly, the industry bodies have done little since the abortive scheme of the early 1990s and until recently to lobby government to enact GI-type legislation (Banks et al. 2007; Murray and Overton 2011; Overton and Murray 2014b). A possible reason for this inaction is the likely difference over the detail and strength of any possible GI scheme for wine in New Zealand. Whilst all producers, arguably, would be in favor of protection of the 'New Zealand' and higher-level territorial names (such as Marlborough), not all might support sub-regional protection or strict regulations concerning place of origin. Under the present sale of goods legislation, it is possible to have up to 15% of the volume of a bottle of wine not conforming to, and not listed on the label, the given variety or region or even year (MPI 2015). Thus, a bottle of Marlborough Sauvignon Blanc can be mixed with other varieties and juice from other regions as long as it contains 85% Sauvignon Blanc and 85% of juice from Marlborough. Given the very high price of sought-after vineyard land in the Wairau Valley in Marlborough—and high demand for Marlborough Sauvignon Blanc—it is therefore advantageous for some producers of this wine to mix their product up to 15% with Sauvignon Blanc from Waipara or Nelson (This can be the case especially if a harvest is lower than expected and forward orders have to be met.) Furthermore, as we will see below, a wide definition of the Marlborough region to include land outside the core Wairau and Awatere valleys helps widen the catchment and potentially lower the cost of grapes for Marlborough wine. On the other hand, other producers may use marketing strategies that emphasize the exclusiveness of their region and seek much tighter rules on specification of origin and smaller and more exclusive regional definition.

One issue that this existing 15% rule has given rise to is the possibility that up to that proportion of a bottle of New Zealand product may contain wine from another country without explicit acknowledgement.[2] A discussion document from the Ministry of Primary Industries in 2015 raised this issue and asked the industry for its views on options to include in the proposed GI legislation and in particular whether to remain silent on the origin of the 15% or exclude the possibility of including overseas wine in any New Zealand wine with a geographical indication). The Ministry's view was that a risk existed where the inclusion of overseas wine could damage New Zealand's reputation as a higher quality producer, particularly from places where the regulatory oversight was weak on, say, environmental matters, use of chemicals etc.) (MPI 2015).

Yet this range of views within the wine industry should not be unexpected if we consider the nature of the wider New Zealand agro-industrial scene. New Zealand is a country that is highly reliant on its agricultural industries and exports. Even though it is now a highly urbanized society with few directly employed on the land, agricultural products still account for around half of the country's exports and agriculture comprised 6% of the country's GDP in 2015. Historically, this dependence has been very strong. From its founding as a British colony in 1840 until Britain's entry to the then European Economic Community in 1973, New Zealand aligned its agricultural sector to produce wool, frozen meat, dairy products and pip fruit for the British market. Although such products were identified as coming from New Zealand, and there were some rare examples of sub-national identification (such as Canterbury lamb), the general approach was 'Fordist': to produce agricultural commodities in large volumes, relatively uniform quality and undifferentiated form (such as bulk butter and cheese).

Production and processing systems (farms, freezing works, dairy factories) sought economies of scale and efficiencies of production rather than a more differentiated niche marketing approach with specialist products with localized branding and high added value. Producer boards (the Wool Board, the Dairy Board, the Apple and Pear Board etc.) acted as monopsonies and encouraged bulk production and marketing.

Following Britain's entry to the EU and the gradual loss of preferential access to its market, New Zealand sought to diversify its export markets. However, diversification was more about finding alternative markets rather than substantially restructuring its production systems to produce more differentiated products. Governments continued to offer incentives to farmers to clear more land, increase

[2]In fact the bottle would have to state that it includes wine from 'New Zealand and another country' (to meet country of origin regulations) but, confusingly, could still call itself 'Marlborough' wine under the proposed GI rules.

stock numbers and produce more basic commodities for exports. Neoliberal reform following 1984 cut the subsidies and other incentives so that the agricultural sector became more open and responsive to global market signals (these reforms were particularly influential for the remaking and development of the modern wine industry—see Barker et al. 2001; Stewart 2010).

In subsequent years, there have been some changes in New Zealand agriculture and there has some deregulation of the producer board system, allowing for independent processing and marketing. Some initiatives have been successful in seeking high value-added production—merino wool production for high quality apparel and specialist meat processing by companies such as Silver Fern Farms are examples (Pawson and Perkins 2016). These operations have sought to link high value products and consumers with the producers, methods of production and locales in ways which leverage claims of high quality, environmental integrity and exclusivity in order to obtain a premium in the market. Production is smaller scale, more intricately monitored and place of origin becomes a more central element in the production profile.

Yet, despite these initiatives, the fundamental structure of New Zealand's agricultural sector and its reliance on bulk commodity production has remained and, in some respects, deepened. Whilst global trends have led to a relative decline in bulk wool production and New Zealand is not able to compete well with large-scale beef exporters, the dairy sector has boomed. In 2001, the New Zealand Dairy Board was transformed into Fonterra, now a global corporation that manages not only most of New Zealand's dairy sector but also is heavily involved in dairy production, processing and value-chain management in many parts of the world. It is the largest company in New Zealand, the globe's largest trader of milk powder and responsible for about 30% of the world's dairy exports. Although dairy prices dipped significantly in 2015, there was much expansion in the preceding twenty years as Fonterra strove to tap into the massive Chinese market for milk powder in particular. New Zealand has seen a very large expansion of dairying as former wool and meat farms were converted, with the help of irrigation schemes, to dairy production. Dairy farms have grown, the sector has expanded into new regions, especially in the South Island, there has been much investment (including FDI) and dairy products have become the country's main export earner. Yet such exports have been of bulk commodities—milk powder especially but also butter and cheese. Here New Zealand is able to compete globally because of the ability to grow grass virtually year round with the help of irrigation.[3]

There have been some attempts to move to higher value-added production both by Fonterra (with products such as spreadable butter) and new smaller enterprises (the Lewis Road company has moved into organic and specialist production of milk

[3]Whilst New Zealand dairy production is very efficient compared to many established smaller-scale farm systems as in Europe, it may be less so as new large dairy units with cheaper land and labour are established in South America, Asia and Africa. Fonterra's long-term strategy seems to be to move into such operations offshore as well as expanding within New Zealand while it can.

and butter products and there has been a growth in small-scale and artisanal cheese production). In addition, New Zealand dairy exports make much of 'brand New Zealand' playing on narratives of 'clean green' production.[4] Yet the old Fordist model of economies of scale, bulk production and relatively undifferentiated products still dominates. Governments have proved reticent to restrict in any way the Fonterra model of growth and commodity production: a monolith committed to globalization and trade liberalization.

8.3 The Early Evolution of GIs for Wine in New Zealand

With these comments on the evolution of New Zealand's agricultural sector, and the wine industry, in mind, we can now proceed to track the way Geographical Indications have been debated, stalled and engaged with in the country over the past 25 years. Although New Zealand has been slow to enact GI legislation (Barker 2004), moves to establish a regional denomination scheme for the wine industry were mooted in the early 1990s. At that time the national industry body for winemakers, the Wine Institute of New Zealand,[5] took steps to put in place a scheme that would identify and monitor the use of regional labels in the wine industry. In this matter, it was advised by Warren Moran, a geographer at the University of Auckland who had conducted extensive research on the wine industry in New Zealand and France (Moran 1988, 1993a, b, 2001; 2001). Moran was familiar with the *Appellation d'origine controllee* (AOC) system of France and also schemes in USA and elsewhere. His approach was pragmatic in that he was aware of how the French system, despite its claims to be based on the concept of *terroir* and sound scientific data concerning soils, climate etc., was in reality highly politicized and a lot more flexible in practice. Recognizing its function in protecting certain interests and regions, he nonetheless saw merit in a possible scheme that would promote localized identification of regions and place names. He advised against using a highly regulatory and prescriptive approach to regional denominations in New Zealand wine that would define styles, varieties, yields and so forth—as followed in France—and he also advised against having fixed boundaries to the regions.

Moran's scheme instead promoted the idea of regional committees being formed that would comprise grape growers and winemakers as well as some external—and neutral—involvement.[6] These committees would then decide which wineries and

---

[4]This environmental narrative has been challenged by some, such as scientist Mike Joy, who point to the serious deterioration of rural water quality in the country as a direct result of the expansion of dairying.

[5]In 2002 the Wine Institute of New Zealand in collaboration with the New Zealand Grape Growers Council formed New Zealand Winegrowers which is now the main industry body.

[6]The early steps to introduce the scheme involved particularly the suggestion that geographers from various universities take on this role.

grape producers could use the regional name on a case-by-case basis, rather than by drawing a line on a map, which delimited certain regions. This method was 'organic' in that Moran favored an approach that would allow regions to evolve and define themselves and attempt to be inclusive, rather than using mapped delimitations to exclude some producers. Initial meetings were held in 1993–94 with the Wine Institute beginning a process to form the various regional committees. At the time, it was believed that the government would proceed to support the scheme with necessary legislation.

However, the scheme faltered then was shelved indefinitely when it was realized that the government (of the National party led by PM Jim Bolger) of the day was not inclined to support such an approach. Significantly, this was at a time when global trade negotiations over the TRIPS agreement on intellectual property were underway (Josling 2006). New Zealand's position at these negotiations tended to side with the USA and against the European Union. In other words, it favored a more liberal regime with regard to what were regarded by the USA as attempts at quasi-protectionism of trade in the form of geographical indications. Although the USA might accept the French argument with regard to protecting names such as Champagne (already established in the courts), it did not want to see a proliferation of GIs worldwide or into other agro-commodity sectors. No New Zealand enabling GI legislation was forthcoming in such circumstances, despite the fact that what the Wine Institute proposed was a far cry from the apparent 'Old World' approach of tightly defined and protected regions and place names. Without government support, the wine industry was left to its own devices with regard to promoting and protecting the use of regional descriptors. By this time, it was clear that New Zealand wine was developing successfully based in part on some clear regional differences in varieties and styles (see below). There was rapid growth in the area planted in grapevines—a six-fold increase in area between 1992 and 2015—and some distinct regional specializations were emerging. Marlborough Sauvignon Blanc was by far the most notable of these regional specializations in the industry but in the smaller regions, producers found both that some varieties seemed to suit their climate and soils more than others and that there were advantages in co-promoting a region with other wineries around a flagship variety. Thus Hawke's Bay (and Auckland/Waiheke Island—Baragwanath and Lewis 2014) came to focus more on red wine varieties—both Bordeaux (Merlot dominant) and Rhone (Syrah) styles—though Chardonnay, Sauvignon Blanc and latterly Pinot Gris continued as white varieties of prominence. The cooler climate regions of Martinborough, Waipara (Schuster et al. 2002; Tipples 2007, Overton et al. 2014) and, most prominently, Central Otago put emphasis on Pinot Noir (Cull 2001) whilst Gisborne (in 1990 it had produced nearly a third of the country's wine) struggled and largely failed to secure a place a 'New Zealand's Chardonnay capital'.

In export markets, the dominance of Marlborough and Sauvignon Blanc meant that there was little place for other regions in terms of global consumer consciousness of New Zealand's wine geography—'New Zealand' and 'Marlborough' were really the only two place brands that had any major traction. However, on the local market greater consumer knowledge and competition meant that producers fought hard to establish their brands. Varietal specialization was one aspect but also the need to establish a regional brand became increasingly of concern. Certain smaller regions (Waiheke, Martinborough, Waipara) were able to exploit their proximity to metropolitan areas (Auckland, Wellington and Christchurch respectively) to build wine tourism and a degree of local/regional brand loyalty (Howland 2008, 2014b) whilst success in wine competitions and gaining favour of international wine writers helped producers in Hawkes Bay, Central Otago, Waiheke and Martinborough build reputations for the quality (and often high price) of their bottles of wine.

8.4 Emergent GIs: Case Studies

In this context, and without any lead from government or even the industry body, groups of producers began to develop local strategies to name, delimit and promote their regions. Three examples are illustrative of the possibilities but also of the limits of such local initiatives: the Martinborough Terrace, Gimblett Gravels and Waipara (see Fig. 8.1).[7]

8.4.1 The Martinborough Terrace

The first notable attempt to establish a wine appellation in New Zealand was in the small settlement of Martinborough, east of the city of Wellington. The first modern vineyards were established on the margins of the town of Martinborough in the late 1970s after the area had been identified as having a suitably dry and sunny climate coupled with good, well-draining soils. The vineyards and winery operations at first were very small and operated by individuals and families (the wineries of Ata Rangi, Dry River and Martinborough Vineyard) but early success in producing well-received wines made from Chardonnay, Pinot Noir, Riesling and Gewurztraminer varieties encouraged others to join them. The best location for grape-growing appeared to be on a gravel terrace that lay between the town and the Huangarua and Ruamahanga Rivers to the north and west; land that had been seen to be too dry and stony for cropping. During the 1980s much of the terrace land was bought and developed into vineyards. The vineyard area was small but relatively contiguous and the community of winemakers was concentrated in and around the town (Howland 2014a, b).

Perhaps drawing on the experience that some of the winemakers had gained from touring and sometimes working in wineries in Europe, some began to talk of

[7]Much of the following discussion of these case studies draws from our earlier paper on the making of wine regions in New Zealand (Murray and Overton 2011).

establishing a regional 'appellation', a defined area in and around Martinborough that could be linked to an environmental narrative (the stony terrace) and used to promote the district's wines. They believed that the apparently unique combination of soils, topography, rainfall and sunshine provided the basis for backing distinctive and high quality wines. Several of the early winemakers had backgrounds in scientific research and they drew on some of their expertise to construct an environmental narrative concerning the wine land:

> "the virtue of this locality was its 30-year record of the lowest rainfall in the North Island and a heat summation appropriate for growing quality cool-climate varieties ... the low rainfall area was limited to a tiny locality roughly 5 km in radius, and a study of soil maps revealed that the deep, free-draining gravels they sought within this were restricted to an even smaller part". (Dry River 2009)

This narrowly defined area required a name—an appellation. The name 'Martinborough' was an obvious choice given the close proximity of the town of that name and to this was added the term 'terrace' to refer to the area on top of a small escarpment that fell away from the level of the town down to the two rivers. Although the early scientific surveys had shown the climate to be critical in favoring grapevine quality, it was soils and topography that proved to be the easiest feature to use to define and delimit the wine district. "This resulted from the belief that the soils on top of the terrace were the free-draining deep beds of gravel that provided the best conditions for vine growth, whilst those on the lower river flats were both frost- and flood-prone and too thin" (Murray and Overton 2011: 425). Topography provided an additional basis for areal definition: the terrace marked a natural boundary to the north, west and northeast, whilst the town happened to mark a barrier to the south and southeast, sitting on the deeper and less free-draining soils that seemed to distinguish the terrace area from farmland to the south. The Martinborough Winemakers Association was formed and in 1991, drawing on an earlier set of defining characteristics of the region, they launched 'The Martinborough Terrace Appellation'. The Association published and promoted the new appellation and administered its use. In all the area of the appellation amounted to some 1000 ha and was defined by both rainfall and soil criteria:

> a) [...] that area receiving an average of less than 800 mm annual rainfall over the period 1940-1980 based on rainfall normally contained within NZ Met. Service Misc. Publ. 145 & 185.
>
> b) [...] which also fell within a particular area defined carefully in terms of trig points etc. and contained 80% or more of the technically defined free-draining soils. (Dry River 2009)

The appellation label was used by most producers within the area on wine bottles and there was growing recognition of the region on the national market at least. The reputation of the new appellation was helped by recognition of and prizes awarded to some key wines from the area—Pinot Noir from Dry River, Ata Rangi, Palliser Estate and Martinborough Vineyards. Expansion continued and the wine industry helped transform the town of Martinborough from a sleepy rural service town to one characterised by restaurants, new lifestyle dwellings, tourist facilities and businesses servicing the growing wine sector.

Yet the very success of the Martinborough Terrace Appellation came to threaten its existence. The Terrace area, as defined, soon filled up almost completely with vineyards, the result of both new entrants and the expansion of existing wineries. Expansion had become necessary. With profitability threatened in a highly competitive market for wine, both domestically and in export markets, Martinborough producers had to hope that their premium wines could continue to command higher than average prices and thus support the relatively small scale of production. Yet whilst the price of wine struggled to rise, the price of vineyard land in Martinborough began to soar as scarcity was perceived and little or no land with the Appellation remained undeveloped.

New vineyard developers soon realized that good land was available nearby but outside the strict Appellation area. Low rainfall areas located on well drained soils lay a short distance to the east along Te Muna Road and these were bought and planted in vines by some relatively large operations (Escarpment and Craggy Range wineries notable amongst them). Some land was also planted in grapes to the south on deeper and heavier soils but still close to the town.

In response to these developments, the Martinborough Winemakers Association could have closed ranks and defended its Terrace Appellation. However, instead it decided to expand and admit the newcomers. It added a third clause in the definition of the Martinborough Terrace Appellation:

> c) Nevertheless, it was also allowed that the area could be expanded to include neighboring areas which yielded fruit similar in quality and style over five successive vintages from 5 + year vines or which corresponded to the rainfall and soil type definitions of the existing Terrace. It is hoped that neighboring areas will examine their situation and join the Appellation as time goes on. (Dry River 2009)

Today the term ‘Martinborough’ is recognized to include a group of wineries around the original terrace area north of the town but also the Te Muna Road producers and others to the south of the town. The term ‘Martinborough Terrace Appellation’ has all but disappeared from the labels of wine bottles from the district. What began as a restrictive form of locally defined geographical indication has now given way to a fluid and expanding form. What is now used as ‘Martinborough’ as a geographical indicator of wine origin is no longer limited by the imagery of a terrace and the environmental discourse that accompanies that, but now seems to be without fixed and impermeable boundaries and with the resultant potential for further expansion and contestation.

“Such flexible geographical delimitation and porous boundaries have clearly yielded benefits for the local industry in its aspirations to build a place brand and to increase the absolute level of production” (Murray and Overton 2011: 428). However, the Martinborough appellation experiment is generally regarded to have failed, at least in its early aim to recreate a tight almost European-style wine region. Interestingly though, the appellation idea has recently been revived.

A newly-named local winegrowers organization, 'Wines From Martinborough Inc' claims on its website that:

> Taking a pro-active approach to enhance and protect their stellar reputation for premium wines, local winemakers have defined the specific geology and geography that underpins their uniqueness and created a Martinborough Appellation … Martinborough was the first wine district to prepare its appellation under New Zealand's Geographical Indications (Wine and Spirits) Registration Act 2006 … This geographical identification protects the use of 'Martinborough' by restricting its use to wines produced from grapes grown within the legally defined area, and containing a minimum of 85 percent of Martinborough grown grapes. The boundaries are defined by a combination of physical points (including the Ruamahanga River to the west and Haurangi State Forest Park to the south) and territorial points between local authorities. (Wines from Martinborough 2016)

The words 'terrace' and 'appellation' have gone, now representing a retreat from the tight environmental narrative and the mimicking of Europe, but instead there is a broader and more inclusive definition that promotes the name 'Martinborough' and attempts to align the rebranded local initiative with a new regulatory environment (see below).

8.4.2 Hawkes Bay: The Gimblett Gravels

Another well-known local initiative to establish a place brand concerned the Gimblett Gravels district in the Hawke's Bay region on the east coast of North Island, north of Martinborough. This is an example of a reactive strategy to repel a threat to potential vineyard land that turned to a successful initiative to establish a reputation as a premium wine-producing place.

The Hawke's Bay region has, by New Zealand standards, a long history of wine production. From origins in the nineteenth century and stalled development through much of the early twentieth century because of a state largely hostile to winemaking—Stewart 2010), the region became one of the centers of the country's wine production (along with Gisborne and Auckland). Much of this production before 1970 was of wine for domestic market. There were a few examples of good quality wine being produced at times but much was poor and both grape varieties and wine-making techniques were focused, as elsewhere in the country, on cheap sweet wine. However, the region proved favorable for ripening red grape varieties and, when the industry began to turn to classical grape varieties, high quality production and, eventually greater export orientation, Hawke's Bay developed rapidly. As cooler climate varieties (aromatic white varieties and Pinot Noir) thrived further south, Hawke's Bay began to focus on white varieties such as Chardonnay and both Bordeaux and Rhône red varieties (Merlot, Cabernet Sauvignon and Syrah). Grapes were grown through the province but several sub-regions emerged exploiting local micro-climates and soil types.

One such area was on the Heretaunga Plains to the west of the city of Hastings. Wineries had been developed just to the south of this area (in what has been variously called the Ngatarawa Triangle, and recently the 'Bridge Pa Triangle') but an area comprised largely of an old riverbed with very thin or non-existent gravelly soils around Gimblett Road seemed to be equally promising. The grapes ripened well and the thin soils encouraged deep root development. Yet because of the very poor soil quality, the area was deemed no good for farming (either the orcharding or sheep farming that were otherwise nearby). Instead it had become a virtual wasteland with very low value peri-urban land uses (such as drag racing strip, a rifle range, and a refuse land fill[8]). There was also a quarry, which extracted and crushed the gravels for construction and roading uses. An application was made by the quarrying company to expand its operations substantially. This led to a reaction from a small number of existing winemakers in the area who objected to the application in 1992, eventually leading the quarrying company to abandon its plans. The action seemed to galvanize the wine makers and grape growers who had said much about the high quality and potential of the area and who formed their own association in 2001.

This association then set about promoting its wines as the area began rapidly to see a transformation from low value land uses to the point where it is now all but planted in grapes—a couple of industrial sites were even transformed back into agriculture in the form of vineyards (Overton and Heitger 2008). In doing so, the association undertook two important measures. Firstly, it created a name for the area: the Gimblett Gravels. This appellation took its name from the previously small unsealed no exit road that ran through the center of the area (of about 800 ha in all)—but the name had no other provenance, certainly not in the soil maps that the association used to draw its boundaries. Secondly, the association sought not only to define the extent of the district but also to protect the name it invented. In this it sought advice from the (then) Wine Institute. By that time (the early 2000s) the earlier Moran appellation scheme had not progressed, nor were there any strong indications that the government would support a national GI set of regulations. As a result, the Gimblett Gravels Winegrowers Association sought and gained trademark status for its logo and the name 'Gimblett Gravels'. Further, it put in place a set of rules for its members regarding the use of the name. It established a 95% rule: 95% of the grapes used to make Gimblett Gravels wine had to come from within the demarcated area—in practice a much stricter restriction than the national 85% policy.

Unlike Martinborough, the boundaries of this appellation have proved to be fixed, a soil map indicating a group of gravelly soils, the extent of which defined the area. With new wineries establishing in the area, including larger enterprises such as Craggy Range and Villa Maria, much capital investment and rapidly escalating land values, together with active promotion of the place brand, wine from the Gimblett Gravels gained a very good reputation and appeared to be able to command a premium in the market. Red varieties in particular seemed to do very well.

[8]The site on Highway 50 was not so much a land 'fill' as a place where refuse was piled on top of the flat plain, forming a low hill still visible today in the middle of valuable vineyards.

The success of this geographical indication in wine, then, resulted not from GI law but through recourse to another form of intellectual property legislation—a trademark. It also demonstrated how collective action of producers and an imaginative use of an invented place name could construct, exploit and protect a place name in the market. The Gimblett Gravels provides an example of a small-scale, well-defined, localized and cooperative GI strategy by producers can succeed.

8.4.3 Marlborough

As we have seen, Marlborough Sauvignon Blanc is New Zealand's iconic wine (Hayward and Lewis 2008). It dominates production and exports and is responsible for establishing a global awareness of the national wine industry, an awareness that other regions and varieties can leverage off. 'Marlborough', it could be argued, is New Zealand's most successful geographical indication, yet it is one that lacks any major official definition or protection and, in practice, it is marked by lack of definition and thus potential free-riding.

Marlborough's wine industry, despite a very few small early efforts, began in effect in the late 1970s with a few growers—Hunters, Te Whare Ra and, most significantly, Montana[9]—discovering that a combination of readily available and affordable land, high sunshine hours and light and gravelly soils made for reliable harvests of, particularly, white grape varieties. Early experiments found that many varieties did well, Chardonnay and Muller Thurgau amongst them, but by the late 1980s the Sauvignon Blanc grape produced juice that, when made into wine using fairly basic techniques (largely without oak aging), produced a very distinctive flavour profile. Marlborough seemed to produce this style like no other, probably a result of particular climatic conditions. The wine grew in popularity with early success in the United Kingdom market. Further success spurred a boom in grape planting in the province. In 1992, a total of 1902 hectares were recorded as being planted in grapes in Marlborough, 22% of this in Sauvignon Blanc (Jordan and Veldhuizen 1992). By 2012 this had risen to a total of 22,587 ha, a remarkable 78% in Sauvignon Blanc. This made Marlborough not only by far the largest wine region in the country (Hawke's Bay as the next largest region had 4841 hectares in 2012) but also the most specialized (only central Otago's 76% coverage in Pinot Noir came close).

This spectacular growth has transformed the landscape and economy of Marlborough. It has also stretched and questioned the idea of 'Marlborough'. The early plantings in the 1970s took place in the lower Wairau Valley, the main valley that runs west-east housing the town of Blenheim and most of the flat and arable land. Vineyards were established particularly on the more stony soils to the west of

---

[9]Montana is now owned by Pernod Ricard and has been largely rebranded as 'Brancott Estate', after the name of one of Montana's early Marlborough properties.

Blenheim and around the town of Renwick. Planting also took place on the floors of tributary valleys to the southern side of the Wairau and some also started in the Awatere Valley, another large valley running south and parallel to the Wairau. Wine brands from the region sometimes took their names from local place names: Cloudy Bay, Mt Riley, Wairau Plains. The main valleys gradually filled up with vineyards, especially during a boom in planting in the late 1990s and early 2000s. This was accompanied by a spread of the wine country and a change in narratives about the region. Away from the main stony valley flats of the Wairau and Awatere Rivers, vineyards were planted on the lower slopes on north facing hills of the valleys and even on the heavier soils to the north and east of Blenheim. Winemakers started talking of the different soil characteristics of these wines, even of different *terroir* in Marlborough.

However, regional expansion was not so much about establishing new and distinctive sub-regions and much more about merely increasing the aggregate supply of grapes that could be turned into Marlborough wine. Planting spread somewhat into more the more humid and heavier soiled valleys to the north of the Wairau, and far up the Wairau Valley itself. Grapes were also planted away from the valleys and on low hills south of the Awatere around the town of Ward. These were districts that were environmentally rather different from the Wairau Valley: heavier and more clay and loess-based soils, equally dry but a little cooler and, arguably, without the distinctive Marlborough flavor 'kick' to the Sauvignon Blanc. A few grapes were also planted further south around the Kaikoura Peninsula—with very different climate and soils but still within the province of Marlborough.

This expansion of the Marlborough vineyard well beyond its original hearth—and the particular conditions that seemed to have given rise to the distinctive character of its wines—reveals how the flexible, liberal and largely unregulated—use of a name can permit industrial growth and profits for those relatively far-removed from the original location to which the name refers.

8.5 Towards a Regulated GI Scheme in New Zealand

Thus the decade of the 1990s saw a brief attempt at a national GI scheme for wine but overall regulatory inaction and apparent legislative disinterest. Instead local schemes (such as Martinborough and the Gimblett Gravels as discussed above) represented efforts largely by the small and medium-sized enterprises and regions to promote their own areal definitions. These certainly helped establish these geographical identities in the New Zealand wine market, if not overseas, but their impact remained limited and localised. The Wine Institute's initiative in the early 1990s, under the advice of Warren Moran, did not come to fruition, stalled by lack of regulatory response. Indeed the later 1990s and early 2000s could be regarded as a hiatus for the industry with regard to GIs. While the industry was booming interest in GIs at the national level appeared to have subsided.

Eventually in 2006 the Geographical Indications (Wine and Spirits) Registration Act was passed into law. The Act provided for the registration of place names associated with wines and spirits in New Zealand and for sanctions against misuse. But although it was enacted, it was not implemented.[10] It seems as if one of the main reasons for creating this legislation at the time was the threat that the European Union would block access to New Zealand wine because did not sufficiently regulate its use of geographical names (New Zealand Cabinet 2015: 2). New Zealand felt that it had to fall into line with the international TRIPs (Trade Related Aspects of International Property Rights) agreement. Thus, the New Zealand GI legislation was largely a defensive move, a reaction to international pressures rather than a proactive strategy to help promote a particular model of agro-commodity exports built on high value added, distinctiveness and place of origin.

The legislation has remained inactive but in 2015 there were strong signals that the government would move in 2016 to amend and fully implement the 2006 Act. The amendments will seek to tidy up some concerns with the original Act, such as the problems relating to the 85% rule (and the use of wine from other countries). It seems that the wine industry is now pushing for implementation: 'They consider that implementation as useful for protecting and promoting their products and reputation in export markets' (New Zealand Cabinet 2015: 1). It is notable that the GI legislation is to apply to wines and spirits only; a more generic GI framework applicable to be used for other commodities (such as cheese or meat) has not been pursued. Again the timing of the legislation seems to be a reaction to the international trading environment: as New Zealand moves to a free trade agreement with Europe it is aligning with European systems so that its wines can continue to have access to markets there. In addition, it has been influenced by an industry report by PricewaterhouseCoopers in 2011 which argued for a strategic push into Asian markets and the need there to develop brand integrity: "The new strategy involved the use of GIs to give added focus to marketing authentic, distinctive, yet evolving, wine stories, and to protect the geographical aspects of "Brand New Zealand" from misappropriation or obstruction by offshore parties" (New Zealand Cabinet 2015: 4).

Although the details of the GI regulations have not been finalized, it seems as if some of the outlines are already in place. These can be found in a New Zealand Winegrowers labeling guide in 2013, which specifies New Zealand wine GIs to be used in the European Union (New Zealand Winegrowers 2013—see Table 8.1).[11] The guide also details a separate list of 'appellations of origin' for use in USA. This is a list of 90 place names varying from the macro level ('New Zealand' and 'North Island') to small areas where few grapes are grown ('Galatea', 'Kapiti'). In many ways this outline of the GIs to be used (and already being employed) in the New Zealand wine industry, reveals how the slow, even reluctant, moves to adopt GIs

[10]A copy of a Cabinet paper on proposals to revise and implement the legislation made public in 2014 has many sections withheld (under the Official Information Act) presumably a reflection of the sensitivity of discussion in terms of the implications for wider trade negotiations.

[11]Table 8.1 available in Appendix 8.1.

have led to a relatively weakly defined scheme that is still largely defensive in nature and which allows a high degree of continued flexibility and 'margin expansion'.

Perhaps the most notable feature of the regions used and defined in this guide is the adoption of broad political (local and regional authority) boundaries and the 'layering' of regions (for the European if not the USA list). These boundaries, particularly those which separate regional councils, commonly use watershed boundaries. This has a solid environmental basis and is useful for, say, separating humid westerly regions from the drier eastern regions (such as Hawke's Bay, Gisborne, Canterbury) where wine is produced. Yet, because these regions are large and cover a very wide range of climates and micro-climates, soil types, topography, underlying geology and aspect, they are very blunt instruments for defining distinctive wine producing areas. Of course, at the broadest level, GIs such as 'New Zealand', 'East Coast' or 'North Island' permit a high degree of mixing of wines from different areas and give little or no indication of the regional character of the product. On the other hand, there is mention of some of the smaller and more tightly defined wine districts, such as Matakana, Waipara and Martinborough and these do give a good degree of protection of local place names. Interestingly, the Gimblett Gravels place name is not included in the list, being a trademark rather than a recognized and gazetted place name. Furthermore, there are several winegrowing areas that have legitimate place names and promote themselves as wine districts (for example locations such as the Awanga, Ngatarawa, Esk Valley, the Awatere Valley) but which are not on the schedule. It will be interesting to see if these local identities and promotion will survive without formal GI recognition, or whether they might in time seek inclusion in the schedule. Overall, it seems that within the industry, some sectors have adopted a strategy of tight place definition and registration (Gladstone, Clevedon etc.) whilst other regions (such as Hawke's Bay) have chosen just to include the broad regional GI but allow sub regions to develop their own branding strategies (at this stage without GI protection).

In terms of layering of place names, the schedule permits a range of options so that wine producers can choose how they scale and identify their wine. A wine produced in, for example, the Waipara valley could legally use any of the following GIs on its label: Waipara, North Canterbury, Canterbury, South Island, East Coast or New Zealand! There are clear advantages of using the smallest scale appellation (in this case 'Waipara') in order to indicate most clearly the favored place it originates from (rather than an amorphous 'South Island' name which would hint at a high degree of mixing from different areas and, it might be assumed, lower quality). Yet some producers might believe it is better to go for a broad scale place name ('Canterbury' for example) because that may have some recognition in overseas markets rather than a smaller locale ('Waitaki') that is relatively unknown. The schedule is not without a number of boundary issues. The Kaikoura District, for example, is part of the Canterbury Regional Council (and therefore within 'Canterbury Region') but is included within the 'Marlborough' wine region.

Overall, it seems as if the New Zealand GI scheme for one will reflect the overall differences and tensions within the industry, whilst allowing for different place-branding strategies to be used by different sectors of the industry. Mostly, the GIs can be seen as reactive and defensive, providing a broad and flexible framework so that large regions can expand and include new areas. Thus the inclusion of Kaikoura District in Marlborough (not without some historical antecedents) allows for possible further southward expansion of vineyards that can be called—and gain 'rent' from—the Marlborough appellation. On the other hand, there is also room for smaller local and more exclusionary initiatives, such as Waiheke Island or Waipara Valley, which are relatively tightly defined, thereby allowing for possible future 'rent' from a scarce resource.

8.6 Conclusion

The formal definition and application of GI legislation has come late to the New Zealand wine sector. Undeniably, in a general sense, New Zealand has not pursued formal GIs in the economy as a whole. This has come about for a range of reasons including a fundamental pursuit since the mid-1980s of neoliberal re-regulation. This has become an important component of the profile of New Zealand in the international political economy as well as part of its self-image. As a trade-dependent nation it has pushed its credentials as a non-protectionist economy. GIs, in wine or any other sector, represent a form of quasi-protectionism in the eyes of neoliberal policy makers. Notwithstanding this, when the political moment has suited, as it has since in the mid-2000s, there have been moves to design and adopt legislation. In other words the creation of GIs, and their formalization has been largely defensive rather than pro-active and has come about as a matter of political necessity rather than economic innovation.

As noted above however, the rules (or perhaps more accurately relative lack of them) have been designed to allow flexibility and growth. This flexibility and growth has been a hallmark of place-branding strategies in a number of cases in New Zealand, as exemplified in two of the three case studies outlined in this chapter. There is the issue of how regions are defined—some attempts at rigid boundaries (Gimblett Gravels, Martinborough)—create scarcity and are driven by local environmental narratives and yet they impose limits; others are broader (e.g. environmental watersheds such as Waipara or political ones—Marlborough) and these have weak environmental narratives but allow greater expansion. In a broad sense, the use of trademark and sales of goods legislation to identify and apply geographic indication without strict rules and simultaneously allow expansion where required suggest that formal laws have not been crucial in the growth of the industry. This has certainly operated with some success on the domestic market, and the creation of 'fictive place' (Overton and Murray 2016b) manufacturing of scarcity and consequent extraction of rents has not required formal delimitations generally. Yet questions regarding future expansion globally may require a re-thinking of this strategy to an extent, as consumers become more demanding with regard to the provenance of the products they purchase. Ensuring the international perception of quality remains constant may require more fine-grained up-scaled place-making strategies.

In a more general sense however, it could be suggested that there are some fundamental tensions and fissures in the country's agricultural sector, which mitigates against developments such as GIs. The national brand is recognized and used by nearly all agricultural producers to sell their products globally. Yet those who wish to promote smaller-scale forms of production and image making, such as artisanal, localized, organic or other high value-added (and often higher-cost) strategies, remain in the minority in terms of political and economic 'weight'. Furthermore, New Zealand, especially since its neoliberal turns in the 1980s has sided with others pushing for global trade liberalization, notably the USA. New Zealand has been a champion of free trade as it has seen its best prospects for economic growth in getting better access to protected markets for its main agricultural exports (meat and dairy). This has been cleared exemplified by both Labour and National governments over recent decades and mostly clearly manifested in the pursuit of the Trans-Pacific Partnership Agreement (TPPA), which has evolved out of the APEC environment and built on the establishment of the so-called P-3 with Singapore and Chile in 2004.

As such, there is general consensus in the New Zealand mainstream political economy against protectionism both direct and indirect. Just as it promotes the idea of a clean, green environment to sell its products overseas, it also plays on its record of ending agricultural subsidies and other forms of trade protection in trade negotiations. Perhaps this is why the country has been very slow in recognizing and adopting largely European initiatives to recognize geographical indications as a form of intellectual property and, thereby, a form of protection of localized production systems. Thus, we would suggest that New Zealand has a dualistic, export strategy: it often seeks to project an image of a small, distinctive, sustainable, smart, agile and clean industry yet the underlying structure is built on mass production exploiting economies of scale, globalization, vertical and horizontal integration and other Fordist imperatives.

The GI legislation currently under discussion is not without significant uncertainties as noted above and it will be interesting to see what scalar strategies different winemakers choose. It will be useful to monitor if wine making places not currently included (such as Gimblett Gravels) will seek such inclusion or continue to differentiate based on other place-branding strategies. Ultimately, there appears to be no clear consensus within the New Zealand wine industry if GI legislation should tighten the rules and move to a more defined and restrictive scheme of regional definition. It would seem, at present, that the interests of larger-scale producers of export wine to keep the regulations flexible is taking precedence over those who would want to move to a more upscale and exclusive GI regime. This strategy is influenced by the structural nature of New Zealand's agro-export sector in general, which is heavily skewed towards large-scale production whilst at the same time seeking to appeal to notions of quality, locality, and environmental integrity. For now at least, this dualistic approach to agro-exports, wittingly pursued or otherwise, is likely to persist in New Zealand in general, and in particular with regard to wine. Whether this disingenuous binary is viable in the longer term remains uncertain.

Appendix 8.1

See Figs. 8.1, 8.2 and Table 8.1.

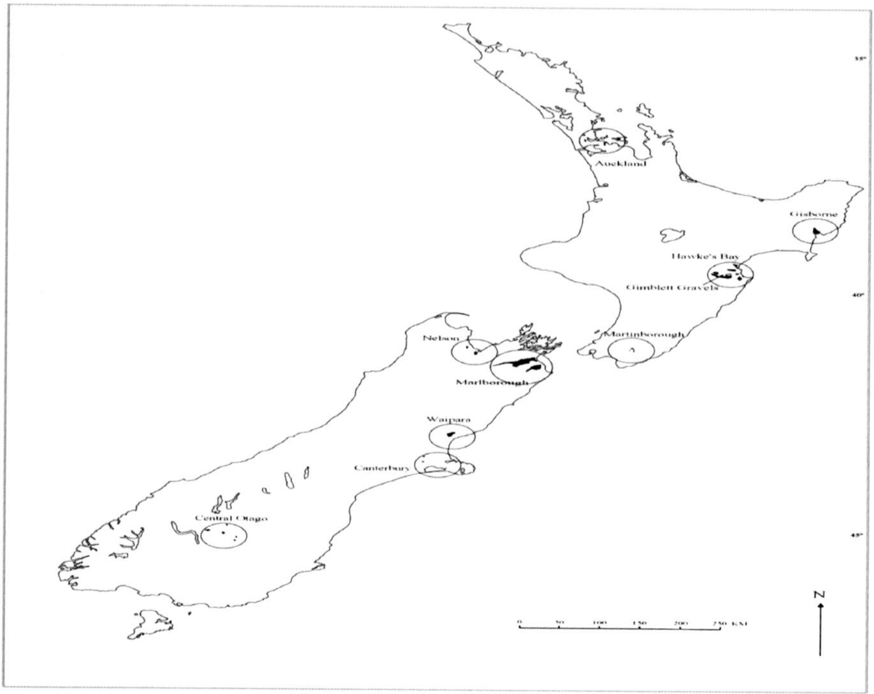

Fig. 8.1 New Zealand wine regions

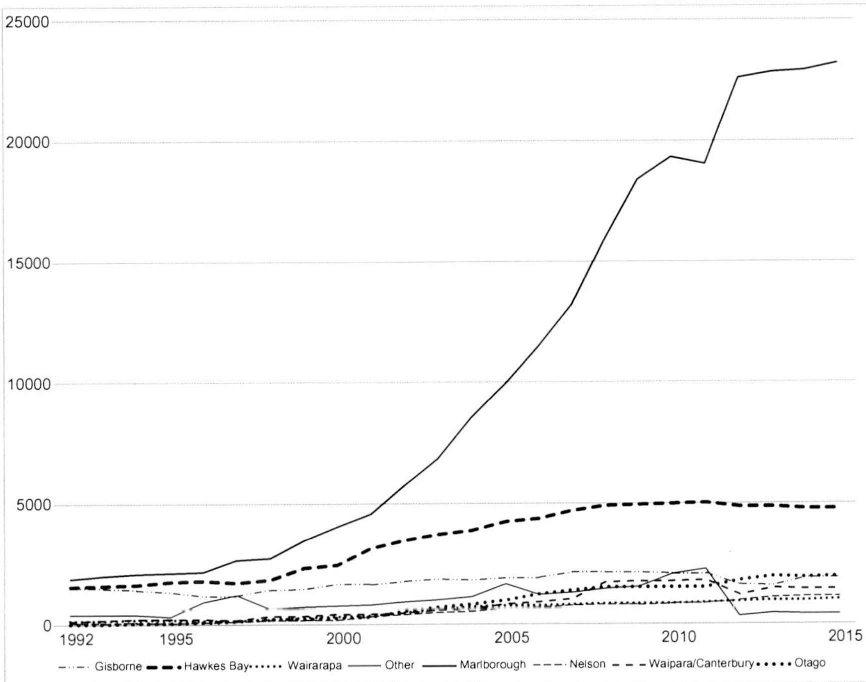

Fig. 8.2 Changes in New Zealand Vineyard area by region 1992–2015 (hectares). *Note* Slight changes in data collection methods in 2010–11 resulted in some anomalies in this data set. The apparent drop in areas in Marlbough and elsewhere at this time may not be accurate. *Source* Wine Institute of New Zealand annual reports 1990–2001; New Zealand Winegrowers annual reports 2002–2016

Table 8.1 New Zealand wine geographical indications as defined for european markets

Geographical indication	Defined area
NEW ZEALAND	The country of New Zealand
EAST COAST	Northland Region, Auckland Region, Bay of Plenty Region, Coromandel District, Hauraki District, Gisborne Region, Hawkes Bay Region, Masterton District, Carterton District, South Wairarapa District, Marlborough District, Canterbury Region
NORTH ISLAND	The North Island of New Zealand
Northland	Northland Region
Auckland	Auckland Region
Clevedon	Winegrowing area in Auckland Region
Western Auckland (comprising Henderson, Huapai, Kumeu, Waimauku)	Winegrowing area in Auckland Region

(continued)

Table 8.1 (continued)

Geographical indication	Defined area
Matakana	Winegrowing area in Auckland Region
Waiheke Island	Waiheke Island
Waikato	Waikato Region
Coromandel Peninsula	Thames-Coromandel District
Lake Taupo	Taupo District
Te Kauwhata	Winegrowing area in Waikato Region
Bay of Plenty	Bay of Plenty Region
Gisborne	Gisborne Region
Hawke's Bay	Hawke's Bay Region
Central Hawke's Bay	Central Hawke's Bay District
Wairarapa	Carterton, Masterton and South Wairarapa Districts
Gladstone	Defined winegrowing area In Carterton District
Martinborough	Defined winegrowing area in Martinborough Ward
Wellington	Greater Wellington Region
SOUTH ISLAND	The South Island of New Zealand
Marlborough	Marlborough and Kaikoura Districts
Nelson	Nelson and Tasman Districts
Canterbury	Canterbury Region
North Canterbury	Waimakariri and Hurunui Districts
Waipara Valley	Winegrowing area in Hurunui District
Waitaki Valley	Winegrowing area in Waitaki and Waimate Districts
Central Otago	Central Otago and Queenstown Lakes Districts

Source New Zealand Winegrowers (2013, p. 30)

Bibliography

Banks, G., S. Kelly, N. Lewis and S. Sharpe. 2007. Place "from one glance": The use of place in the marketing of New Zealand and Australian wines. *Australian Geographer* 38,1: 15–35.

Baragwanath, L. and N. Lewis. 2014. Waiheke Island. In *Social, Cultural and Economic Impacts of Wine in Zealand*, ed. P. Howland, 211–242. Abingdon & New York: Routledge.

Barker, J., N. Lewis and W. Moran. 2001. Reregulation and the development of the New Zealand wine industry. *Journal of Wine Research* 12,3: 199–221.

Barker, J.P.H., 2004. Different worlds: Law and the changing geographies of wine in France and New Zealand. Ph.D thesis, University of Auckland.

Cull, D. 2001. *Vineyards on the Edge: The Story of Central Otago Wine*. Longacre Press: Dunedin.

Dry River. 2009. The Martinborough terrace terroir. http://www.dryriver.co.nz/Terroir/index/Terroir/Martinborough%20Terrace/ (accessed 9 November 2009).

Gwynne, R.N. 2006. Export-orientation and enterprise development: A comparison of New Zealand and Chilean wine production. *Tidjschrift voor Economische en Sociale Geografie* 97,2: 138–156.

Gwynne, R.N. 2006. Governance and the wine commodity chain: Upstream and downstream strategies in New Zealand and Chilean wine firms. *Asia Pacific Viewpoint* 47,3: 381–395.

Hayward, D. and N. Lewis. 2008. Regional dynamics in the globalising wine industry: The case of Marlborough, New Zealand. *The Geographical Journal* 174,2: 124–137.
Howland, P. 2008. Martinborough's wine tourists and the metro-rural idyll. *Journal of New Zealand Studies* 6–7: 77–100.
Howland, P. 2014a. From 'civilising' Maori to fruit-driven exuberance: An introduction to wine in New Zealand. In *Social, Cultural and Economic Impacts of Wine in Zealand*, ed. P. Howland, 1–21. Abingdon & New York: Routledge.
Howland, P. 2014b. Martinborough: A tourist idyll. In *Social, Cultural and Economic Impacts of Wine in Zealand*, ed. P. Howland, 227–242. Abingdon & New York: Routledge.
John Barker Law. 2016. Wine and spirits to get GI registration. *The Food and Beverage Agenda* edition no.1 p. 4 http://www.johnbarkerlaw.com (accessed 11 March 2016).
Jordan, D. and S. Veldhuizen, S. 1992. *New Zealand Vineyard Survey*. The Horticulture and Food Research Institute of New Zealand: Hamilton.
Josling, T. 2006. The war on *terroir*: Geographical indications as a transatlantic trade conflict. *Journal of Agricultural Economics* 57,3: 337–363.
Lewis, N. 2014. Beyond the flawed narratives of a crisis of oversupply: A conceptual fix for New Zealand wine. In *Social, Cultural and Economic Impacts of Wine in Zealand*, ed. P. Howland, 86–102. Abingdon & New York: Routledge.
Moran, W. 1988. The wine appellation: environmental description or economic device? In *Proceedings of the Second International Cool Climate Viticulture and Oenology Symposium New Zealand Society for Viticulture and Oenology*, eds. R.E. Smart, R.J. Thornton, S.B. Rodriguez and J.E. Young, 356–360. New Zealand Society for Viticulture and Oenology: Auckland.
Moran, W. 1993a. Rural space as intellectual property. *Political Geography* 12,3: 263–277.
Moran, W. 1993b. The wine appellation as territory in France and California. *Annals of the Association of American Geographers* 83,4: 694–717.
Moran, W. 2001. Terroir – the human factor. *Australian and New Zealand Wine Industry Journal* 16,2: 32–51.
MPI. 2015. Wine Geographical Indications. Proposed amendment to the 85% rule in section 21 of the Geographical Indications (Wine and Spirits) Registration Act 2006. *MPI Discussion Paper* No: 2015/27. Ministry of Primary Industries: Wellington. http://www.mpi.govt.nz/ (accessed 11 March 2016).
Murray, W.E. and J. Overton. 2011. Defining regions: The making of places in the New Zealand wine industry. *Australian Geographer* 42,4: 419–433.
New Zealand Cabinet. 2015. Geographical Indications (Wine and Spirits) Registration Act 2006. http://www.mbie.govt.nz/publications-research/publications/intellectual-property/gi-cabinet-paper-and-ris.pdf (accessed 13 January 2016).
New Zealand Winegrowers. 2013. *New Zealand Winegrowers Labelling Guide 22nd Edition August 2013*. http://www.nzwine.com/assets/sm/upload/5l/ep/ut/un/NZW_labelling_guide_Aug2013.pdf (accessed 11 March 2016).
New Zealand Winegrowers. 2015. *New Zealand Winegrowers Annual Report 2015*. http://www.nzwine.com/new-zealand-winegrowers-annual-report-2015/ (accessed 14 January 2016).
New Zealand Winegrowers. 2014. *Vineyard Register Report 2013*. New Zealand Winegrowers: Auckland.
Overton, J. and J. Heitger. 2008. Maps, markets and merlot: The making of an Antipodean regional wine appellation. *Journal of Rural Studies* 24,4: 440–449.
Overton, J. and W.E. Murray. 2014a. Boutiques and behemoths: The transformation of the New Zealand wine industry 1990–2012. In *Social, Cultural and Economic Impacts of Wine in Zealand*, ed. P. Howland, 25—40. Abingdon & New York: Routledge.
Overton, J. and W.E Murray. 2014b. Finding a place for New Zealand Wine: *Terroir* and regional denominations. In *Social, Cultural and Economic Impacts of Wine in Zealand*, ed. P. Howland, 41–57. Abingdon & New York: Routledge.
Overton J. and W.E. Murray. 2016a. Investing in place: Articulations and congregations of capital in the wine industry. *Geographical Journal* 182,1: 49–58.
Overton J. and W.E. Murray. 2016b. Fictive Place. *Progress in Human Geography* 40,6: 794–809.

Overton, J., G.A. Banks and W.E. Murray. 2014. Waipara. In *Social, Cultural and Economic Impacts of Wine in Zealand*, ed. P. Howland, 243–252.

Overton, J., W.E. Murray and G. Banks. 2012. The race to the bottom of the glass? Wine, geography and globalisation. *Globalizations* 9,2: 273–287.

Pawson E. and & Perkins H. 2016. New Zealand 'going global': The tensions of rapidly shifting external relationships and the remaking of domestic rural landscapes. Keynote address, New Zealand Geographical Society Conference, Dunedin 3 February 2016.

Schuster, D., D. Jackson and R. Tipples. 2002. *Canterbury Grapes and Wines 1840–2002*. Shoal Bay Press: Christchurch.

Stewart, K. 2010. *Chancers and Visionaries: A History of Wine in New Zealand*. Godwit: Auckland.

Tipples, R. 2007. Wines of the farthest promised land from Waipara, Canterbury, New Zealand. In *Wine, Society and Globalization: Multidisciplinary Perspectives on the Wine Industry*, eds. G. Campbell and N. Guibert, pp. 241–254. Palgrave MacMillan: New York and Basingstoke.

Wines from Martinborough. 2016. Martinborough Geographic Indication (GI). http://www.winesfrommartinborough.com/about_martinborough/martinborough_geographic_indication_9gi0/index.htm (accessed 14 January 2016).

Author Biographies

John Overton is a Professor at the School of Geography, Environment and Earth Sciences at the University of Wellington, NZ. His main research interests are aid and development, with particular regard to New Zealand and the Pacific Islands region; and wine, place and capital with regard to New Zealand and Chile.

Warwick Murray is a Professor at the School of Geography, Environment and Earth Sciences at the University of Wellington, NZ. His main research interests are development geography and economic and rural geography with a Major Focus on Latin America and the Pacific Islands.

Chapter 9
Do Geographical Indications for Handicrafts Deserve a Special Regime? Insights from Worldwide Law and Practice

Delphine Marie-Vivien

Abstract According to the TRIPs Agreement, Geographical Indications (GIs) identify a good as originating in a region where a given quality, reputation or other characteristics of the good is essentially attributable to its geographical origin, be it an agricultural or non-agricultural product such as a handicraft. However, in Europe, and until recently in France, GIs are limited to agricultural products and foodstuffs along with wines and spirits, in contrast to the flourishing registration of GIs for handicrafts in other WTO member States, especially in Asia. Due to the absence of a uniform framework for all types of goods, negotiations in bilateral agreements are less smooth and international trade is hampered while European artisans cannot obtain protection for their handicraft's GIs. The purpose of this chapter is to analyze the legal basis for the protection of GIs over time and over practice in several countries worldwide and whatever the product is, to question the relevance of a product-based approach. The key findings of this theoretical and empirical analysis suggest that the discrimination between products is not justified, including the existence of a special regime for wines and spirits. The chapter recommends the use of the same criteria of the link to the origin to be applied to GIs for all kinds of goods. Whatever the product is, it is recommended to apply a two-level GI reference according to the presence or absence of natural factors. Such insights cast new light on the existing legal regimes for GIs and suggest potential avenues for reform: in France, which has just passed a new law on GIs for handicrafts; in the EU where the protection of GIs for handicraft is a debated issue; as well as in international treaties on GIs, be that the TRIPs or the Geneva Act of the

---

This chapter draws largely on a previous publication by the same author.

Marie-Vivien (2016).

D. Marie-Vivien (✉)
CIRAD, UMR Innovation, MALICA, Montpellier, France
e-mail: delphine.marie-vivien@cirad.fr

D. Marie-Vivien
CIRAD, UMR Innovation, MALICA, Hanoi, Vietnam

W. van Caenegem and J. Cleary (eds.), *The Importance of Place: Geographical Indications as a Tool for Local and Regional Development*, Ius Gentium: Comparative Perspectives on Law and Justice 58, DOI 10.1007/978-3-319-53073-4_9

Lisbon Agreement for the international registration of appellations of origin and geographical indications.

9.1 Introduction

A Geographical Indication (GI) identifies a good as originating in a region, where a given quality, reputation or other characteristics of the good is essentially attributable to its geographical origin.[1] This definition from the WTO's TRIPS Agreement is based on an old concept of products with reputation attached to their place, for example to minerals (marble), art objects (bronze, ceramics, pottery or terracotta), textiles (silk), perfumes (incense) and processed agricultural products (honey) (1985). These examples from ancient times show that the kind of products notorious for their place of origin are limitless. Nowadays, however, in some countries and more particularly in Europe, GIs are still limited to agricultural products and foodstuffs,[2] as part of the Common agricultural policy, along with wines and spirits.[3]

The European vision can be explained by the interpretation of the link to a given territory primarily through the concept of *terroir*, which emphasises the land and the soil. This vision results from using the concept of appellation of origin that was first applied to wines in France, the birthplace of the modern appellation of origin, in the early 20th century. Such a vision influenced the TRIPS Agreement, whereby wines and spirits benefit from a higher level of protection compared to other goods (Addor and Grazzioli 2002). However the TRIPS protects GIs for all kinds of goods, beyond agricultural products, thus offering many WTO State members the opportunity to protect non-agricultural goods such as handicrafts (Audier 2008; O'Connor 2004). In India, for example, GIs are predominantly registered for non-agricultural and non-foodstuff products. These made up two-thirds of the 238 registered GIs as of January 2016 (Marie-Vivien 2010).[4] In Colombia, almost half of the recognized Denominations of Origin are for handicraft goods (10 out of 22 in December 2014).[5] In Switzerland an ordinance for GIs for non-agricultural goods is

---

[1]The Agreement on Trade-Related Aspects of Intellectual Property Rights (TRIPS), Art 22.1.

[2]Products are listed in Annex I of the European Community Treaty and in Annexes I and II of the European Regulation No. 510/2006, replaced since 2012 by Regulation No. 1151/2012 of the European Parliament and of the Council of 21 November 2012 on Quality Schemes for Agricultural Products and Foodstuffs [2012] OJ L323/1. These Annexes may be amended in order to extend or reduce the list of products eligible for the registration of PDOs and PGIs as long as they constitute agricultural products or foodstuffs.

[3]In 2008 wines and spirits were integrated into the PDO and PGI categories; see Council Regulation (EC) No 479/2008 of 29 April 2008 on the Common Organization of the Market in Wine [2008] OJ L148/1. Previously, the production of wines and spirits was regulated by a system of specific rules, due to the then extant structure of the European market in wine.

[4]*See* http://www.ipindia.nic.in/girindia/ as consulted on February 29, 2016 at 16:46 pm.

[5]*See* http://www.sic.gov.co/drupal/productos-con-denominacion-de-origen as consulted on February 29, 2016 at 16:48 pm.

entering into force on January 1st 2017.[6] In Thailand, nine GIs for handicraft goods are registered, out of which four are for silk, an emblematic product of the country.[7]

The TRIPS framework not only led to new domestic laws protecting GIs but also provided for mandatory protection of foreign GIs in abroad jurisdictions beyond the country of origin. In the EU, whether it is achieved via the registration of a non-Member's individual GI or through free trade agreements including a list of GIs to be protected by the signatory States, the practices of countries protecting handicraft GIs is challenging and influencing the current EU vision. Countries such as India, Brazil, Vietnam and Malaysia are requesting GIs for their handicrafts to be protected in the EU, be it for the famous Kashmir Pashmina[8] from India, the conical hat from Hué in Vietnam (which could not be included in the list of GIs of the EU-Vietnam bilateral agreement signed in 2015[9]) or for the two non-agricultural GI products, Guacamayas (spiral basketwork) and Chulucanas (pottery) not to be protected in Europe despite the bilateral agreement with Peru and Colombia in 2012.[10] In general, the preference is for protection within the harmonized sui generis EU GI regime,[11] and not according to a trade mark regime (Gangjee 2007). Due to the absence of a uniform framework for all types of goods, negotiations in bilateral agreement are difficult and international trade is hampered.

European producers themselves are now demanding a unitary protection framework at the EU level, according to a recent study on Geographical Indications protection for non-agricultural products in the internal market. The study, ordered by the European Commission, identified several candidates within the EU. For example in France alone around 100 possible candidates, such as Stone from Bourgogne or Lace from Calais were identified.[12] Subsequently in 2014, France passed a law recognizing Geographical Indications for non-agricultural goods, which is enforceable since June 4, 2015,[13] with the first GI registered the

---

[6]*See* art. 50a of the Trademark and Indication of Source Law of Switzerland. Plutot Ordonnance sur les AOP et les IGP non agricoles (https://www.admin.ch/opc/fr/official-compilation/2015/3669.pdf) (date le 2 septembre 2015) (as consulted on February 29, 2016).

[7]*See* www.ipthailand.go.th/en/index.php?option=com_docman&view=docman&Itemid=434 as consulted on February 29, 2016 at 17:48 pm.

[8]Indian GI application No. 46, filed on 9 December 2005, Geographical Indication Journal No. 13.

[9]The negotiations of the FTA concluded in December 2, 2015, with agreement published on February 1 2016, available at http://trade.ec.europa.eu/doclib/press/index.cfm?id=1437.

[10]*See* Trade Agreement between the European Union and Colombia and Peru at: http://trade.ec.europa.eu/doclib/docs/2011/march/tradoc_147704.pdf.

[11]No foreign GIs for non-agricultural goods have been protected in the Member States of the EU. See Insight Consulting, REDD & OriGIn, *Study on Geographical Indications for Non-Agricultural Products in the Internal Market* (Final Report, 18 Feb 2013) (Hereafter, *EU Study*).

[12]GI 15-002, BOPI Marques 2015-44 du 30/10/2015.

[13]10 new articles L.721-2 to L. 721-10 were introduced into the Intellectual Property Code, by the Law on Consomption of 17 March 2014, and its regulation of application "Décret 2015-595 du 2 juin 2015 relatif aux indications géographiques protégeant les produits industriels et artisanaux et portant diverses dispositions relatives aux marques; JORF 3 juin 2015".

2 December 2016 for "Siège de Lifol" and 3 GI applications pending. [14] Indeed, France is facing a strong deindustrialization process, with many French geographical denominations of manufactured and handicraft goods becoming generic leading to an increase in imported goods of low quality but also low price, using the same name. The risk is then the loss of local know-how and disappearance of local producers and artisans.[15]

In Europe and France, the discussions around the extension of the protection of GIs for non-agricultural and non-food products have raised many questions regarding the legal basis for GI protection in general, as well as whether GIs should be treated differently according to the kind of goods in question. Indeed, in general, for most of the handicraft goods, the divide between agricultural and non-agricultural products resides in the absence of physical elements linking non-agricultural goods to the soil. While natural factors besides the soil such as the climate, the origin of raw materials, or environmental elements such as water can indeed influence product quality, the territorial link for most non-agricultural products or handicraft goods is based above all on the producers' know-how, skills and practices – that is, on human factors. The issue is whether geographical names designating goods linked to their place of origin essentially via human practices can be registered as GIs. Can GIs be considered to be applicable to non-agricultural as much as agricultural products, as much to culture as they apply to nature (Gangjee 2015a)? And what about the non-agricultural products linked to their geographical origin through natural factors? Nature is not exclusive to agriculture. Finally, an underlying question is how to assess the link between the quality, reputation or characteristics of a product and its place of origin.

This chapter sets out to analyze the validity of such a link to origin, since this is the central legal criterion setting GIs apart as a distinct intellectual property right and to assess whether this discrimination based on the nature of a product is justified. Section 9.1 explores the concepts underpinning GIs, with a historical discussion of the evolving legal definitions of GIs over time. The chapter then proposes an original framework for better scrutinizing the different types of links with the origin, drawing from different jurisdictions on empirically informed case studies for both agricultural and non-agricultural GIs. It extensively reviews GI specifications resulting from the practices both of producers/applicants and of examining authorities (that are registrars), drawing on the concepts of natural and human factors introduced by the Lisbon Agreement for the Protection of Appellations of Origin and their International Registration.[16] In order to take into consideration the looser definition of GIs in TRIPs now also introduced in the Lisbon Agreement, the analysis uses these two concepts

---

[14]https://www.inpi.fr/sites/default/files/02122016_cp_inpi_1ere_indicationgeographique.pdf

[15]Art. L.721-6 of the French Code of Intellectual Property provides for "the development of territories, of local traditions and know-how as well as products derived therefrom".

[16]Lisbon Agreement for the Protection of Appellations of Origin and their International Registration of 31 October 1958.

both as cumulative and alternative criteria. The selected cases highlight the various possible combinations of natural and/or human factors in linking products to their origin, beginning with products linked to the territory exclusively through human factors in Sect. 9.2 and then with products linked to the *terroir* through human factors combined with natural factors in Sect. 9.3. Section 9.4 discusses the key findings of this theoretical and empirical analysis. First, one finding is that the existing categorization of goods in the EU does not allow for clear distinction between products that deserve GI protection and those that do not, or between different types of GIs, as the link to the territory can widely differ within the same category of products, while there may be similarities in the nature of the link across different categories of products. Second, it is proposed for the international and national GI legal frameworks to consider the concepts of natural and/or human factors in determining the strength of the link to the origin, regardless of whether the product is classified as agricultural, a foodstuff, handicraft or industrial. Such insights cast new light on the existing legal regimes for GIs and suggest potential avenues for reform. This chapter argues that a link to the origin based solely on human *or* natural factors might be qualified as weaker than a link based on the combination of natural *and* human factors. It consequently develops the rationale for establishing two levels of GI protection based on the strength of that link. The argument contributes to the implementation of GIs for non-agricultural goods introduced in French law, to the evaluation of the EU Commission on the feasibility of recognizing GIs for non-agricultural goods as well as, providing guidance for GI examination throughout the world.

9.2 History of GIs and Appellations of Origin Concepts for All Goods

A look back at the history of GIs reveals that, in France (Olszak 2001), the first modern law on the protection of appellations of origin was in 1905, designed to combat fraud in the sale of goods and adulteration,[17] and was created to take account of and protect place names for qualifying natural products whose specificity was dependent on natural elements. Prior to this, place names were protected only for 'manufactured' goods (Plaisant and Fernand-Jacq 1921).[18] However, the subsequent law of 1919,[19] still in force, gave the Court the task of defining

[17]*Loi du 1er août 1905 sur la repression des fraudes dans la vente des marchandises et des falsifications des denrées alimentaires et des produits agricoles* (Law of 1 August 1905 on the repression of fraud in the sale of goods and adulteration of foodstuff and agricultural products) (5 Août 1905) Journal Officiel No. 210.

[18]*Loi du 28 juillet 1824 Relative aux Altérations ou Suppositions de Noms dans les Produits Fabriques* (1825) 7 *Bulletin des Lois* no 19, 65 (Law of 28 July 1824 on the Misuse of Names for Manufactured Products).

[19]*Loi du 6 Mai 1919 Relative à la Protection des Appellations d'Origine* 8 Mai 1919 Journal Officiel 4726 (Law of 6 May 1919 on the Protection of Appellations of Origin).

appellations of origin in case of conflicts between users and did not distinguish between various kinds of products, allowing for handicraft appellations to be protected. Most important was the consideration of the concept of '*terroir*' (Audier 1993), which ultimately included, in addition to the natural environment, human skills, know-how, practices and the knowledge of producers. Appellations designating non-agricultural goods were recognized by the judiciary, in accordance with local, fair and constant use of appellations of origin. Finally, in 1935, a law was passed and it provided for the *appellation d'origine controlée* (AOC) and the creation of a national committee comprised of representatives of ministries and professionals for their recognition, initially only for wines and spirits, but subsequently extended to cheeses in 1990, and then to agricultural, forestry and food products (Marie-Vivien et al. 2015). The French system of AOC evaluated by a national committee of experts has therefore never been applied to non-agricultural goods.

Recently, following the misappropriation of the name of the city Laguiole, famous for its knives, the French Law on consumption was passed on 17 March 2014 providing for the creation of GIs for non-agricultural products[20] defined as "the name of a region or a specific place used to describe a product, other than agriculture, forestry, food or marine products, originating from that place and which has a specific quality, a reputation or other characteristics that can be essentially attributed to its geographical origin; [t]he conditions of production or processing of this product, such as the cutting, extraction, or manufacture must comply with a specification".[21] The specification shall describe "the process of preparation, production and of processing, elaboration, including the steps of production of processing which shall take place in the geographical area or the specific place as well as the steps which ensure the specific characteristics".[22]

At the international level, the Lisbon Agreement, signed in 1958 within the framework of the Paris Convention (Gangjee 2012) provides a very precise definition of appellations of origin which has to be applied by all State members (presently 28): "appellation of origin means the geographical name of a country, region or locality, which serves to designate a product originating therein, the quality and characteristics of which are due exclusively or essentially to the geographical environment, including natural and human factors".[23]

The Lisbon Agreement clearly points to the link between the quality and characteristics of the product and its geographical environment. The *travaux préparatoires* of the Lisbon Agreement reveal that the first proposal did not mention the geographical environment, referring rather to "the place and method of production,

---

[20]Loi No. 2014-344 du 17 mars 2014 relative à la consommation, Journal Officiel de la République Française No 65 du 18 mars 2014, p. 5400.

[21]Art L. 721–2 of the Intellectual Property Code.

[22]Ibid.

[23]Art 2 of the Lisbon Agreement. The same definition was introduced in France in 1966, via Art 1 of the Law of 6 July 1966 amending and completing the Law of 6 May 1919, JO 7 July 1966, p. 5781, which is now Art L. 115-1 of the Consumer Code.

manufacturing, extracting or assembling of products".[24] This development in the text makes it possible to interpret the concept of human factors: the practices, skills and know-how of the producers incorporated in obtaining the product. The term "the place" has been replaced by "natural factors", which is more restrictive, focusing on nature, which comprises the climate and the elements of the natural environment (soil, water, clay, caves, and so on) that can influence the quality of the raw material or processing for processed goods. Based on a mandatory combination of human and natural factors, the Lisbon Agreement is very restrictive regarding the strength of the link, but does not, however, distinguish according to the kind of products, which can be agricultural or non-agricultural.[25] The Lisbon Agreement provides a very high level of protection against any usurpation or imitation, even if the true origin of the product is indicated or if the appellation is used in translated form or accompanied by terms such as "kind", "type", "make", "imitation", etc.[26] The system provides for a notification and opposition procedure within its membership, and has led to the international registration of more than 800 appellations of origin.

This definition of the appellation of origin was introduced in European countries which were signatories to Lisbon and later Europeanised with the introduction of the harmonized European regime for the protection of GIs for agricultural products and foodstuffs in 1992, subsequently replaced by a new regulation in 2006[27] and more recently by Regulation (EU) No 1151/2012 in 2012.[28] Wines are regulated by Council Regulation (EC) No 1308/2013 while spirits are regulated by Regulation (EC) No. 110/2008 of 15 January 2008, and aromatized wines by Regulation (EU) No. 251/2014.

The EU's Protected Designation of Origin (PDO),[29] is very similar to the AO present in the Lisbon Agreement, albeit with greater precision on the origin of raw

[24]As proposed unanimously by the 4th Committee of the work and reports of the Lisbon Conference in 1956.

[25]Art 2.1 of the Lisbon Agreement refers to product, with no further details. Consequently in practice, non-agricultural products have been registered under the Agreement.

[26]Art 3 of the Lisbon Agreement.

[27]Council Regulation 510/2006 of 20 March 2006 on the Protection of Geographical Indications and Designations of Origin for Agricultural Products and Foodstuffs [2006] OJ L93/12, which replaces Council Regulation 2081/92 of July 14, 1992 on the Protection of Geographical Indications and Designations of Origin for Agricultural Products and Foodstuffs [1992] OJ L208/1.

[28]Regulation No 1151/2012 of the European Parliament and of the Council of 21 November 2012 on Quality Schemes for Agricultural Products and Foodstuffs [2012] OJ L323/1.

[29]See Art 2(2)(a) of Regulation 510/2006; similar wording is evident in Art 5(1) of Regulation 1151/2012. For wines, the designation of origin is defined in Art 93.1(a) of Regulation 1308/2013 as:

> the name of a region, a specific place or, in exceptional cases, a country used to describe a product referred to in Art 92.(1) that complies with the following requirements: (i) its quality and characteristics are essentially or exclusively due to a particular geographical environment with its inherent natural and human factors; (ii) the grapes from which it is produced come exclusively from this geographical area; (iii) its production takes place in this geographical area; (iv) it is obtained from vine varieties belonging to *Vitis vinifera*.

materials that must originate in the geographical area where the production, processing and preparation occur.[30] In order to take into account the tradition of Northern Europe countries, the Court of Justice of the European Union (CJEU) recognized protection for geographical names for products "which cannot be shown to derive a particular flavor from the land, but which may nevertheless enjoy a high reputation amongst consumers and constitute for producers established in the places to which they refer an essential means of attracting custom".[31] This distinction has led to the creation of the Protected Geographical Indication (PGI), for "a name which identifies a product (a) originating in a specific place, region or country; (b) whose given quality, reputation or other characteristic is essentially attributable to its geographical origin; and (c) at least one of the production steps of which take place in the defined geographical area".[32] Regarding the link to the territory, the PGI definition does not expressly require that natural factors such as the raw materials be sourced in the defined geographical area, as only the step of processing/manufacturing can be localized. From legal practice and jurisprudence, however, it appears that the demarcation of the geographical area of origin of raw material is authorized only if the source of the raw material influences the quality of the final product (Evans 2012).[33] In addition to product qualities being causally attributable to origin, characteristic and reputation are also recognized criteria. As a result, there are three alternative criteria for qualifying for recognition and protection as a PGI.

Thus for a PGI, unlike for a PDO, the strength of the link to the origin is weak, less stringent and less exclusive (Bérard and Marchenay 2008). This weaker conception of the link to origin is a result of the lower importance attached to natural factors. This flexibility in the European regulation regarding the link to the origin contrasts with the strict conception regarding the kind of products that may benefit from GI protection, which is restricted to a list of agricultural products and foodstuffs, as part of the Common Agricultural Policy of the EU. Such a list can be amended, but additions must fit into the general definition of agricultural products

---

[30]See Art 3 of Regulation 510/206; Arts 5(1) and 5(3) of Regulation 1151/2012.

[31]*Exportur SA v LOR SA and Confiserie du Tech SA* (C-3/91) [1992] ECR I-5529 at [28].

[32]Art 5(2) of Regulation 1151/2012. Regarding wines, according to Art 93.1(b) of Regulation 1308/2013, a GI means 'an indication referring to a region, a specific place or, in exceptional cases, a country, used to describe a product…' which complies with the following requirements: '(i) it possesses a specific quality, reputation or other characteristics attributable to that geographical origin; (ii) at least 85% of the grapes used for its production comes exclusively from this geographical area; (iii) its production takes place in this geographical area; (iv) it is obtained from vine varieties belonging to *Vitis vinifera or* a cross between the *Vitis vinifera* species and other species of the *Vitis* genus'. Regarding spirits, according to Art 15(1) of Regulation 110/2008, a geographical indication shall be 'an indication which identifies a spirit drink as originating in the territory of a country, or a region or locality in that territory, where a given quality, reputation or other characteristic of that spirit drink is essentially attributable to its geographical origin'.

[33]*Carl Kühne & Others v Jütro Konservenfabrik GmbH* (C-269/99) [2001] ECR-I 9517. .

or foodstuffs.[34] Regarding the level of protection, Europe has kept on with a high protection of GIs, with the prohibition of any misuse, imitation or evocation, even if the true origin of the products or services is indicated or if the protected name is translated or accompanied by an expression such as 'style', 'type', 'method', 'as produced in', 'imitation' or similar, including when those products are used as an ingredient.[35]

The introduction of the PGI in the EU framework was concomitant with the negotiations of the TRIPS Agreement which has introduced the same concept but with a more flexible link to the origin into its framework, with the definition of GI being very similar to the PGI used by EU legislation. By introducing an autonomous criterion of reputation to the TRIPS and the EU legislation, the definition of GI and the EU PGI provides greater opportunities of products linked to the origin while opening the door to innovative experiences (Gervais 2009).[36]

When implementing TRIPS, many countries introduced a broad GI definition into their domestic legal framework. For example in India, the Geographical Indications of Goods (Registration and Protection) Act (1999) defines GI in the same manner as the TRIPS. Furthermore, the Indian GI Act defines 'goods' as any agricultural, natural or manufactured good or any handicraft of industry goods and includes foodstuffs.[37] It thus formalizes the validity of GIs for handicraft goods in India.

While not discriminating against non-agricultural goods, the TRIPs Agreement separates wines and spirits that benefit of a higher level of protection than other goods (Thévenod-Mottet and Marie-Vivien 2011). The level of protection for all products (Article 22) represents a minimum *standard* of protection for all products. It prohibits any use that constitutes an act of unfair competition in the sense of Art 10 *bis* of the Paris Convention, which misleads the public as to the geographical origin of the goods. It also prohibits the registration of a trade mark that would contain or consist of a GI for goods not originating in the territory indicated but only when a mislead as to the true place of origin also exists. Article 22 also applies in situations where the sign in question is accurate as to the territory in which the goods originate, but would falsely indicate to the public that the goods originate in an alternative territory. The minimum standards in this provision focus on preventing consumers from being misled and in case of unfair competition, it allows for judicial consideration. Such a case-by-case application of protection through judicial decisions often generates problems.

---

[34]An example is the addition of salt, whereas mineral waters have been withdrawn. Council Regulation (EC) No 692/2003 of 8 April 2003 Amending Regulation (EEC) No.2081/92 [2003] OJ L99/1.

[35]Art 13 of Regulation 1151/12.

[36]According to this author, the difference between the Lisbon Appellation of Origin and the TRIPS GI is mainly focused on the additional criterion of reputation found in the latter. However, the mandatory existence of both natural and human factors in Lisbon is underestimated.

[37]S2 of the Indian Geographical Indications of Goods Act, 1999.

The level of protection for wines and spirits (Article 23) strictly prohibits the use of a false GI, even if it is used in translation or accompanied by a qualifying expression such as 'kind', 'type', 'imitation', and so on. The registration of a trade mark containing a GI for wines or spirits not having this origin is prohibited, even if the public is not misled as to the true origin of the product. Moreover, Article 24.1 mentions that members will enter into negotiations aimed at increasing the protection of individual GIs for wines and spirits.

The successful dissemination of the concept of GIs present in the TRIPS has consequences for the Lisbon Agreement and its restrictive definition of appellation of origin, considered by many as the reason for the limited number of signatory States. In 2009, a working group on the development of the Lisbon system was established and it has since worked extensively on a draft for a new Act of the Lisbon Agreement, finally adopted on May 20, 2015 in Geneva, with the objective of improving the system, *inter alia*, by incorporating the TRIPS definition of the GI, more flexible than the appellation of origin. The purpose of a new Act is to attract new signatory States thanks to the flexibilities in the definition of GIs and to the possibility for GIs beneficiaries to directly file an application for international registration without going through their national authorities.

Nevertheless, in the absence of any detailed provisions – in TRIPS, in the Lisbon Agreement or in national adaptations of these provisions – in the substantive examination of the link to the origin, practice and case law provide valuable guidance when attempting to define the basis for qualifying a GI, both for non-agricultural goods as well as for agricultural goods.

9.3 The Human Factors in Linking the Product to the Place

Most of non-agricultural goods and particularly handicrafts are usually linked to a given place via human factors such as the practices and know-how of the producers, with relatively few natural factors. Unexpectedly, however, an emphasis on human factors in the absence of natural factors also occurs with foodstuffs in Europe.

9.3.1 Handicraft Goods

9.3.1.1 Sophisticated Know-How

For many GIs in the handicraft sector, the method of production is highly intricate, based on the consideration that only manual methods lead to desired results, whilst machine-made copies are of inferior quality as illustrated by the Indian GI Kancheepuram Silk, for silk woven in the ancient, royal town of Kancheepuram, famous for its temples.[38] The method of production is characterized by the use of thick silk yarn which gives it its heavy weight and bright colors as well as by the use of two extra shuttles on each side of the loom, besides the shuttle used for the main body of the sari, to weave the borders in contrasting colors. The reputation of Kancheepuram saris also lies in the use of silver, gold and red silk threads known as "zari". It is generally accepted that counterfeit Kancheepuram saris are made of thinner silk yarn and have only one border and do not contain gold.[39] The method of production includes dressing the yarn, sizing, the degumming of the yarn, including the number of times it is rinsed and the time required to do so, as well as a meticulously detailed dyeing process. The intricate details of the description demonstrate the sophistication of the skills involved. In France, meanwhile, the Dentelle du Puy (lace) appellation of origin has been recognized by the Court according to the law of 1919. The numerous certificates offering guarantees of the independence, impartiality and sincerity that proved the existence of traditional, local, constant and fair usage were the motivations for the Court to reserve the appellation exclusively for handmade lace by artisans whilst excluding machine-made lace.[40] In India, woodcraft products are also protected by GIs, with the product being entirely hand-chiseled then painted. Take, for example, the GI Kondapalli Bommalu for painted wooden figurines.[41] Their production requires know-how about selection of wood, cutting and seasoning, woodcarving and painting.

Besides the know-how involved in making an object, assembling it and treating the raw material, GIs are also registered on the grounds of traditional designs and drawings. While the method of production may be widespread, motifs and drawings are often specific to a region. Such GIs raise issues about their 'uniqueness'—embedded essentially in the designs—which, according to intellectual property laws, are normally protected as 'models and designs'.[42] Nevertheless, exclusive rights on designs and models are only granted for a limited period of time and only for *new* models and designs, which does not apply to designs in other traditional products.

The sophistication of the design may ensue from the technique used and vary according to the degree of its mastery, as illustrated by the Indian GIs Pochampally Ikat[43] and Orissa Ikat.[44] Ikat—a Malaysian-Indonesian word that means tie-dye—

---

[38]GI application No. 14, filed on 7 October 2004, Geographical Indication Journal No. 4.

[39]Interview with a manager of the weaving service centre of Kancheepuram, December 2006. On file with the author.

[40]*Chambre Syndicale des Fabricants de Dentelles et Passementeries de la Haute-Loire v Gouteyron et Jérôme* [1931] Propriété Industrielle 188 (Le Puy-en-Velay Civil Court, 19 February 1931).

[41]GI application No. 44, filed on 10 November 2005, Geographical Indication Journal No. 13.

[42]See e.g. Section IV of TRIPS and the Hague Agreement Concerning the International Registration of Industrial Designs of 1925.

[43]GI application No. 4, filed on 15 December 2003, Geographical Indication Journal No. 13.

[44]GI application No. 22, filed on 1 February 2005, Geographical Indication Journal No. 12.

involves the sequence of tying or wrapping and dyeing sections of bundled yarn to a predetermined color scheme prior to weaving. Two elements distinguish imitations from the originals: they are machine-woven and above all the motifs are printed after weaving and not created during the weaving process with pre-dyed yarn.[45] The GI Pochampally Ikat is entirely based on the Ikat technique and is described in a very detailed manner. The diamond shaped motifs, or 'chowka', which, because of the recent history of Ikat production in Pochampally, are simpler than Ikat made in other regions of India, are only briefly mentioned in the specification. However, such motifs distinguish Pochampally Ikat from Ikat produced in other places,[46] for example, the GI Orissa Ikat[47] is characterized by floral motifs with sophisticated shaded effects. Pochampally Ikat and Orissa Ikat GIs highlight the situation of creative know-how intertwined with the Ikat technique. The creation of designs cannot therefore be dissociated from the mastery of the technique. Their *uniqueness* lies in the particular way the technique is applied in a specific place and not only in the type of design.

9.3.1.2 Absence of a Specific Origin Requirement for Raw Material

In many cases the raw materials used for non-agricultural GI products are not sourced locally, but must nevertheless be of high quality, known as *generic quality* (Allaire 2004). In Europe, for example, out of 127 products considered in the EU Study, 46 non-agricultural GI products do not exclusively source their raw materials from within the defined geographical area (Insight_Consulting et al. 2013). In India quality raw material implies traditional raw material, such as natural dyes[48]; or, for woodcrafts, it is timber quality.[49] The use of *forged* raw material, such as synthetic thread, helps identify counterfeits. However, the quality of the raw material is not attributed to local natural factors. For Indian handicraft GIs, either the geographical origin of raw material is not described, or this origin is far removed from the product's manufacturing zone and indicated on a purely documentary basis without being mandatory. The silk yarn used for Kancheepuram Silk according to GI specification is bought from Gujarat, situated in North India. The *uniqueness* of the Indian Konark Stone Carving[50] GI registered for sculptures of traditional dancers

[45]A case between the owner of the Pochampally Ikat IG and an infringer who manufactures saris with machine printed pattern. Complaint to the High Court of Delhi, 887/2005.

[46]These simplified modern designs have proved more popular with new generations of Indian consumers.

[47]Applicant's reply, dated 3 November, 2005 to the letter from the GI Registry dated 21 October 2005, accessed in the file available at the GI Registry, Chennai.

[48]Presentation of the Assistant Registrar of the GI Registry, Mr Natarajan, 17–18 September 2008, Delhi. The six criteria are: 'Quality of raw fibre, Natural dye, Quality of water, Colour fastness, Durability, Professional skill.'

[49]E.g. the 'Sankheda Furniture' GI specification insists on the use of 100 per cent teak wood.

[50]GI application No. 87, filed on 9 April 2007, Geographical Indication Journal No. 15.

from Orissa is due to the raw material that is the snake stone. The mineral composition of the stone is described but its geographical origin is not documented, demonstrating the extent to which this aspect is considered superfluous.

For many cases, the source of raw material explains the localization of production but local sourcing is not mandatory, as it is not seen as a condition for obtaining a quality product. For example, the Indian GI Kondapalli Bommalu, designating wooden figurines, indicates that the wood comes from the surrounding region of Kondappalli, but the area identified corresponds to the area where the figurines are sculpted—a village of 1.5 km^2—and does not include the nearby forests. Similarly, the reputation of GI Mysore Silk[51] is mainly attributable to the silk yarn produced in the ancient kingdom of Mysore where there is a tradition of silk worm farming.[52] The specification emphasizes its *uniqueness*, attributed to the superior quality of the silk yarn used, a zari consisting of 65% silver and 0.65% gold and a special process of twisting the yarn which gives the fabric its wavy effect. However, there are no provisions specifying the source of the silk.

While such handicrafts also exist in Europe, they may not be granted protection according to EU GI law and according to French GI law, at least not before June 2015. In France, in the 1990s the 'faïence de Moustiers' producers association drafted an appellation of origin to protect the revival of earthenware production in Moustiers, which attracted many avid prospectors. But since the raw material (fuller's earth, enamel, and so on) was no longer directly sourced from Moustiers, as it had been in the 18th century, the application was not accepted (Pollaud-Dulian 1999).[53] Indeed, French law has, since 1966, incorporated the definition of the appellation of origin of the Lisbon Agreement, which provides for a combination of natural and human factors. For the same reasons, Porcelaine de Limoges and Porcelaine de Nevers applications were rejected. This is why thanks to the new GI for non-agricultural goods available in France since June 2015, all those products that gained their reputation at a time when the raw material was available locally and was of particular quality, will be able to apply for GI protection.

What is now possible in France is impossible in the EU, due to the absence of a GI legal framework. For example, Harris Tweed from Scotland is only protected as a certification trademark in the absence of a mandatory local wool requirement. The certification trademark relies on article 7(a) of the Harris Tweed Act of 1993 as 'a tweed which has been hand-woven by the islanders at their homes in the Outer Hebrides, finished in the Outer Hebrides, and made from pure virgin wool dyed and spun in the Outer Hebrides", but the trademark does not protect the name as such (Anthony 2013). In Colombia as well, only the trademark system is available, despite the existence of local raw material. Indeed, Colombia protects GIs through

[51]GI application No.11, filed on 22 July 2004, Geographical Indication Journal No. 3.

[52]Personal interview with Mr Vijayan, the General Director of KSIC (Karnataka Silk Industries Corporation), the GI applicant. On file with the author.

[53]See 'Faïences de Moustiers' PIBD (1992) No. 509.I.85, Ministerial Answer No. 15479 of 6 June 1991 and official journal of the Senate debates, 26 September 1991, p. 2088.

the denomination of origin, where natural factors are mandatory, accompanied by a mandatory sourcing of raw material from the geographical area. Therefore, the *Mochila Arhuaca*, a handwoven bag, that has dyed wool originating from Peru is only protected as a collective trademark (Totem 2012), despite of the existence of human factors.

9.3.2 Agricultural Products and Foodstuffs

Worldwide, most GIs for agricultural goods and foodstuffs are widely perceived to satisfy the link requirement on the basis that natural factors determine the specificity of the product. According to WIPO: ‘agricultural products typically have qualities that derive from their place of production and are influenced by specific local, geographical factors such as climate and soil’. It is therefore not surprising that a majority of GIs throughout the world are applied to agricultural products, foodstuffs, wine and spirit drinks’ (WIPO_Publication_No.952(E)). However, a careful analysis of European registered PGIs and Indian agricultural GIs does not support such an argument. Many agricultural products and foodstuffs are in fact linked to their origin mainly through human factors, even if in France, the birthplace of the concepts of appellation of origin and *terroir*, this acknowledgement has been relatively ignored.

In India, the first GI registered for a liqueur, Feni[54] illustrates a GI based on traditional know-how in the absence of mandatory sourcing of raw material locally. The cashew tree was introduced to Goa by the Portuguese in the 16th century and the use of cashew apples for the production of the liquor only exists in this Indian state. The apples are collected and crushed to extract the juice, which is then fermented and distilled without the addition of any foreign ingredient, using traditional tools (Rangnekar 2009). The GI specification does not highlight natural factors, mentioning only that the quality of apples varies, depending on soil characteristics and the place of cultivation, which is not demarcated. In contrast, the distillation zone is restricted to Goa. Even though Feni is undoubtedly an agricultural product, the choice of the GI applicant was to refer only on the human factor of the distillation techniques that should take place within Goa.

The UK’s ‘Melton Mowbray Pork Pie’ PGI provides another illustration where the emphasis relies on the quality of the raw materials rather than on their origin within the defined region. The pie was made in Melton Mowbray and its surrounding areas, being historically linked to the established practice of hunting in this area. In the specification, the emphasis is on traditional methods of production of the pies having a traditional bow walled pastry giving them their characteristic shape. The meat content of the whole product must be at least 30%. The pies must

[54]GI application No. 20, filed on 19 December 2007, Geographical Indication Journal No. 27.

be free of artificial colours, flavours and preservatives (Tregear and Giraud 2011).[55] A preliminary question referred to the ECJ by the High Court in London regarding the validity of this PGI application and thus the absence of local raw material was filed but eventually withdrawn. Consequently, there will be no further clarification by the ECJ on the requirements for registration of PGIs without natural factors, an option considered valid by the European Commission.[56]

In France, where GIs are used for rural development, the requirement for locally sourced raw material in order to support their local producers, often disadvantaged, is a hotly debated issue. Few GIs are linked to origin solely through human factors, even though the situation is changing. One case is the PGI Bergamote de Nancy, for sweets made traditionally with sugar, glucose syrup and natural bergamot essence, a recipe used in Lorraine cuisine since the 18th century, which are renowned in the region.[57] Yet it is difficult to assess the conformity to European criteria for this PGI, since it was registered via the fast-track procedure for incorporating lists of pre-existing national GIs within the newly harmonised EU-wide GI regime.[58] The PGI Pâtes d'Alsace represents another controversial example due to the mechanisation of the processing and the consequently weak human factor, while no local natural factors are involved as the raw materials are not sourced locally. The GI specification explains that since ancient times, Pâtes d'Alsace[59] have been produced from flour and eggs. This distinctively Alsatian tradition is based on a production method that, until the 19th century, was essentially domestic and rural, the special domain of the housewife, with recipes and know-how handed down from mother to daughter, using eggs which were easily available from farms. Such pasta is eaten with traditional dishes such as jugged hare, fish matelote or Rhine salmon. According to Professor Olszak, the application was made largely because a small Italian producer of pasta had begun to produce an imitation that did not contain the same ingredients and involved only two producers. Yet he suggests that 'Pâtes d'Alsace' do not deserve GI registration.[60] The PGI specification essentially indicates that pasta produced on an industrial scale in the region respects the traditional balance of ingredients: 'This organoleptically optimal mix [of eggs and durum-wheat semolina] is today unanimously accepted by Alsatian producers,

[55]See Dossier No (UK/PGI/0005/0335) at http://ec.europa.eu/agriculture/quality/door/list.html as consulted on March 1, 2016 at 14:58 pm.

[56]*Northern Foods Plc v The Department for Environment, Food and Rural Affairs & Anor* [2005] EWHC 2971 (Admin); see also D Gangjee, 'Melton Mowbray and the GI Pie in the Sky' (2006) 3 IPQ 291.

[57]Dossier No (FR/PGI/0017/0195).

[58]Established in Commission Regulation (EC) No 1107/96 of 12 June 1996 on the Registration of Geographical Indications and Designations of Origin under the Procedure Laid Down in Art 17 of Council Regulation (EEC) No. 2081/92 [1996] OJ L148/1. The PGI application was based on the specifications in the French Dossier LA/19/90 (Red Label).

[59]Dossier No (FR/PGI/0005/0324).

[60]Interview with Professor Olszak, on file with the author.

enabling them to carry on the tradition of eggpasta'. Indeed, in this case the link to origin does seem too weak to be eligible for the rights conferred to GIs.

Apart from these rare examples of dubious validity, French tradition is reluctant to protect GIs for products where raw materials are not sourced locally. One case is the PGI Calisson d'Aix (lozenge shaped sweets made out of ground almonds), where 70–80% of these sweets are made by eight calisson makers. The local public authorities wanted to boost almond production in the Aix region by introducing into the PGI a large but localized almond supplying zone around Aix, whereas calisson makers at the time were using almonds coming from California. The European Commission objected to the localization of almonds in the Aix area as there was no justification for a link between the quality of the almonds and their place of cultivation.[61] The final specification attributed the link to origin entirely to the know-how of the calisson makers.[62] This case is still under examination at the French level. Another controversial case concerns the Alsace geranium. Horticulturists wanted to protect the traditional geranium from Alsace, which flourished in the 1950s and is a resistant type which flowers quickly and abundantly. The planting of cuttings in pots and the monitoring of their growth until they are ready to be sold takes place in Alsace. The know-how of local horticulturalists is essential in adapting the cuttings to the cold climate. However, the application was rejected by the French authorities because cuttings that previously originated from Alsace now come from Kenya and are chosen according to generic quality criteria such as good health.

Yet, in a similar case of a product's specificity based on human factors influenced by the local climate, the Rillette de Tours, a PGI was granted. This region of Touraine is characterized by a mild but relatively wet climate, which prevented the development of naturally dried meats as seen in southern France. This climate is conducive to the vine, resulting in a significant production of wines that influenced, too, the flavoring of "Rillettes de Tours".[63]

In conclusion, it appears that the French tradition of GI protection does not easily promote products linked to their origin primarily via human factors, whereas European legislation (in terms of recognizing the validity of PGIs) allows for this, suggesting that there are different approaches to assessing the link to origin for protecting GIs even within the European Union.

---

[61] Interview with E Monticelli, on file with the author.

[62] Specifications for the PGI 'Calissons Aix', Union of Manufacturers of Calissons Aix en Provence, version of 2 June 2009, http://www2.inao.gouv.fr/repository/editeur/pdf/PNO-IGP2009/CDCCalissonDAix.pdf, where are visible the amendments between the two application with the withdrawal of the local sourcing of the almonds.

[63] Specification of the PGI « Rillettes de Tours » , https://info.agriculture.gouv.fr/gedei/site/bo-agri/document_administratif-7b236d55-dae1-42bc-ab4a-8c05c7c9846b/ as consulted 29 March 2016.

9.4 The Link Between the Product and Its Origin via a Combination of Natural and Human Factors

Most agricultural and foodstuff products are linked to their origin through both natural and human factors. The importance of natural factors mainly depends on whether the product is raw or a processed good. Practice shows that handicraft goods can be linked to a given territory in the same way.

9.4.1 *Agricultural Goods and Foodstuffs*

(a) Agricultural raw products obtained from a local variety

When the final product is subject to relatively little processing, as in the case of horticultural goods and cereals, the natural factors influencing the quality of the product are the soil and the climate, while the human factors consist of cultivation methods. Products resulting from the cultivation of an ancient local variety are strongly anchored in their environment, as those varieties are particularly well adapted to their surroundings. GIs can also help maintain a diversity of varieties as illustrated by the Indian Navara Rice GI,[64] which designates rice resulting from the crossbreeding of two indigenous varieties from Kerala in South India. Navara, like Basmati (Marie-Vivien 2008) or Rooibos (Biénabe 2009), is not a geographical name but the name of a rice variety, named after its short cultivation cycle, endemic to Kerala according to the GI application. Historical sources from 2500 BC testify to the Ayurvedic medicinal properties of this rice, which is used in Ayurvedic formulations.[65] The cultivation process is exclusively organic, since it is used for medicinal purposes, thus involving certain know-how and a specific *modus operandi*. This rice is slowly becoming extinct because of the difficulty of cultivation and low yield.[66] *Imitation* Navara rice is grown in the same geographical zone but from other varieties.

Similarly, the French appellation of origin Châtaigne d'Ardèche is characterised by the cultivation of local varieties. For centuries local society was organized around chestnut groves, which began to deteriorate. New varieties resulting from the hybridization of plants from different regions, more suited to certain technical and marketing criteria, were proposed. Since such an innovation would have radically changed the cultivation of chestnut groves, shifting it from the domain of agroforestry to intensive orchard farming, producers rejected it and applied for an appellation of origin to protect local varieties, a mode of cultivation and the landscape. From a census of 66 local varieties, the Châtaigne d'Ardèche appellation

[64]GI application No. 17, filed on 25 November 2005, Geographical Indication Journal No. 17.

[65]'Susruta Samhita', Susrutacharya, 2500 BC.

[66]See http://njavara.org; 'Njavara Facing Extinction', The Hindu, 21 December 2007.

selected 19 main varieties, specifically ruling out hybrid varieties.[67] To counter the European commission's argument that a common characteristic between the different varieties of chestnuts had not been established, the producers decided to use the 'local ecotype concept' found in all 66 varieties.[68] In such cases, the GI specificity is strongly related to its natural features based on local varieties and shaped by human practices.

(b) Processed products obtained from local raw material

The Darjeeling Tea GI is a significant Indian example of a processed product linked to origin through human and natural factors,[69] now also registered in Europe, despite some EU members opposing the prohibition of the use of the indication for blended tea.[70] Its qualities are attributed to an exceptional natural environment, including the soil, but above all from the altitude of between 600 and 2000 m at which it is cultivated, the steep slopes and very specific weather conditions: wet and cool in summer and dry and cold in winter. Yields are much lower than in non-Darjeeling districts. The species *Camellia sinensis*, is used for cultivating all Indian tea. Darjeeling tea leaves are processed in the production area, in the *conventional* way initially invented by the British, who adapted the processes from China. In France, processed GI products obtained from local raw material are common features. Such is the case of the olive oil from Nyons processed from olives of the Tanche variety grown in the departments of Drôme and Vaucluse. The 'tanche' is a variety typical of this region, particularly well adapted to its mixed climate. The smallest olives are crushed and mixed using traditional methods; the paste is either pressed or centrifuged to extract the oil.[71]

(c) Processing under special environmental factors

A famous French case is the Roquefort appellation of origin, protected since the 15th century through a royal charter. Protection was granted to the inhabitants of Roquefort along with a monopoly over the maturing process, carried out in well-guarded cellars (Bienaymé 1995). Being the first appellation of origin for cheese in France, it contains provisions on production and refining conditions, as well as 'natural factors' referring to both the source of raw materials and the

---

[67]See Art 4 of the Decree of 28 June 2006 concerning the 'Chestnut Ardèche' appellation of origin: 'Chestnuts from local varieties of the species *Castanea sativa* Miller listed in the technical regulations are provided for in Art 1 of this Decree. Hybrids are prohibited.'

[68]Letter from the European Commission, General directorate for agriculture and rural development, 9 October 2009 and interview April 2010. The PDO was registered in 1994. See Dossier No (FR/PDO/0005/00874).

[69]GI application No. 1, filed on 27 October 2003, Geographical Indication Journal No. 1.

[70]See Dossier No (IN/PGI/0005/0659). The objections are recorded in Commission Implementing Regulation (EU) No. 1050/2011 of 20 October 2011.

[71]For Huile d'olive de Nyons, see Dossier No (FR/PDO/0117/0142).

processing environment.[72] The particular character of Roquefort lies, on one hand, in the characteristics of the milk from traditional breeds of sheep fed according to traditional farming, and on the other, in the originality of the natural caves of Roquefort sur Soulzon, entirely dug from fallen rocks at the foot of Combalou's limestone cliffs 'where a miracle of nature takes place that gives Roquefort its unique flavour'.[73] The raw material was initially sourced from a very large area but subsequent to the EU definition of the appellation of origin that provides for mandatory local sourcing of raw material, the milk-supplying region was limited to the mid-mountain ranges south of the Massif Central, where there has been a long and particular tradition of sheep farming.[74]

In the absence of local raw material, quality might be related to environmental factors during processing. The Indian GI Monsooned Malabar Coffee[75] is processed following a technique directly linked to climatic conditions. The coffee was once stored for a long time before being shipped, but then underwent a transformation that was initially unexpected because of the weather, which is hot and wet on the Malabar Coast during the monsoon season. This process was called 'monsooning' and the resultant coffee became famous. The geographical origin of the coffee is non-specific and includes coffee cultivated anywhere in India. It can be compared to the French PGI 'Saucisse de Morteau', linked to origin by the slow smoking practice and know-how, with wood obtained from softwood forests found throughout Franche-Comté, in particular from the mountainous areas. This practice gives the Morteau sausage its amber color and smoky taste and is inextricably linked to its manufacturing region.[76] This example is quite unique in France, where the strategy generally pursued for specifying the product is to localize the source of the raw material. For the Saucisse de Morteau, the producers and the French government initially sought to do so, but such a limitation was rejected by the European Commission on the grounds that the zones supplying pork are different from the sausage production zones and it is not possible to establish a link between the source of the pork and the quality or reputation of the Saucisse de Morteau.[77] Yet Saucisse de Morteau is linked to its origin via natural factors such as the local wood used for the smoking of pork.

---

[72]*Loi du 26 juillet 1925 Ayant pour but de Garantir l'Appellation d'Origine du Fromage de Roquefort* (30 Juillet 1925) JO 7190 (Law on the Appellation of Origin Roquefort, 26 July 1925).

[73]Ibid.

[74]Dossier No (FR/PDO/0217/0131).

[75]GI application No. 85, filed on 5 April 2007, Geographical Indication Journal No. 21.

[76]Dossier No (FR/PGI/0005/0556).

[77]INAO, Comité National des IGP, Labels rouges et STG. PGI application 'Saucisse de Morteau ou Jésus de Morteau' request for additional information from the European Commission. Answer from the ODG, File No. 2008-414, 27 November 2008. The first letter of the Commission is dated 24 August 2007, the response of the applicant group is of 13 August 2008 and the response of the Commission is dated 3 October 2008.

9.4.2 *Non-Agricultural Goods—Handicraft Goods*

(a) Local source of raw material

Under the Lisbon Agreement which requires the presence of both natural and human factors, 28 appellations of origin for handicrafts based on human and natural factors such as raw materials have been registered at the time of this writing. Examples include Emaux de Limoges from France, Marbre Lepenica for natural products, or Trojanska Keramika for processed products from Bulgaria.[78] However, these represent a small number compared to wines and spirits. In France, the Poterie de Valauris appellation was recognized through a judgment of the Court of Cassation on 18 November 1930 for pottery made with local clay.[79] In Colombia, the preparation of the spiral woven handicrafts designated by the denomination of origin 'Guacayamas', is from a fiber extracted from the fique, a plant cultivated in the geographical area by the specific indigenous community of Laches.[80] In Vietnam, one GI designates a conical hat from Hué, prepared from local variety of leaves.[81] Within the potential GIs identified in France which are in the process of filing an application, Stone from Bourgogne, Stone from Britany are per se made out of local raw material.

In India, a rare handicraft GI example specifying local raw materials is the Kashmir Pashmina[82] for shawls made in Kashmir from the undergrowth of fleece from the mountain goat Capra Hiracus with a fineness of 12–16 microns. The specification outlines the following steps: procurement of the raw material, combing, cleaning; soaking in clean cold water, draining of the water and mixing with rice flour; the storage of the delicate pads of wool in deep stone pots, spinning of the yarn on the traditional 'chakra'; warping, dressing and reeling the yarn; weaving; clipping the loose threads; dyeing; washing and packing. The GI specification retraces the geographical route of the different phases of production and identifies three sub-regions: Ladakh, the grazing ground of the goats where the wool is collected, the outskirts of Srinagar, the capital of Jammu and Kashmir, where the wool is spun and the entire region of Jammu and Kashmir, where the wool is woven. The link to territory is characterized by different types of know-how, including weaving, and by local raw materials. However, the Ladakh zone is not demarcated in the same level of detail, especially in terms of latitude and

---

[78]See the Lisbon Express Registry database at http://www.wipo.int/ipdl/en/search/lisbon/search-struct.jsp as consulted on March 1, 2016 at 15:39 pm.

[79]In France, following the implementation of the 1919 law which allowed for the recognition by the Court, some appellations of origin for handicraft were recognised by the Courts, e.g., the 'Emaux de Limoges' appellation through a judgment of the Court of Appeals of Limoges on 18 February 1946 and the 'Cholet' appellation through a judgment of ruling of the Angers Court of Appeals on 17 November 1936.

[80]See Cestería Guacamaya, File no. 06-85475, registered on 19 June 2009.

[81]GI Certificate No. 00020, National Office of Intellectual Property, Vietnam (19 July 2010).

[82]GI application No. 46, filed on 9 December 2005, Geographical Indication Journal No. 13.

longitude, as the overall GI zone. It suggests that collecting the wool is accorded less importance, combined with a certain disregard for this activity compared to spinning and weaving.

India may have a different perspective from that of Europe, where a PDO was registered for Native Shetland Wool,[83] which designates only the wool, collected and spun in a specific area according to traditional methods. The weaving phase is not included in the PDO, so the fabric can be produced anywhere, according to any method.

(b) Natural factors influence the processing

As with agricultural goods, non-agricultural goods can derive their reputation from natural factors which impact upon the processing. In France, the appellation of origin Toiles de Cholet has been protected since 1936 by judicial recognition,[84] which noted that the reputation of Cholet sheets and textiles was owed to the bleaching techniques, consisting of stretching the textiles out on green, wet and clayey meadows. Textiles are bleached with water pumped directly from the ground, the quality of which is seen as exceptional for bleaching and not available anywhere else in the region. Their special weight and strength also depends on natural elements from the soil and specific local climate conditions.[85]

9.5 Replacing the Categorization of Products by the Criteria of Human and/or Natural Factors

9.5.1 No Justification for Discriminating on the Basis of Subject Matter Categories

The practice of GI protection in many countries including the EU and France described above shows that the link with the place of origin can be analyzed on the basis of human factors, individually or combined with natural factors, not only for agricultural products and foodstuffs but also for non-agricultural goods. The review of existing practice demonstrates that, whatever the nature of the product is, GI links have been recognized in the same way.

[83]Dossier No (UK/PDO/0005/0737).

[84]Commercial Court of Cholet, 8 January 1936, Etablissements Béra c./ Syndicat patronal des industries textiles de la région de Cholet, regional archives, Conseil général, Département Maine et Loire, côte 143a63.

[85]1992. *La Blanchisserie de la rivière Sauvagean et le blanchiment des toiles à Cholet,* Cahors, Association des Amis du Musée du Textile Cholerais, REMPART. The protection of this appellation is still effective because the same Angers Court of Appeal decided in 1992 that the trademark 'Création Maret Cholet France' used to designate woven textile was misleading since it could be confused with the 'Cholet' appellation of origin, if the methods of manufacturing did not meet the appellation criteria. See Court of Appeal of Angers, Chamber 1B, 17 February 1992.

	Natural factors + human factors	Human factors
Agri-food products	Roquefort Nyons Olive Oil Darjeeling Tea Monsoned Malabar Coffee	Melton Mowbray Pork Pie Feni alcohol Calisson d'Aix Rillettes de Tours
Handicraft products	Pashmina shawls Toile Cholet Conical Hat Hué Guacayamas Stone of Bourgogne (possible)	Kancheepuram silk Pochampally Ikat Dentelle du Puy Faience Moustiers Marseille Soap (possible)

For any kind of goods, the analysis of the existence of human and/or natural GI factors provides a useful way to approach and assess the link to origin. As a result, there is no reason to treat categories of products differently.

9.5.2 The Validity of GIs Based on Human Factors

In practice, GIs based on human factors are widely acknowledged and registered. However, this requires further assessment. Is a link to geographical origin via know-how alone legally valid with respect to the definition of a GI? Can know-how be rooted in an area—essentially attributable to origin in the language of TRIPS—and can it confer a product a quality or reputation linked to its geographical origin? Can this know-how be influenced by the place? Indeed, this link to territory based on human factors is often questioned, with know-how and methods of production seen as easily transmissible beyond the region of origin. Then there is the well-documented migration of artisan communities, whose history suggests that their roots in a given area can prove to be transient.[86] For example, the Kancheepuram GI application states that the weavers were originally from another state, Andhra Pradesh, and migrated four hundred years ago when their village was swept away by the sea. In the event of artisans migrating outside their area of origin, would a reservation of the name only to goods produced in the demarcated area be an unfair restriction? Can an artisan, or a group of artisans who migrate, produce an identical product in their new environment? Will the know-how evolve in the course of this migration? How can this be assessed?

Even if such an assessment of the human factors is also needed for GIs based on a combination of natural and human factors, a link to the origin essentially through human factors needs specific appraisal. Several criteria can be used to assess the specificity of the know-how of producers from a place. The hypothesis is that know-how is influenced by surroundings, including both elements of the nature and human interaction.

[86] Amongst the Indian GIs, see in particular Orissa pattachitra, Nirmal paintings and Kota doria.

Along with the natural environment, climate (a natural factor) influences the type of produce, for example light cotton saris in South India, shawls in the mountainous regions of the Himalayas, specific types of geranium flowers in Alsace, Rillettes de Tours. For our purposes it can be accepted that, though exceptions might still exist, the know-how of producers is adapted and specific to their natural environment. When migration occurs, know-how is not blindly transferred as such, but adapted when used in new surroundings.

Second, the definition of geographical origin requires some elaboration: origin refers to a place having a certain reputation thanks to history and shared know-how (Bérard and Marchenay 2008). The collective nature of know-how that results from it being shared within a community located for a long time in a specific area ensures its continued existence in this place and ensures the existence of a certain reputation. GI products are not only local, but part of the local culture due to the existence of a group of producers exploiting the same product in the same place for a certain time. Individuals isolated from the community, or even small groups cut off from the main community, will not be able to execute this know-how with equal proficiency and with the same reputation. This underlies the prominent collective dimension of GIs, resulting from a group of producers sharing their know-how and cross-controlling the quality of the product. Beyond GIs, in any kind of human activity, human environment is important. In India, as a consequence of the caste system that associates a given community with a specific activity, know-how linked to handicrafts is held by specific groups and passed down from generation to generation (Mahias 1997). For example, the GI Pipli Appliqué Work[87] is made by artisans from the caste of Darjis, who hand down their knowledge and titles in this way. The idea of a 'basin of skills' is introduced when skills are developed over a large area (Bérard and Marchenay 2008). The know-how might as well spread among the locals. In the Kancheepuram Silk case, the silk was originally woven only by the Salia community, but now all the communities within eight kilometres of Kancheepuram, representing 75% of the population of Kancheepuram, weave silk (Gopalakrishnan et al. 2007). The historical depth of the localization of the community of producers is the criterion of validity for such GIs. Indeed, the Indian GI Rules state that the link to origin must be demonstrated through the history, in a section specially designed for this purpose: 'Proof of Origin (Historical Records)'. This principle has been expressly implemented in India for the GI Chanderi sari: producers who want to use the GI must have resided in Chanderi for at least 15 years.[88]

In France and at the EU level, proof of historically verifiable human factors or pre-existing know-how is required to objectivize the link to territory (Bérard and Marchenay 2004). Such criteria are provided in the new French GI regime for non-agricultural goods "the specification […] gives details about the quality, the

[87]GI application No. 86, filed on 9 April 2007, Geographical Indication Journal No. 15.

[88]Rules of the Chanderi Foundation; interview with the GI applicant, on file with the author.

reputation, *the traditional know-how* or other characteristics".[89] The criteria of history and tradition balances in favor of granting a GI for traditional methods of production, whether there are manual or mechanized, to balance the absence of natural factors. For example, the conflict over the Marseille soap GI, opposed producers willing to maintain the traditional method of production of the soap without any additives and produced in caldron by soap masters, while others want to authorize the use of additives such as perfumes.

Similarly, for human factors it may be worth considering a level of sophistication or refinement as an additional requirement, which would then impart real uniqueness to the product and compensate for the absence of natural factors. In conclusion, GI goods linked to origin through mainly human factors such as handicraft goods or foodstuffs cannot be disqualified a priori, when know-how is shared by a community and has existed for a certain period of time, as part of the local culture and tradition.

For GIs using raw materials which are not exclusively sourced from the area demarcated by the Geographical Indication, any risk of consumer confusion on the origin of the raw materials should be avoided. The mandatory mention of this *external* source in the labelling of the processed good is recommended, whether for agricultural or non agricultural goods (Gangjee 2015b). Besides, the link to origin of products characterized by specific designs and resulting from creative know-how seems rather fragile, and subject to the risk of changing patterns, following fast-changing fashion in the textiles domain.[90] However, for traditional designs, GIs are a more desirable tool of protection than the mere protection of design and models, since GIs are protected without time limitation.[91] GIs are also much more relevant when it is the protection of a product's name and reputation that is sought. For example, the Pochampally designation is famous for its Ikat, and it seems very unlikely at present that the weavers of Pochampally will decide to weave other motifs which are typical of Orissa: they will not abandon their own motifs and above all their name, because the Pochampally appellation is as well-known as the Orissa appellation.

9.5.3 Same Legal Regime?

The question remains of whether a link to origin solely via human factors should be governed by the same legal regime as the rules applying to the more robust link to *terroir*, which involves a combination of natural and human factors. It is argued that the concept of *terroir* shall not be applied to GIs where both soil and nature are not cumulatively influencing the quality of the product. In such cases, the more

[89]Art. L. 721-7 of the Code of Intellectual Property.

[90]Statement of A Mohamed Jamuluddin regarding Kancheepuram Silk GI, on file with the author.

[91]Subodh Kumar, CII Personal interview, April 2010.

restricted concept of origin is arguably more appropriate. In the absence of natural factors, the link to origin is undoubtedly weaker. Maintaining the principle of two legal options is thus proposed, as it is the case for Europe with coexisting categories of PDO and PGI, based on the criteria of human and natural factors.

9.6 The Implications for France and the EU

9.6.1 France

(a) The lost opportunity of a unique Institution for GIs for all goods

The recommendation of a consistent and uniform legal regime for GIs regardless of the type of goods, with the criteria of validity of GIs being assessed by the existence of natural and/or human factors would be rather useful to France which is experiencing a difficult start for implementing the new law providing for GI for non-agricultural goods. Indeed, the French Law of 2014 recognizing GIs for non-agricultural goods maintains a product-based approach at the institutional level. GIs for non-agricultural goods will be scrutinized and approved by the National Institute of Intellectual Property (INPI), alongside other intellectual property rights, whereas PDOs and PGIs for agricultural goods and foodstuff continue to be scrutinized and regulated by a dedicated body, the National Institute of Quality and Origin (INAO) under the aegis of the Ministry of Agriculture. For assessing the validity of agricultural goods and foodstuff PDO/PGIs, INAO has 21 local branches working with producers and decision-making bodies, as well as five national committees in charge of the approval of labels of quality and PDO/PGIs.[92] All national committees bring together representatives of producers/processors engaged in the relevant products and benefiting from PDO/PGI (or other quality labels), who account for at least 50% of the members of the committee, qualified experts and members of Ministries and governmental agencies. INAO is also responsible for checking the representative nature of the applicant and the plan of inspection/control prepared by the certification body together with the producers (Marie-Vivien et al. 2015).

In contrast, for deciding on approving or rejecting GI specification applications for handicrafts and manufactured goods, at INPI, at the time of this writing, only one examiner has been appointed to deal with all GI applications. A patent examiner can be also consulted in the area concerned. After an official examination, INPI shall conduct the public inquiry, open to any person and shall conduct the consultation of the national organizations representing the enterprises and craftsmen and professional organizations representing the conformity assessment of industrial

[92]See Article R642-6 of the Rural Code; see also http://www.inao.gouv.fr/Institut-national-de-l-origine-et-de-la-qualite-INAO/L-INAO-sur-le-territoire as consulted on March 1, 2016 at 16: 13 pm.

products and handicrafts bodies.[93] But failure to receive a response within a period of two months suggests opinions submitted are deemed to be favorable.[94] The consultation of INAO in the process of registering GIs for non-agricultural goods is provided for by the law, but only in cases of a risk of confusion with a PDO/PGI for agricultural goods.[95] There is no requirement for a general collaboration between the two institutions regarding the examination of the link with the origin and thus the validity of the GI. The objective of the procedure at INPI is to be fast-tracked, with short deadlines and no visit to the field before the application is published for public inspection. Indeed, professionals often complained that the recognition of PDO and PGIs for agricultural goods is too slow. Reaching an agreement between producers or manufacturers around a common definition of the GI specification is always time consuming, either after or before the publication of the GI. The first application for Marseille Soap filed on June 16, 2015 was opposed by another group of manufacturers during the public inspection procedure which closed on the 21 December 2015.[96] Such opponents filed their own application late December 2015, with the public inquiry probably open in March 2016.[97] Had a mechanism of substantive examination on field existed for non-agricultural GIs as for agricultural GIs, following the experience of INAO, this problem would have been potentially avoided. Such a dichotomy in the recognition of GIs according to the kind of good is all the more regrettable if we consider that in 2006 INAO expanded its mandate to include all voluntary quality labels in agriculture and foodstuffs such as organic farming, guaranteed traditional specialties, Label Rouge, with the risk of diluting INAO's specific skills for appellation of origin and Geographical Indication that have been acquired progressively throughout its long history. INAO could have profited from its unique skills concerning the link to the origin by extending its activities to GIs for non-agricultural products, transcending the Ministries of Agriculture and Industry, and thus promoting the consistency and visibility of GIs.

(b) The abandon of two levels of geographical reference for handicraft

Another problem regarding the French Law is that it only provided for the registration of a GI and not for an appellation of origin. It is true that recognition and protection as an appellation of origin is already available for non-agricultural goods, but only according to the judicial procedure provided by the Law of 1919, which is far more complicated in terms of implementation. A GI ready to be protected, the 'Pierre de Bourgogne' is a stone extracted from Bourgogne whose soil is distinctive for wine-making, qualifying for an appellation of origin and it is thus unfortunate

---

[93] Décret 2015-595 du 2 juin 2015 relatif aux indications géographiques protégeant les produits industriels et artisanaux et portant diverses dispositions relatives aux marques; JORF 3 juin 2015.

[94] Art. L 721-9(3) of the Intellectual Property Code.

[95] Art. L 721-9(3) of the Intellectual Property Code.

[96] https://www.inpi.fr/fr/la-base-des-indications-geographiques-de-l-inpi-est-en-cours-de-maintenance as consulted on March 1, 2016 at 17:19 pm.

[97] Ibid note 55.

not to have created the possibility of registering appellations of origin alongside geographical indications. It also marginalizes handicrafts as not being the equivalent to agricultural and foodstuff goods, not benefiting from the same regime. Moreover, if France does not defend the concept of appellation of origin it conceived, it might jeopardize the ongoing relevance of appellations of origin at the EU level, where it may be overtaken by subsequent developments.

9.6.2 *Europe*

(a) The proposal of a specific legal regime

The possibility of extending GI protection to non-agricultural goods was identified by the European Commission in its 2011 Communication on a single market for intellectual property rights,[98] which proposed a review of the existing legal framework for GI protection of non-agricultural products across the EU, and its implications for the internal market. Member States offer different legal solutions for the protection of non-agricultural products: for instance, through unfair competition or consumer protection law, or via collective or certification marks. The Commission considered that this fragmentation of the legal framework to grant protection for GIs for non-agricultural products may negatively affect the functioning of the internal market and trade negotiations with third countries.

The ensuing Study on Geographical Indications protection for non-agricultural products in 2013[99] recommended a fourth legal instrument for non-agricultural goods, besides those for agricultural products/foodstuffs, wines and spirits. The Commission eventually decided to pursue its analytical work through a Green Paper published in July 2014.[100] It aims to consult with all stakeholders in the broadest possible manner on whether there is a need in the EU to increase GI protection for non-agricultural products, and if so what approach should therefore be taken. 136 responses were collected originating from 18 Member States (Koenig et al. 2015) and the European Parliament adopted in October 2015 a resolution on the possible extension of GI protection of the EU to non-agricultural products, in response to the Commission Green Paper.[101]

In the Green paper, the Commission was open to exploring the possibility of a supervisory, operational and administrative role taken by the Office for

[98]European Commission 'A Single Market for Intellectual Property Rights' COM (2011) (24.05.2011).

[99]*EU Study* (n 15).

[100]European Commission, 'Making the Most out of Europe's Traditional Know-How: A Possible Extension of Geographical Indication Protection of the European Union to Non-Agricultural Products' COM(2014) 469 final (15.07.2014).

[101]2015/2053(INI), Possible extension of geographical indication protection of the European Union to non-agricultural products.

Harmonization in the Internal Market (OHIM). Indeed, it would require new human and financial resources, which can be difficult to secure. The Study considers that an office could be better positioned to manage a system of GIs and, as the main EU-wide intellectual property office, the OHIM could be considered as a suitable candidate in that respect, having legal, administrative and financial independence. The Study justifies this by mentioning that the establishment of a new EU-wide system will not be free of costs. Moreover, as conflicts or potential conflicts between GIs and other IP rights often involve trademarks, there could be an interest in having the trade mark register and the GI register managed by the same office. The EU Parliament supports this proposal. Yet the Study reports the disadvantages of the OHIM, which lacks specific expertise when it comes to Geographical Indications.[102] Undeniably, the examination of the link to the region of origin is completely different from the examination of the grounds of validity of trademarks, such as distinctiveness.

(b) The issue of the two levels of geographical reference

The Study recommends a single definition for a geographical designation, and, therefore, when considering which GI definition(s) should be used, the options offered do not include both PDOs and PGIs.[103] The Study also considers that a double definition approach could pose problems in its implementation. For instance, the regulating body would have to verify the strength of the link between the product and its geographical area to define whether it can be registered as a PDO or a PGI. Addressing precisely this issue, the recommendations provided in the present chapter would be to use the criteria of natural and/or human factors in examining the strength of the link with the origin, even if ‘each protected product is to be seen as an original with its own history, composition and specific quality’ (Profeta et al. 2009). At present, it remains unclear on what grounds is the distinction between PDO and PGI based, in the present EU regime under Regulation 1151/2012 (Geiger et al. 2010).

The notion of two distinct types of geographical designations is supported by the majority of producers in Europe,[104] who have reaffirmed their commitment to maintaining this distinction, whereas the EU Commission had proposed merging the two categories of GIs when drafting Regulation 1151/2012,[105] following a report from the European Court of Auditors, 2011,[106] which considered that two different links might confuse consumers. The preference for two types of

---

[102]*EU Study* (n 15) 318.

[103]Ibid., 298.

[104]Proposal for a Regulation of the European Parliament and of the Council on Agricultural Product Quality Schemes COM(2010) 733 final (10.12.2010). This also reiterates the position of maintaining the concept of the appellation of origin.

[105]See European Commission, Green Paper on Agricultural Product Quality: Product Standards, Farming Requirements and Quality Schemes COM(2008) 0641.

[106]European Court of Auditors, *Do the Design and Management of the Geographical Indications Scheme Allow it to be Effective?* (Special Report No 11, 2011).

geographical designation continues to be evident as illustrated by the responses of the stakeholders to the Green Paper of 2014.[107]

The author also wishes to emphasize that with PDOs and PGIs enjoying the same level of protection at the EU level, the attractiveness of PDOs is diminishing. One way of avoiding this would be to create a meaningful difference between the two types of GIs, by providing a different level of protection in accordance with the strength of the link.

Finally, having two distinct link requirements which vary according to the strength of the connection to origin is not entirely new in the broader intellectual property context. If we look at patent law, in many countries alongside the regular inventive patent there is the utility model patent (also known as the petty patent or innovation patent), which benefits from a shorter period of protection.[108] With GIs claiming to support cultural and biological diversity, it would be regrettable not to provide for a diversity of legal references and to abandon the originality of the South European countries' vision, where the appellation of origin was born.

9.6.3 International Agreements

At the WTO, in the TRIPS Agreement, there is no discrimination between agricultural goods and handicrafts but there is a distinction between wines and spirits and other goods. Such discernment should clearly be abandoned and replaced by a distinction based on the strength of the link. Such a distinction for wines and spirits reflects a history that has been largely modified with the implementation of GI laws in non-wine producing countries and is no longer justified. The proposal to base the protection on the strength of the link to territory and not on the kind of goods sheds new light on the current WTO negotiations around the extension of additional protection, currently granted only to wines and spirits, to all products (Evans and Blakeney 2006; Kongolo 2011; Thevenod-Mottet 2009). This distinction could be implemented by WTO State members. For example, looking at Colombia, which uses only the concept of Denomination of Origin, providing two levels of geographical references would allow the protection of GIs for products which have raw material coming from outside the area and that otherwise can only be protected as trademarks under current regimes.

Finally, the potential revitalization of the Lisbon Agreement (Geiger et al. 2010), which for the first time introduced these criteria of human and natural factors to define appellations of origin, suggests an optimal path towards a global system for the protection of GIs. First, let's remember that the Lisbon Agreement is open to

[107]The figures being: 33.3 per cent are in favour of two links with the origin; 26.5 per cent are against and 42.2 per cent do not have an opinion.

[108]WIPO, 'Protecting Innovations by Utility Models' at: http://www.wipo.int/sme/en/ip_business/utility_models/utility_models.htm, as consulted on March 12th 2016 at 12:38 pm.

any kind of goods. Second, the Geneva Act of the Lisbon Agreement on appellations of origin and geographical indication, adopted at a Diplomatic Conference on May 20, 2015,[109] introduces the definition of GIs as provided for in the TRIPS Agreement alongside the definition of the appellation of origin, thus confirming that a two-level approach is appropriate. Regarding the level of protection conferred, the new Act provides for a uniform protection for AOs and GIs. However, the new Act of the Lisbon Agreement still does not address the issue of how to assess the link with the place of origin for GIs and AOs, which is unfortunate since it is the main international reference point for using the concepts of natural and human factors. The aim of this chapter was to develop the potential for interpreting the link requirement by drawing on a combination of these two elements, a methodology that can be used in any jurisdiction, for any kind of goods, giving GIs their full opportunities as intellectual property rights recognized internationally.

Bibliography

Addor, F. & Grazzioli, A. 2002. Geographical Indications beyond wines and spirits, a roadmap for a better protection for Geographical Indications in WTO/TRIPS Agreement. *The Journal of World Intellectual Property,* 5, 865–897.

Allaire, G. 2004. Quality in economics: a cognitive perspective *In:* Harvey, M., Mcmeekin, A. & Warde, A. (eds.) *Qualities of food.* Manchester University Press.

Anthony, D. 2013. From Croft to Catwalk: The Harris Tweed Certification Mark. *WIPO Magazine,* 24.

Audier, J. 1993. Réflexions juridiques sur la notion de terroir. *Bulletin de l'O.I.V.*, 423–437.

Audier, J. 2008. Passé, présent et avenir des appellations d'origine dans le monde: vers la globalisation. *Bulletin de l'O.I.V.*, 405–435.

Bérard, L. & Marchenay, P. 2004. *Les produits de terroir, entre cultures et règlements,* Paris, CNRS Edition.

Bérard, L. & Marchenay, P. 2008. *From Localized products to Geographical Indications, awareness and action,* Bourg-en-Bresse, CNRS.

Biénabe, E. 2009. Le rooibos d'Afrique du Sud : comment la biodiversité s'invite dans la construction d'une indication géographique. *Autrepart,* 2, 117–134.

Bienaymé, M.-H. 1995. L'appellation d'origine contrôlée. *Revue de Droit Rural*, 419–424.

Caroline_Le_Goffic 2015. La saga « Laguiole » , illustration du conflit entre marques et indications géographiques. *Revue Francophone de la Propriété Intellectuelle,* Septembre.

Evans, G. E. 2012. The Strategic Exploitation of Geographical Indications and Community Trade Marks for the Marketing of Agricultural Products in the European Union. *W.I.P.O.J.*, 1, 159–179.

Evans, G. E. & Blakeney, M. 2006. The Protection of Geographical Indications After Doha: Quo Vadis? *Journal of International Economic Law,* 9, 575–614.

Gangjee, D. 2007. Quibbling Siblings: Conflicts between Trademarks and Geographical Indications. *Chicago-Kent Law Review,* 82, 46.

Gangjee, D. 2012. *Relocating the law of Geographical Indications,* Cambridge, Cambridge University Press.

[109]Geneva, May 20, 2015, PR/2015/779.

Gangjee, D. 2015a. Geographical Indications and Cultural Rights: The Intangible Cultural Heritage Connection? *In:* GEIGER, C. (ed.) *Research Handbook on Human Rights and Intellectual Property.* Edward Elgar Publishing.

Gangjee, D. 2015b. Proving Provenance? Geographical Indications Certification and Its Ambiguities. *World Development,* Forthcoming.

Geiger, C., Gervais, D., Olszak, N. & Ruzek, V. 2010. Towards a Flexible International Framework for the Protection of Geographical Indications. *W.I.P.O.J.,* 1, 147–158.

Gervais, D. J. 2009. Traditional Knowledge: Are We Closer to the Answers? The Potential Role of Geographical Indications. *ILSA Journal of International and Comparative Law* 15, 551–567.

Gopalakrishnan, N. S., Nair, P. S. & Babu, A. K. 2007. *Exploring the Relationship between Geographical Indications and Traditional Knowledge, An Analysis of the Legal Tools for the Protection of Geographical Indications in Asia,* Geneva, ICTSD Program me on IPRs and Sustainable Development, International Centre for Trade and Sustainable Development.

Insight_Consulting, Origin & Redd 2013. Study on geographical indications protection for non-agricultural products in the internal market.

Koenig, M., Czerska, S. & Marie_D'avigneau, V. 2015. Public Consultations on Geographical Indication Protection for Non-agricultural Products in the EU: Summary of Results. Industrial Property Unit J1, DG Growth

Kongolo, T. 2011. Any new developments with regard to GIs issues debated under WTO? *E.I.P. R.,* 33, 83–90.

La Blanchisserie de la rivière Sauvagean et le blanchiment des toiles à Cholet, Cahors, Association des Amis du Musée du Textile Cholerais, REMPART (1992).

Mahias, M.-C. 1997. Les sciences et les techniques traditionnelles en Inde *L'Homme,* 37, 105–114.

Marie-Vivien, D. 2008. From Plant Variety Definition to Geographical Indication Protection: A Search for the Link Between Basmati Rice and India/Pakistan. *The Journal of World Intellectual Property,* 11, 321–344.

Marie-Vivien, D. 2010. *Le droit des indications géographiques en Inde: un pays de l'Ancien monde face aux droits français, européen et international,* Paris, Ecole des Hautes Etudes en Sciences Sociales, Paris.

Marie-Vivien, D. 2016. A comparative analysis of GIs for handicrafts: the link to origin in culture as well as nature? *In:* Dev_S.Gangjee *Research Handbook on Intellectual Property and Geographical Indications.* (pp. 292-326). UK: Edward Elgar Publishing.

Marie-Vivien, D., Bérard, L., Boutonnet, J.-P. & Casabianca, F. 2015. Are French Geographical Indications losing their soul? Analyzing recent developments in the governance of the link to the origin in France. *World Development,* XX.

O'Connor, B. 2004. *The Law of Geographical Indications,* London, Cameron May.

Olszak, N. 2001. *Le droit des appellations d'origine et des indications de provenance,* Paris, Tec&Doc.

Plaisant, M. & Fernand-Jacq 1921. *Traité des noms et appellations d'origine,* Paris, Librairie Arthur Rousseau.

Pollaud-Dulian, F. 1999. *Droit de la propriété industrielle,* Paris, Montchrestien.

Profeta, A., Balling, R., Schoene, V. & Wirsig, A. 2009. The Protection of Origins for Agricultural Products and Foods in Europe: Status Quo, Problems and Policy Recommendations for the Green Book. *The Journal of World Intellectual Property,* 12, 622–648.

Rangnekar, D. 2009. Geographical Indications and Localisation: A Case Study of Feni. CSGR Report.

Thevenod-Mottet, E. 2009. Avenir des indications géographiques dans le contexte international. *Revue suisse Agric.,* 41, 331–335.

Thevenod-Mottet, E. & Marie-Vivien, D. 2011. Legal debates surrounding geographical indications. *In:* Barham, E. & Sylvander, B. (eds.) *Labels of origin for food: local development, global recognition.* CABI.

Totem 2012. Protecting Traditional Knowledge and Culture: Colombia Gives an Example. 9 Jan.

Tregear, A. & Giraud, G. 2011. Geographical indications, consumers and citizens. *In:* Barham, E. & Sylvander, B. (eds.) *Labels of origin for food: local development, global recognition.* CABI.
Une réussite française: l'appellation d'origine contrôlée, INAO (1985).
WIPO_Publication_No.952(E) Geographical Indications: An Introduction.

Author Biography

Delphine Marie-Vivien Researcher in Law, CIRAD, UMR innovation, Montpellier, f-34398 France, based in Hanoi, Vietnam. She is a graduate in chemistry and intellectual property law. After 2 years as patent engineer, she joined CIRAD, a French public research institute dedicated to agricultural research for developing countries. After 5 years working on intellectual property and genetic resources issues, she was a visiting researcher at NLSIU, Bangalore, India from 2005–2008. She is since 2012 based in Vietnam, working on research projects regarding the protection of geographical indications in south-east Asia and more particularly in Vietnam.

Chapter 10
Old World Case Study: The Role of Protected Geographical Indications to Foster Rural Development Dynamics: The Case of Sorana Bean PGI

Giovanni Belletti, Andrea Marescotti and Alessandro Brazzini

Abstract The protection of Geographical Indications is an issue of growing importance all over the world, as it offers local producers a tool to differentiate their products on the market and escape price competition. In the European Union the legal protection of Geographical Indications dates back to 1992, and aims at both preventing misuses and abuses of brand names on the market fostering fair competition among producers and transparent and complete information to consumers, and supporting rural development dynamics, especially in marginal areas. In this chapter, after describing the many and multifaceted effects the protection of the Geographical Indications may exert on the economic, social, and environmental spheres, the case study of the Sorana Bean PGI in Tuscany (Italy) will be analyzed. The case is related to a very small production system, where a few small farms are using the protected Geographical Indication to market their product. The case shows that the protection granted by the European Union, besides supporting farmers' income, exerts important economic and social effects on the territory, thus supporting rural development in a marginal area.

10.1 Introduction

The legal protection granted to Geographical Indications (GIs) is an issue of growing interest and concern worldwide. From an economic and social standpoint, interest in this topic is rising due to an increasing international competition on the

G. Belletti (✉) · A. Marescotti · A. Brazzini
Department of Economics and Management, University of Firenze, Via delle Pandette 9 - 50127, Firenze, Italy
e-mail: giovanni.belletti@unifi.it

A. Marescotti
e-mail: andrea.marescotti@unifi.it

A. Brazzini
e-mail: brazzini.alessandro@gmail.com

W. van Caenegem and J. Cleary (eds.), *The Importance of Place: Geographical Indications as a Tool for Local and Regional Development*, Ius Gentium: Comparative Perspectives on Law and Justice 58, DOI 10.1007/978-3-319-53073-4_10

level of product quality differentiation, which helps products to stand out and avoid competing purely based on prices. As a consequence, many public and private stakeholders at both local and global levels have fostered this new turn to the element of quality. GIs protection appears to be one of the most interesting and *locally manageable* tools for attaining this aim. At the same time, GIs protection can help safeguard traditional knowledge and cultural heritage, and generate other territorial externalities, supporting rural development dynamics. This potential stems from the strong link these products have with their territories, and from the specificity of the local human and physical resources used in the production process, that give these products unique quality features. Furthermore, consumers are looking for a reconnection to the locality where food is being produced, sometimes for reasons of identity, in other cases for food safety and quality reasons. Given this conjuncture, GI protection is expected to exert positive rural development effects: economic effects (both inside and outside the supply-chain, at a local level), as well as social, cultural, and environmental ones.

The European Union has a longstanding tradition in supporting and regulating these products and the GIs they bear. In 2015 in the European Union, 1244 food products and 1579 wines are registered as Protected Denomination of Origin (PDO) or Protected Geographical Indication (PGI). Italy is the leader in the EU, with 278 food products and 523 wines. Empirical evidence on Italian protected GIs shows that over 80% of the turnover of PDO-PGI products in 2014 is generated by the big ten protected GIs, among which the famous Grana Padano PDO, Parmigiano-Reggiano PDO, Prosciutto di Parma PDO.

Thus, two main research questions emerge. The first question is related to the scope of the GI protection, which is presumed to be wider than any other kind of collective intellectual property right. In many places around the world, protected GIs are the pivot, or one of the main pillars, of territorial development strategies. However, academic (and non-academic) literature focuses mainly on marketing issues of GI products, and there is not sufficient awareness about the "side-effects" of the GIs protection (Belletti et al. 2015a, b). So the first question is the following: is the protection of GIs just a tool for fighting abuses and misuses of the GIs on the market, helping producers to skip unfair competition and consumers to access transparent and complete information? Can GI protection be also capable of activating more general economic and social development trajectories inside specific territories? What effects can be expected in that sense?

The second question is partly related to the first, and deals with the economic size of GI production systems. In many countries as in the EU, many GIs that have been protected show a huge diversity in terms of geographical scale of the production area and production volumes: some refer to a small village and its surroundings, while others refer to an entire region or even to a State. The production volume greatly varies across different cases. The question posed by many policy-makers is if only the *big* GIs deserve protection, considering a rough cost-benefit analysis that balances the efforts made by public bodies for the registration and enforcement, and the value of sales of the certified product.

We will try to answer these questions by drawing on the case of Tuscany, one of the most reputed Italian regions for agri-food quality; in particular, a small PGI related to a high-quality bean will be analysed. The structure of the chapter is the following. First of all, in the following section, we will introduce the concept of origin product and two ideal typical strategic orientations for its valorisation. In Sect. 10.3, we will discuss how Geographical Indications are protected in the European Union, with a special reference to food products (wines having a specific regulation in the EU). Section 10.4 will shortly describe the many effects that the GI protection can exert. Section 10.5 will be devoted to introducing the situation of protected GIs in Tuscany, while Sect. 10.6 will be entirely devoted to describing and analysing the case study. The final section will shortly resume and draw particular conclusions.

10.2 The Valorisation of Origin Products as a Tool for Rural Development

Origin products (called also typical products) are products whose specific quality comes from a strong link to their territory of origin (Delfosse 1996). A Geographical Indication often identifies them on the market. According to a number of scholars (Bérard and Marchenay 1995, 2004; Casabianca et al. 2005; Vandecandelaere et al. 2009) three relevant dimensions determine the special nature of an origin product (OP): the specificity of local resources used in the production process; the history of the product, included its production and consumption tradition; and the collective dimension, including the presence of a shared production and consumption knowledge at the local level. This strong link to the territory originates a rural development potential of OPs, which are expected to exert not only economic effects in the local supply chain but also other economic effects at local level, as well as social, cultural, and environmental ones.

The positive attitude of many consumers towards a reconnection to the locality (both for identity and for safety and environmental reasons) allows for a strengthening of their effects on local economy and communities (Weatherell et al. 2003), even though behavioural patterns are not univocally consistent with attitudes (Vermeir and Verbeke 2006). Thanks to their multifunctional character, OPs are often the object of valorisation strategies developed by actors belonging to the supply chain and to the territory they come from, with the frequent support of local public bodies, such as Municipalities, Provinces, Local development agencies, Chambers of Commerce. Valorisation strategies of OPs can be oriented by different logics, which depend on the category of actors involved in defining the strategy. Based on empirical evidence, two main approaches that actors may adopt with regard to the economic role of the OP can be identified: a supply chain strategy and a territorial strategy (Pacciani et al. 2001; Tregear et al. 2007). The supply chain strategy aims at building a network of actors in the production and processing of the OP, with a specific focus on the improvement of product quality (also by defining common rules), the management of quantities to be produced, and the implementation of effective collective marketing strategies also thanks to labelling initiatives. The initiators of a supply chain strategy are normally the enterprises belonging to (one or more phases of) the supply chain, with the aim of increasing the degree of differentiation of the product in order to improve the product added value, opening new marketing channels or penetrating new markets. By this approach, the OP contributes to socio-economic well being through the strengthening of the local production network, and increases employment and revenues from the effective management of the supply chain and marketing of the product.

The territorial strategy conceptualizes the OP as an asset for rural development paths. Here, actors perceive such products as offering a series of related resources including environmental (e.g. distinctive landscapes, local animal breeds and plant varieties), cultural (e.g. techniques, know-how, local folklore and heritage), and economic (e.g. skilled employment). In this strategy, OPs are seen as having the potential to contribute to a wide range of initiatives that encourage diverse activities and novel interactions between multiple types of actors (e.g. tourist trails, markets, festivals, educational initiatives, community events). What is important in this strategy is not only the physical output of the supply chain (the volume of OP produced and sold), but also the territorial identity of the product that can be integrated in other value creation processes as an immaterial asset. A range of local actors much broader than the supply chain can use OPs: citizens' associations and local development agencies are normally part of this type of strategy. These actors may develop strategies of 'basket' goods and services, trying to integrate different activities and resources in the territory by developing synergies both in the production and consumption phases; these strategies can also result in a wide distribution of economic rent (Pecqueur 2001; Hirczak et al. 2008). In some specific cases, the valorisation of OPs can be directly aimed at enhancing non-economic positive contributions to rural development, especially for the conservation of culture and the environment, for example when the identity of the OP is rooted in a local specific breed or a vegetal variety.

The valorisation of an OP asks for many tools, but the pillars of such a strategy are the following (Vandecandelaere et al. 2009): an agreement between local producers on the key-features of the product and of its production process, a control system able to guarantee both producers and consumers the real quality and identity of the product, and a collective organization that encompasses producers and other stakeholders in the product and support communication with consumers. The legal protection of geographical indications (GIs) is one of the most used tools for OPs valorisation, and has the potential to strengthen the above-mentioned effects (Treagar 2003; Rangnekar 2004; Giovannucci et al. 2009; Bowen 2010). GI protection allows fighting against imitations and misuse of the name used to market a product, and makes it possible to strengthen the territorial anchorage of the more relevant phases of the value chain. Therefore, GI protection can generate a better remuneration of efforts made by local producers and preserve local specific resources, according to the "GI quality virtuous circle" model (Belletti and Marescotti 2011; Vandecandelaere et al. 2009).

10.3 The Protection of Geographical Indications in the European Union

The protection of GIs is a tool of growing importance all over the world (Arfini et al. 2011). Following the TRIPS agreement (1994), that defines GIs as "indications which identify a good as originating in the territory of a Member, or a region or locality in that territory, where a given quality, reputation or other characteristic of the good is essentially attributable to its geographical origin" (art. 22), all WTO member States are obliged to provide the legal means for interested parties to obtain the protection of GIs. From an economic point of view, interest in GIs protection is directly related to the need to escape the increasing competition on global markets, GIs being perceived as a useful tool to signal specific quality characteristics and avoid competing purely based on prices. The protection of GIs is advocated to offer opportunities to support local agri-food systems and sustainable development (Belletti and Marescotti 2011). Companies using protected GI are expected to observe a reduction of unfair competition due to abuses or misuses of a GI, and have the opportunity to differentiate their product on the market, thus gaining higher prices, higher sales volumes, and/or access to some marketing channels. Moreover, the protection of GIs is often linked to the production of public goods, such as biodiversity preservation, cultural heritage protection, sociocultural development and rural poverty reduction (Vandecandelaere et al. 2009; Belletti et al. 2015a, b).

Starting with the '80s, the EU agricultural policy seeks to shift the leading paradigm from a quantitative to a qualitative perspective. For this reason, according to several reforms that involved the whole Common Agricultural Policy (CAP), in 1992 the UE introduced some instruments to protect and promote GI food products (regulation n.2081/92, now substituted by the UE regulation n.1151/2012): Protected Designation of Origin (PDO) and Protected Geographical Indication (PGI). These tools were the first attempt to introduce a common way to govern the use of geographical names linked to food products, following the Italian and French pioneering experiences. Both PDO and PGI focus on the relationship that links a product with a specific geographical area. Specifically, the strong connection between producers and their territory contributed to the progressive evolution of the product, developing its unique characteristics and determining its quality. PDO and PGI products have developed a reputation associated with a specific production place, which has become a brand, recognized and valued by consumers.

PDO is used for agricultural products and foodstuffs, which are produced, processed and prepared in a specified area using recognized know-how. Instead, PGI covers agricultural products and foodstuffs closely linked to the geographical area and one, at least, of the stages of production, processing or preparation has to take place in that area. Both PDO and PGI are characterized by a set of rules, which define the production area where the process must take place, the characteristics of the product, and its production process. These rules are codified in a document, called Product Specification (PS). The main difference between PDO and PGI consists in the different intensity of the link between the quality of the product and the geographical area where it is produced. The whole production process has to take place in the area allowed by the PS for PDO products, while in the case of PGI, the geographical link must occur in at least one of the stages.

Finally, the ownership of these two signs is collective and all producers that belong to the geographical area in question and respect PS rules have the right to use it as a GI. An independent third party firm has the task of evaluating the correct respect of the PS by producers. PDO and PGI can be included in the agri-food standard plethora, which has quickly increased from the end of the '90s and has become a leader in coordinating relationships among stakeholders. PDO and PGI have some specific characteristics as compared to other standards, in particular, they posses a hybrid nature. Certainly, they are created by private initiative, following a procedure defined by public rules and approved by public authorities. Another relevant characteristic is their Business-to-Consumer (B2C) nature, which offers several marketing opportunities, specifically taking advantage of two reputational levels: first, the PDO/PGI reputation and second, the product one. The PS represents the core of designations and, due to its structure, a fully-fledged standard. For that reason firms that use PDO or PGI, have to comply with every norm established in the PS. The PS is the result of a complex process of negotiation, which involves a great number of stakeholders, from the firms involved in the different stages of the supply chain to public authorities; therefore, it reflects different point of views and heterogeneous interests (Dentoni et al. 2012). Usually, the debate among producers is based on the definition of the characteristics of three main elements: product, production process, and production area. This decision-making process influences the PS structure and its rules, as the effects on rural development trajectories (Tregear et al. 2007).

10.4 The Effects of the Protection of Geographical Indications

The effects originating from the registration of a GI as a PDO/PGI may cover different aspects and dimensions. As pointed out by Belletti and Marescotti (2011), the multiplicity and complexity of the effects originated by GI protection are linked to the complexity of OP production systems. In fact, they are strictly interrelated to many typologies of local resources; therefore, they have a multidimensional and very strict link (*ceteris paribus*, stricter than other kind of products) with the territory they originate from. GI use and GI effects are therefore very complex, and they are subject to much *exogenous interference*. The analysis of the chains of causality helps to understand the effects of GI protection. Furthermore, GI effects depend strongly on the characteristics of the general framework, legal and institutional, for the recognition, protection and management of all GIs in a given territory. Although EU countries have a single general legal framework, there are many differences between different EU countries in regards to legal and administrative frameworks, the presence and the effectiveness of accompanying polices and the effectiveness of the protection on the market (London Economics 2008).

When a GI obtains legal recognition and protection, a sub-system of firms using the protected GI according to the rules defined in PS can be identified inside the OP production system. Not all the enterprises belonging to the OP system are able to make use of the protected GI (Barjolle and Jeanneaux 2012). Therefore, the effects of GI protection should be evaluated not only with regards to the enterprises using them, but also to those excluded from the use of a geographical name. Positive effects for the first category can correspond to negative ones in the latter category.

Enterprises able to comply with the PS choose whether or not to use the protected GI on their products when they find it profitable according to their global strategy, depending on the marketing channels and customers preferences and knowledge. The boundaries between the use and non-use of the protected GI are fleeting (the same firm can make use of the registered GI only for a part of its OP production) and vary in time. GI protection can affect many aspects of both the OP system and the single enterprises belonging to it, not only in a marketing perspective (quantities sold, prices, marketing channels, etc.) but also in regards to coordination and governance mechanisms inside the local production system and the supply chain. The protection of GIs can modify the competitive equilibrium inside the supply chain, both at a horizontal (competition between firms at the same production stage) and at a vertical level (competition between different stages of the supply chain, typically farmers and processors). Besides, the protection can exert indirect effects on local economy (for example increasing tourist inflows), on local society (increasing social cohesion and identity), and the environment. The contents of the PS (especially the delimitation of the production area, the characteristics of the production process, and the quality of the product) are key-factors in determining the effects of the protection, both inside and outside the production system. Stricter requirements in the PS guarantee high level of product reputation and recognisability among consumers, but small or poorly equipped producers may be excluded, because of their inability to bear the implementation costs and comply with the rules (Galtier et al. 2013). Moreover, even big enterprises oriented towards mass markets may find it too costly to insert a so-specialized and different production line in their activity. Consequently, the total amount of production under PDO or PGI may not reach significant levels, relegating the PDO/PGI product to niche markets and/or impeding appropriate collective action (Barjolle and Sylvander 2002). On the contrary, weaker rules in the PS simplify the implementation process and increase companies' possibility to use PDO and PGI. This situation strengthens both the number of enterprises using PDO/PGI and total amount of certified product quantity, increasing the opportunity of reaching supermarket and international channels. At the same time, weaker PS' reduce product standardization and preserve variations of the product (i.e. under the same protected GI many different kinds of product may co-exist), but menacing product identity and reputation, as well as confidence among buyers and final consumers. Therefore, big enterprises are normally much more interested in having looser rules for their production, as to capture the benefits from scale economies. Dentoni et al. (2012) recently explored the impact of individual group members' heterogeneous characteristics, resources and strategies on their level of cooperation on defining the future regulation of GIs. Higher heterogeneity negatively affects members' agreements on the future level of restrictiveness of "Prosciutto di Parma" PDO as a GI and therefore the effectiveness of the collective action.

The analysis of potential impacts derived from GI protection should proceed from a classification and typology of possible effects (see in particular: Belletti and Marescotti 2011). Five main categories of effects can be identified to assess the performance of the GI protection:

(a) **Effects on the structure of the GI system**

GI registration can affect the structure of a GI system. The rules set in the PS normally lead to some kind of exclusion of some businesses (firms located outside the boundaries of the identified production area, firms that do not have the capacity to comply with the specifications, etc.). GI registration may also affect the organization of the production system, the degree of horizontal and vertical coordination between firms along the supply chain (Réviron and Chappuis 2011), and the level of investment and innovation dynamics. In addition, it can exert effects on the relocation of economic activities and the maintenance of more added values to local producers, thus ensuring a positive effect on employment and income in the local economy of the region.

(b) **Effects on the economic performance of the GI system**

Normally, the most important expected effect is an increase of income, that results from the difference between the turnover (price x quantity) of products sold under protected GI and the costs to produce and certify it. On the other hand, the negative effects on enterprises not able to use the protected GI should be taken into account in the analysis. One should not focus only on the price level as an indicator of the *success* of the GI protection, because higher prices do not necessarily lead to an increase in income for single firms and for the whole production system (aggregate). Enterprises often gain premium prices, but they should be compared to prices of similar products, that is to say the price of other similar GI products and/or the price of the product not using the protected GI. Production costs may increase due to the use of the protected GI, not only because of the higher costs of inspection and certification, but also because of the need to comply with the contents of the PS: adaptation of the production process to the new requirements, implementation of the certification system (acquisition of new skills, change in administrative routines), administrative expenses (time to fill out forms), plus mandatory participation in collective organizations managing the protected GI.

Therefore, the final effect on both enterprises' and system profitability is uncertain, and requires a careful evaluation of many aspects. For example, the protection of the GI can also open new markets and/or new marketing channels (such as mass distribution, export), allowing firms to attain a greater diversification and risk reduction. Moreover, the stability of sales could be improved through the use of the protected GI, or again an increase in firm's reputation benefitting the whole assortment. Generally speaking, the horizontal and vertical distribution of these benefits and additional costs should be carefully analysed, also given the fact that the registered GI can be used more easily by large companies than small producers, or enterprises operating in the downstream sectors of the supply-chain than farmers.

(c) **Effects on consumers and markets**

The performance of the GI protection on the market (final and intermediate) is strictly related to the capacity to controlling abuses and imitations of the GI, especially when the product displays a strong reputation and is widely imitated. The number of imitations and abuses of the GI in both the home country and abroad, and the number of imitations and abuses sanctioned are key indicators that could be used. The GI registration can increase consumers' willingness to pay by increasing the perception of the product's quality. This assessment should be made not only on the final market (consumer level), but also on intermediate markets, such as retailers, restaurants, mass-distribution firms. The PS can change the quality and the identity of the protected GI product, establishing a higher level of quality of the raw material and/or the final product, requiring traceability systems (which are usually appreciated by customers on modern marketing channels). At the same time, the rules of the PS may lead to a standardization of the product, with a consequential loss of specific qualities of the product.

(d) **Economic effects outside the protected GI production system**

The registration of a GI and its effective use by enterprises may engage third order effects from outside the GI production system. The value and reputation associated with the GI may act as a lever by other subjects to enable or reinforce other economic activities. Local actors can use the GI product, its reputation and the specific resources that are linked to it (local gastronomy, traditions, landscape, etc.) as tools to enhance the competitiveness of the entire local economic and social system taking advantage of its ability to attract customers and tourists in the production area. As a result, other economic activities can be developed by both GI producers and by other local businesses (hotels, restaurants, museums, company visits, etc.).

(e) **Effects on other elements of the territorial capital**

GI systems are often strictly related to many local resources as components of biodiversity and other local environment assets (soil, landscapes, etc.), as well as cultural and social capitals. The preliminary step to be made in the analysis of the effects exerted on the territorial capital is the analysis of the relevance between the protected GI and the various components of territorial capital. This analysis should start from the identification of the specific local resources used in the production process (such as local breeds or local vegetal varieties). The amount and ratio of the agricultural area interested by the GI production is also relevant (a very low ratio means that the potential impact of the protected GI on certain resources—such as water—is probably limited). There can be a distinctive relationship between a GI product and other local material or immaterial resources (e.g. local traditions, fairs, specific habitats).

10.5 PDO and PGI Food Products in Tuscany

In 2015 there were 1244 PDO/PGI registered products in the European Union, most of which related to "fruit, vegetables and cereals", "cheeses" and "meat products" categories (Chart 10.1). Moreover, a relevant difference between Northern and Southern European countries exists. In fact, more than 70% of PDO/PGI products are produced in Mediterranean countries, in particular Italy (22%), France (18%) and Spain (14%).

Different motivations can explain this wide difference. Climate conditions and the relevance of agriculture for the economy of Mediterranean area, as well as cultural habits and social perception of food can be considered among the most important factors. Still in 2015, the amount of Italian PDO/PGI products was 280 (164 PDO, 116 PGI), elevated from 81 in 1997. Italy is the leader for PDO/PGI number of products, even though the Italian growth rate is lower than the EU rate. "Fruit, Vegetables and cereals" category represents more than 38% of the whole number of Italian PDO/PGI products, followed by "Cheeses" (18%), "Oils and Fats" (15%) and "Meat Products" (14%) categories. More than 45% of PDO/PGI products are concentrated in Northern regions, while in Central and Southern regions these products represent 26 and 29%, respectively. According to the Italian National Institute of Statistics (ISTAT), in 2013 (latest available data) PDO/PGI

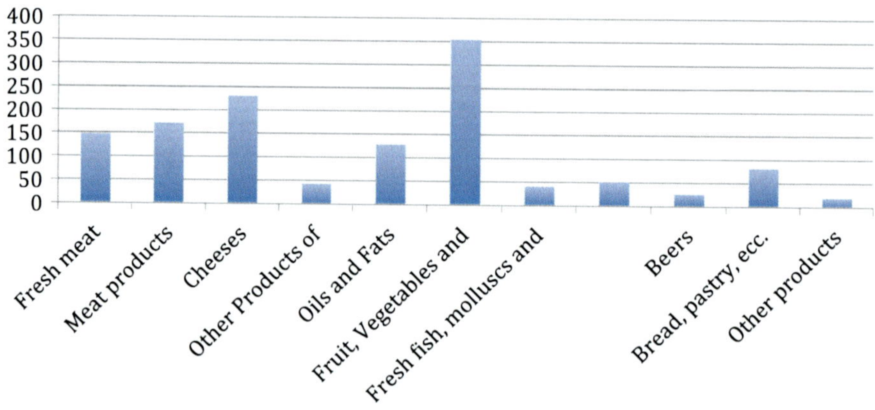

Chart 10.1 Number of registered PDO and PGI products (excluded wines) in the European Union, 2015

productions involved more than 75,000 producers and 7000 processors. Producers worked especially in "Cheeses" category (37%), "Oils and Fats" (25%) and "Fruit, Vegetables and cereals" (22%), and most of them were located in Northern part of Italy (45%). Agricultural land involved in PDO/PGI productions was more than 162,000 ha, 67% of which for "Oils and Fats" category and 32% for "Fruit, Vegetables and Cereals". Breeding farms involved in the production of PDO/PGI products were around 42.000, most of them concentrated in the North, in particular in Lombardy, Veneto and Emilia Romagna (Chart 10.2).

Tuscany is one of the most noteworthy Italian regions, according to the amount and the relevance of PDO/PGI products. Tuscany is a quite small region in the northwest coast of Italy. Due to the morphological and pedoclimatic characteristics, Tuscany has developed a long and important tradition in agri-food productions. Despite a high level of production costs due to the high fragmentation of farming activity that reduces the opportunity to reach scale economies and to introduce innovations, the relevant reputation achieved all over the world has contributed to the development of international trade. Wine, extra-virgin olive-oil and cheese have gained substantial market shares, thanks to the "bucolic" perception of Tuscany, always related to good quality and safe lifestyle. The importance of the brand "Tuscany" gives a plus to food products, but misuse of the name risks spoiling its collective image. The PDO/PGI represents a way to both exploit this brand name capital and to protect it.

In 2015, the amount of Tuscan PDO/PGI products was up to 30 (15 PDO, 15 PGI). Most of these products can be considered *entirely Tuscan*, because the whole production process has to take place in that area. According to the Italian National Institute of Statistics (ISTAT), in 2013 PDO/PGI productions involved around 12,800 producers in Tuscany (around 17% of Italian producers) and 1.000 processors (around 14% of Italian processors). Tuscany is the region characterized by the wider PDO/PGI agricultural area, around 67.000 ha, thanks to the relevant production of extra-virgin olive oils.

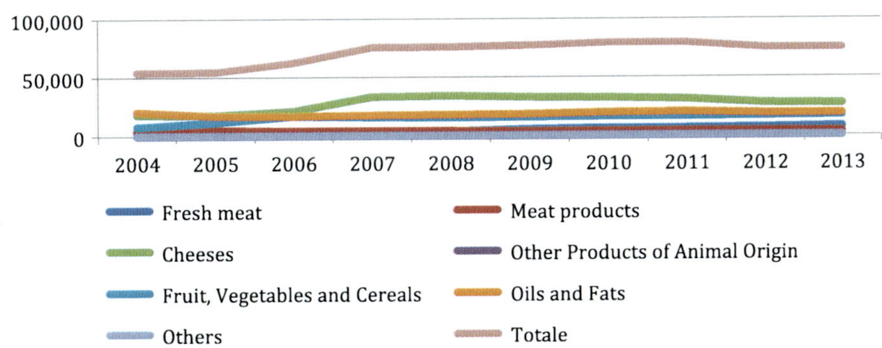

Chart 10.2 Italy: number of producers involved in PDO-PGI productions, 2004–2013. *Source* ISTAT

PDO/PGI production systems in Tuscany can be divided in two main groups (Table 10.1): major PDO/PGI systems and minor ones. The first group is composed by few products (Tuscan Pecorino-cheese PDO, Tuscan Extra virgin Olive oil PGI and Tuscan Ham PDO), characterized by large aggregate volumes (relatively speaking), the presence of some large companies, and regular access to supermarket channels and to foreign markets. The application for the PDO/PGI protection was mainly meant to exploit the brand name reputation associated to "Tuscany" and to clear the market from imitation products or misuses of the name. As a consequence, the contents of the PSs were drawn in a way as to include nearly all the typologies of the product previously produced, without setting too high quality standards. This decision, coupled with the wide dimension of the territory where the production can take place, allows many producers to be able to use the PDO/PGI, but at the same time has some drawbacks. First, these products can be produced in the whole territory of Tuscany, where a high variability in the products typology and quality can be observed, thus causing a reduction in the effectiveness of the message to customers and final consumers. Second, if on one hand the opportunity to access to supermarket channels thanks to the high "appeal" of the brand Tuscany can guarantee quite stable sales to producers, on the other hand it exposes them to the risk of losing control over the brand, due to the unbalanced bargaining power on the two sides of the relationship. Third, the impact these products exert on local development is more "diluted" on the territory.

The second group is composed by most of the PDO/PGI products. These products are characterized by small quantities, traditional/artisan production process, and the preference for direct sales and short supply-chains. Usually, their production area is very small and the number of producers and processors is limited. For instance, in 2013 San Gimignano Saffron PDO and Caprese Michelangelo Chestnut PDO counted only on 3 and 1 producers. The protection granted by the EU represents a tool to protect their products against misuses and frauds, but it is mainly meant to increase (and not only to defend) product reputation and communicate product peculiarities to consumers. Moreover, being locally focused, the effects of the protection are more concentrated on small territories, thus being able to better activate and reinforce rural development dynamics.

10.6 The Fagiolo Di Sorana IGP (Sorana Bean, PGI)

10.6.1 Aims and Methodology of the Study

The Sorana Bean PGI (*Fagiolo di Sorana IGP*) is one of the smallest production systems, which gained PGI protection in Tuscany. Nonetheless, the Sorana Bean PGI can be considered an interesting example of success. Truly, it played an important role for this geographical area, not only from an economic point of view, but also from a wider rural development perspective. Below, the effects related to the GI protection on farmers' activity will be analysed, especially focusing on how

Table 10.1 PDO/PGI products in Tuscany, 2012

		Producers	Processors/bottlers	Quantity (kg)	Turnover at consumer level
Smaller PDO/PGI production	Chianti Classico Extra Virgin Olive Oil PDO	238	79	119,000	1,680,800
	Lucca Extra Virgin Olive Oil PDO	12	7	7200	n.a.
	Seggiano Extra Virgin Olive Oil PDO	29	14	4352	n.a.
	Caprese Michelangelo chestnut PDO	9	1	1700 (2011)	9200 (2011)
	Cinta Senese PDO	85	27	494,228	n.a.
	Colonnata lard PGI	–	17	165,220 (2011)	377,300
	Garfagnana neccio flour PDO	12	5	1700	14,200
	Garfagnana spelt PGI	32	3	118,400	592,000
	Lunigiana chestnut flour PDO	23	2	1100	17,300 (2011)
	Lunigiana Honey PDO	41	12	44,428	1,074,300
	Monte Amiata chestnut PGI	164	4	400	n.a.
	Mugello chestnut PGI	93	5	35,600	224,900
	San Gimignano saffron PDO	4	4	n.a.	n.a.
	Sorana Bean PGI	22	22	7500	155,000
	Terre di Siena Extra Virgin Olive Oil PDO	114	59	36,000	679,700
Bigger PDO/PGI production	Tuscan Extra Virgin Olive Oil PGI	10,825	706	3,583,871	43,006,452
	Tuscan Pecorino-cheese PDO	832	22	3,067,190	46,315,100
	Tuscan Ham PDO	2306	122	2,927,900	65,456,600

Source: Authors elaboration on data by Qualivita, Ismea, Istat, Consorzi di tutela

farmers decide whether and to what extent to make use of the protected GI, and the benefits they obtain. Moreover, the effects induced by the PGI product on the whole rural economy and community will be analysed.

The research methodology consisted in an analysis of the *logic* followed by local stakeholders during the process that led to the application for the PGI recognition, by examining PS contents (also in relation to similar products) and other official and informal documents. Secondly, some semi-structured interviews were conducted with a representative group of Sorana Bean PGI producers (8 out of the 23 registered farmers), in addition to the director of the Consortium. The first aim of these interviews was to understand the motivations underpinning the choice of firms of using the PGI in marketing their products, underlining strengths and weaknesses. The second goal was to verify the existence and the strength of the different effects on the territory. In particular, we analysed the effects from an economic, social and environmental point of view. The questionnaire was divided into six main sections:

- Firm's characteristics. History and evolution of the firm, type of products (assortment), turn-over, marketing channels importance and evolution, quality certification schemes, investments, etc.;
- Implementation of the PGI standard. Quantity produced, PGI marketing channels as compared to other firm's products, geographical markets, etc.;
- Costs of compliance to PS. Implementation costs (administration), raw material costs, production costs, control and certification costs, participation fee to consortium, etc.);
- Direct benefits from PGI. Prices and incomes, turnover by marketing channel and geographical market;
- Other benefits related to PGI use (protection from imitations and abuses, firm's reputation, assortment, access to specific marketing channels, etc.);
- Effects related to PGI use for the rural community (reduction in depopulation rate, increase in tourism, biodiversity protection, etc.).

10.6.2 Basic Characteristics of the Sorana Bean PGI Production System

Sorana is a small village situated at the bottom of the Appennino Mountains, half way between the cities of Florence and Pisa. The Sorana Bean is a niche product, which can boast ancient origins, chronicled by old documents. The production area includes no more than 660 ha in this marginal valley, crossed by a little torrent, and characterized by low level of urbanization, industrialization and infrastructures, and for the presence of very small farms where farmers are often non-professional (retired, hobby or part-time farmers). The biggest part of the valley is in fact covered with woods, while open fields are cultivated with olive-trees and some horticultural products, and in particular beans, using traditional methods. These cultivated fields play an important role for both landscape and habitat functions. High humidity levels and low temperatures ranges characterize the area. Mountains, protect it from cold wind in winter and from sun exposure in the summer. These particular pedo-climatic features contribute to the distinctiveness of these beans: small dimension, pearly white colour with pink veins, and a very thin skin. Traditionally, producers cultivate this bean on the sand lands of the torrent banks (area named *Ghiareto*, traditional production area where yields are lower and product quality higher due to special pedo-climatic characteristics that seem to give the bean a particular texture and flavour), even though the production area has been more recently extended to the surrounding hills (area named *Poggio*). With time, production methods have not been subject to relevant modification: seeds are selected from the last crop, harvest time is from half August to half September and this phase is manual, beans are exposed to sun for 3–4 days and packed in small plastic bags. The Sorana Bean is a climbing pulse bean, which belongs to the *Phaseolus vulgaris L.* species and can reach more than five metres. The seeds used today are obtained from the plants that have been on site for many generations, thus creating an ecotype entirely adapted to the local environment. Firms produce two different varieties of Sorana bean: white Sorana bean and red Sorana bean. While the white bean is the most famous and it benefits of a high recognisability and reputation, the red one is quite unknown for consumers and less appreciated by local population, and, in the last decades it risked extinction. Farmers in Sorana valley are very few, and most of them are non-specialised, being retired or undertaking other work. Traditionally, farmers directly manage all the phases of the production process up to the drying and packaging; very often they also sell the product directly on the final market or to groceries. Consumers pay a price six to seven times higher (20–2 euros/kg) than for a *standard* bean.

10.6.3 The Application of the PGI and the Drawing of the Product Specification

For a long time, the reputation of Sorana Bean did not exceed the surrounding areas and trading was a marginal activity. In time, the high reputation of Sorana bean, coupled with unclear rules on the use of its name, brought problems of misuse and abuse of the name on the market. Following the approval of the EU regulation 2081/92, Sorana bean producers started considering the PGI-application. The small number and physical proximity of producers favoured a direct, informal interaction, in particular in the Ghiareto area. The Associazione dei piccoli produttori del Fagiolo di Sorana Il Ghiareto (Association of small farmers of the Sorana Bean), founded in 1994, managed the PGI-application process, in order to reach an agreement between farmers about the rules of the PS. Some conflicts emerged between farmers located in Ghiareto and Poggio. The latter benefited from higher yields per hectare and thus lower production costs. However, by extending the historical production area to Poggio one has also increased the quantity produced, the visibility in the market and the possibility to carry out collective promotional activities. In the end, the PS has been drawn taking into consideration more traditional production techniques, common to non-professional farmers with less productive land in the area of Ghiareto, which historically gave the reputation of the product. Producers agreed on the PS rules, deciding to extend the PGI production area to the entire Sorana valley, so to include the Poggio area. The municipality of Pescia and regional public authorities supported this solution, given the narrowness of the area and the limited amount of production. At the same time, Ghiareto producers were allowed to add a special mention on the label to highlight the most reputed sub-area of production.

The extension of the production area to other historically less traditional areas, like Poggio, has increased the quantity produced by the whole system, the visibility of Sorana bean on the market, and the possibility to carry out collective promotional activities. Other important points agreed in the PS were the ban of chemical herbicides to grow the beans, and a relatively low maximum yield per hectare (20 quintals), lower than the average yield in Poggio area, but higher than the Ghiareto one. Despite the conflicts between upper and lower areas, the application process allowed the cooperation between farmers to achieve a common aim. Moreover, the process fostered the motivation to produce high quality beans and increased the local pride of producers. The very PGI application process, thanks to local and national media news about the PGI was reported frequently by media - boosted the reputation of Sorana bean favouring its marketing. Traceability and quality control enhanced product quality on the final market and the access to new markets and marketing channels (Quiñones-Ruiz et al. 2016).

In conclusion, the rules included in the PS were the result of a complex process of negotiation, which involved heterogeneous stakeholders, from farmers to local public authorities. It is worth underlining the presence of local public authorities in the process (Municipality, Tuscan Regional Administration, Chamber of Commerce), that highlights how the protection of the geographical name and the support to local farmers is not only a matter of the specific supply-chains and its enterprises, but involves the whole territory, as it both shapes its identity and cultural values and traditions, and may contribute to the whole economy of the area by inducing positive effects on touristic activity and preventing local people to emigrate to other areas.

10.6.4 The Use of the PGI

From a total number of 40 bean producers in Sorana valley, aproximately 20–22 actually use the GI certification. The other mainly non-professional producers do not need the GI certification, as many of them sell small quantities of beans to friends and relatives. Today's production is characterised by very low quantities and high sale prices (22, 00 euro/kg on average, compared to 3–4 euros/kg for

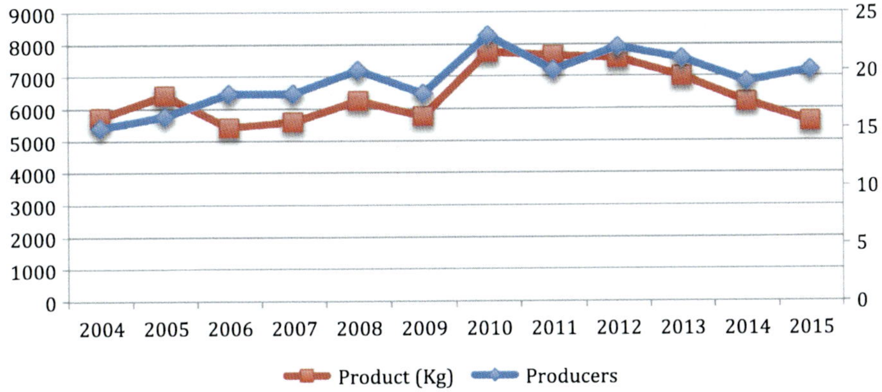

Chart 10.3 Sorana Bean PGI: number of producers (*right axis*) and production quantity (*left axis*, kg). *Source* ICEA (inspection body of Sorana Bean PGI)

conventional beans). Mostly, Sorana beans are sold through direct marketing. Tuscany is the prevalent consumption market, although a small share of beans is sold to restaurants and agri-food shops in Northern Italy. The PGI certified production has grown from 57 quintals of certified beans in 2004 (first year of PGI implementation) up to more than 80 q. in 2010, before a fall in the following years due to adverse weather conditions (Chart 10.3). Most production is carried out by small farms often managed by non-professional producers (retired, hobby, or part-time). The number of producers which use the PGI has slightly grown over the years, but still remains very small (15 in 2004, 20 in 2015). In fact despite the high price level, it is difficult to extend the area under cultivation, as available land is scarce and highly fragmented. On the other hand the slope of fields and their small size do not allow to mechanize the production, which remains thus highly labour-intensive.

Moreover, farmers produce on average 280 kg of dry bean, which means approximately an average turnover of 6000 euro (value at final consumption). Farmers' production ranges from a maximum of 1905 kg to a minimum of 45 kg (2015), highlighting a high heterogeneity of producers. Indeed, the production system is composed by a few large (relatively speaking) professional farms, where the production of Sorana Bean PGI accounts for a high percentage of total farmer's income, flanked by many small farmers, often non-professional, who keep on producing the bean for income integration.

10.6.5 Effects from the Use of the PGI

The results of the research show an interest in using the PGI by market-oriented producers, essentially due to the fact that market price is by far higher as compared to conventional beans, while additional production costs (included inspection and certification costs) are minor. The choice to use the PGI is almost uniform among producers; farms opt for using the PGI certification for the whole production. Specifically, the level of use of PGI is the highest according to both the number of users, compared to potential ones, and the certified quantity, compared to the whole production. Indeed, only a marginal amount of Sorana beans are not certified as PGI, since they are used for personal consumption, to obtain seeds for the following year, or simply they do not comply with the PS. The morphological characteristics of the production area (in particular the cultivation carried out on the torrent banks) and the limited extension of available fields, coupled with the fact that most farmers are pensioners or part-time farmers, do not allow the use of less expensive production methods. Therefore, producing a different bean variety with *free* cultivation practices costs as much as producing Sorana Bean PGI, but the sale price is undoubtedly lower. The banning of the use of chemical herbicides, the product selection and the packing phase are the main aspects that create higher costs to farmers, due to additional work.

In addition to open interviews, a specific enquiry was made in 2014 to a sample of local actors about their perception about ex-post effects of the PGI. A total of 8 people – 3 public actors and 5 producers – were asked to express their level of agreement on a list of 19 potential effects on a scale from 1 (strongly disagree) to 5 (strongly agree). Table 10.2 reports the results of this enquiry.

Improved product identity and perceived quality in the final market, and related satisfactory price premium, are the most perceived benefits from stakeholders, with an average score of 4.5 and 4.4 over 5.0 respectively. The high market price is the main element that encourages producers to use the PGI. The morphological characteristics of this valley restrict the opportunity to introduce innovations or new techniques, affecting producers' activities and reducing their options. Therefore, Sorana Bean PGI is the best and unique product they can cultivate for economic results. Farmers perceive the importance of this product and they pay specific attention to keep the high quality level. As previously mentioned, the strict rules defined in the PS have contributed to increase the product reputation and to give a strong characterization, which has made Sorana Bean PGI easily recognisable among consumers. PS has been built on the basis of traditional modes of production (banning the use of chemical herbicides, setting a maximum yield), typical of non-professional farmers with less productive land, but that has historically given the reputation of the product. Whereas most farmers are pensioners or part-time farmers, Sorana Bean PGI represents an important opportunity to increase their traditional incomes, and therefore they prefer to consolidate the exclusive product image, through a limited supply. This choice strengthens the product position on the market, but it represents a huge limit to innovations. On the other hand, professional farms have an opposite point of view. They perceive the maximum yield rule as an important restriction, which limits the experimentation of technical innovations, the recovery of new lands, the access to new market channels, in particular supermarket ones. The access to supermarket channels is the most criticized element, and represents a daily topic of debate. Professional farmers would increase the production

Table 10.2 Benefits perceived by stakeholders from registration of Sorana bean PGI

Benefits after registration	Average	Coefficient of variation (%)	Public	Private	Difference in %
Improved product identity and perceived quality in the final market (both customers and consumers)	4.5	11.1	4.3	4.6	6.2
Satisfactory price premia of the Sorana bean on the final market	4.4	11.1	4.0	4.6	15.0
Positive effects on the environment and landscape.	4.4	11.1	4.0	4.6	15.0
Retaining of revenues and employment in the GI geographical area.	4.3	10.2	4.3	4.2	−3.1
Improved regional knowledge. self-confidence and identity of producers	4.3	10.2	4.0	4.4	10.0
Introduction of a traceability and quality control logic in firms	4.3	15.6	3.7	4.6	25.5
Improved access to new marketing channels (e.g. supermarket chains. HO.RE.CA.)	4.1	8.0	4.0	4.2	5.0
Positive effects on rural development (e.g. agro-tourism)	4.1	8.0	4.0	4.2	5.0
Development of collective marketing activities	4.1	8.0	4.0	4.2	5.0
Price premia sufficiently transferred to farmers.	4.0	12.5	4.0	4.0	0.0
Improved access to local/regional market	4.0	12.5	4.0	4.0	0.0
Increase of firm reputation and support in selling other firm's products	3.9	23.9	3.0	4.4	46.7
Increase of the market power of farmers	3.8	22.1	3.7	3.8	3.6
Improved access to new markets in EU countries	3.5	24.7	3.3	3.6	8.0
Improved access to new markets outside the EU	3.0	23.6	3.3	2.8	−16.0
Reduction/discouraging of free-riding (decrease of abuse and imitations)	3.0	44.1	4.0	2.4	−40.0
Increase of horizontal coordination between farmers	2.9	32.2	3.7	2.4	−34.5
Strengthening of Interprofessional organization	2.6	18.4	2.7	2.6	−2.5
Improved long-term relations/communication and vertical coordination between farmers. packers and retailers	2.4	20.4	2.7	2.2	−17.5

Source Authors

to become a stable supplier and to gain quite constant incomes. Even though this strategy may add new opportunities for the whole production system, the exclusive image of Sorana Bean PGI would be compromised due to the access to mass distribution. Therefore, some conflicts emerge between the two different production areas, "Poggio" and "Ghiareto", due to the higher average production per hectare in Poggio and consequently different production costs, allowing producers from "Poggio" to sell their product at a lower price. The extension of the production area to other historically less traditional areas (the "Poggio" ones), while it has helped to strengthen the system by increasing the quantity produced, the visibility in the market, and the possibility to carry out collective promotional initiatives, on the other hand introduced tensions among producers: the professional farmers, who are located outside the more traditional production area, ask for the increase of the maximum yield, also at least to partly compensate for the lower price they get on the market. But Ghiareto farmers, underlining the higher quality of their product as compared to those from Poggio, complain about the excessively low prices set by professional farmers, who are likely to confuse the average consumer and to reduce the reputation of the Sorana Bean IGP on the market. Producers underline that PGI certification plays an important role in reducing imitations and misuses.

The high price differential between Sorana Bean PGI and other traditional beans has encouraged frauds, decreased due to the growing controls and packaging rules, in particular the ban of unpacked trade. Moreover, the PGI has helped producers to strengthen marketing relationships, in particular with unconventional clients. PGI logo represents insurance for consumers that are more confident in purchasing this product, despite its high price. The PGI increased the fame of this particular bean and, consequently, favoured the strengthening of direct marketing. Producers assert that, at the beginning, their own firm brand had no importance in direct sales as personal knowledge and reputation was at the basis of market relationships on short supply-chains and direct sales, while PGI mark was crucial to capture new customers, especially on intermediate markets and distant consumers. On the contrary, frequent consumers evaluate much more the firm brand than the EU logo, because of the organoleptic differences between beans coming from different producers. Moreover, consumers perceive the PGI mark as a sign of high quality and particular characteristics of the bean. It is for that reason that producers normally succeed in selling the whole production of the year in 5 months despite the high price. Secondly, Sorana Bean PGI is an important retention tool for consumers. After the first consumption, most customers keep buying year after year. Furthermore, this bean is the driving force for other farmers' products. Producers underline that the excellent impact of Sorana Bean PGI on consumers increases their confidence and, consequently, the chance of also selling other products raises. Farmers underlined that Sorana Bean PGI plays an important role in the economy of the area. The low level of urbanization, industrialization and infrastructures of this valley influenced its economic development and the social fabric dynamics, characterized by a progressive reduction and ageing of the population. The opportunities related to PGI have partially mitigated this phenomenon, thanks to the positive effects registered on the economic sphere. The profitability of the PGI bean allowed for the survival of agriculture in this valley - which otherwise would have been abandoned due to difficult growing conditions - thus preventing for abandonment of land and contributing to the preservation of landscape and habitat equilibrium in this delicate area. Stakeholders in fact cite positive effects on the environment and landscape as the third benefit with an average score of 4.4 (Table 10.2).

Immaterial effects of GI protection are also quoted by producers and public actors as highly relevant. Certainly, the GI registration improved self-confidence and identity of producers and other local people, thus contributing to the strengthening of local social capital. For this reason GI protection acted as an incentive to restore some agricultural productions (especially reclaiming part of the torrent banks), and offered new chances to youth. Besides, the whole territory benefited from the Sorana bean's notoriety gained through the PGI, which supported the valorisation of other products such as local extra-virgin olive oil or by promoting rural tourism (some restaurants offer special bean menus) (Belletti et al. 2014). In fact, the reputation and success of the PGI increased the exposure of the valley in the media, also due to the intense promotional activity carried out by the producers' association, raising opportunities for local tourist industry. Agri-tourisms and restaurants have gained the most from the PGI success. This marginal valley, cut off from the main touristic itineraries and attractions of Tuscany, has succeeded in developing its own touristic inflows, in particular thanks to the relevant communicative effort of Sorana bean PGI producers. Additionally, the producers' association promotes annual festivals, to celebrate the seeding and harvesting time. These festivals are important opportunities for producers to increase Sorana Beean PGI sales, as well as for other stakeholders, such as restaurants, agritourisms, etc., to take advantage of tourism flows.

10.7 Conclusion

The peculiarities of Origin Products – the strong link between quality and territory, the use of traditional production process, their collective and cultural dimension—offer new opportunities to producers to escape price competition in a globalizing world, and are coherent with increased attention towards multi-functionality and diversification of agricultural and rural activities. The vitality and in some cases the survival of rural regions are occasionally endangered, particularly where local human and natural resources cannot easily be managed so as to find new bases of competitiveness. This is particularly true in marginalized rural areas where, due to a complex set of reasons - infrastructural, structural, geographical, cultural—it is not always easy to achieve price competitiveness, and other forms of leverage are not available. The crisis that many rural areas are facing may endanger their local economies, exert negative effects on the quality of the environment, threaten social relationships and livelihoods, and cause a loss of culture and traditions. The valorisation of Origin Products can prove to be an important device to activate and consolidate the dynamics of sustainable rural development. In particular, the legal protection granted to Geographical Indications can reveal to be an effective tool to foster rural development, provided that it is inserted in a wider and coherent network of actors and actions. By means of the case study of the Sorana Bean PGI, we showed how a marginalized territory could benefit from the legal protection of a Geographical Indications. Indeed, the protection of Sorana Bean as PGI exerted positive effects both for producers and for local development. Despite some relevant limits for the production system and its development (non-professional farmers, small production level, preference for direct sales, etc.), the PGI has played a crucial role in protecting the name of Sorana beans against misuses and frauds, safeguarding its strong image, which is the basis of its success among consumers. The decision to include some strict rules in the PS has contributed to forming a strong identity of the product and has encouraged the adoption of a PGI scheme by farmers. In particular, imposing a relatively low maximum yield per hectare has moderate the competition among *big and small* producers. Moreover, these rules have reduced the possibility to obtain different quality levels of the Sorana Bean PGI. Therefore, the image and reputation of the product have been well defined and made it easily recognizable by consumers. Moreover, the very process of application for the PGI has delivered numerous benefits (Casabianca 2003), reinforcing cohesion among farmers by bringing producers together when no association was previously active in the area. Through the producers' association, the interests of producers are now represented in negotiations with agencies and institutions. Finally, by encouraging the defence and promotion of the bean, the association has acted as a catalyst for the involvement of other local and non-local actors. The high reputation and the exposure achieved by Sorana Bean PGI have generated positive effects not only for producers, but also for local stakeholders. For example, local public institutions have funded educational initiatives with local schools related to the bean and the valley history, and the producers' association has developed scientific research projects with the University of Pisa and the University of Florence. Moreover, tourism and local industries have exploited this opportunity, trying to find new businesses and fostering the social environment. Therefore, the initiatives taken by the producers' association, prompted and supported by an *external* network of local public and private organisations show how the PGI helped to activate new connections by capturing the new social demand for multi-*functionality* (product quality, environment, traditions and culture, ethics, social relationships). It is also for that reason that the engagement of public bodies in GI registration processes and GI protection can be justified, also for the *small* products whose turnovers are not significant but which are able to generate other positive effects at local level, in both economic and non-economic spheres. In conclusion, the legal protection of Geographical Indications should not only be considered as a *legal* tool to defend a particular intellectual property right and modify the markets, but also as leverage to foster rural development paths even when the economic dimension of the production system of the protected product is apparently *small*.

Bibliography

Arfini, F., Albisu, L.M., and Giacomini, C. 2011. Current situation and potential development of Geographical Indications in Europe. In Barham, E., Sylvander, B. (Eds.), *Labels of origin for food. Local development, global recognition*, 29-44. Cambridge (USA): CABI International.

Barjolle, D., Sylvander, B. 2002. Some Factors of Success for "Origin Labelled Products" in Agro-Food Suppply Chains in Europe: Market, Internal Resources and Institutions. *Économies et Sociétés* n° 25, 9–10:1441–1461.

Barjolle, D., Jeanneaux, P. 2012. Raising Rivals Costs Strategy and Localised Agro-Food Systems in Europe. *International Journal on Food System Dynamics* 3(1):11–21.

Belletti, G., Marescotti, A. 2011. Origin products, geographical indications and rural development. In Barham, E., Sylvander, B. (Eds.), *Labels of origin for food. Local development, global recognition*, 75–91. Cambridge (USA): CABI International.

Belletti, G., Marescotti, A., Brazzini, A. 2014. Collective rules and the use of protected geographical indications by firms. *International Agricultural Policy*. 1: 11–20

Belletti, G., Canada, J.S., Marescotti, A., Vakoufaris, H. 2015. Linking Protection of Geographical Indications to the environment: Evidence from the European Union olive-oil sector. *Land Use Policy* 48:94–106.

Belletti, G., Marescotti, A., Touzard, J.M. 2015. Geographical Indications, Public Goods and Sustainable Development: The roles of actors' strategies and public policies. *World Development*. doi:10.1016/j.worlddev.2015.05.004.

Bérard, L., Marchenay, P. 1995. Lieux, temps et preuves: la construction sociale des produits de terroir. Terrain. *Carnets du patrimoine ethnologique* 24:153–164.

Bérard, L., Marchenay, P. 2004. *Les produits de terroir: entre cultures et règlements*. Paris: CNRS Editions.

Bowen, S. 2010, Embedding Local Places in Global Spaces: Geographical Indications as a Territorial Development Strategy. *Rural Sociology* 75(2): 209–243.

Casabianca, F. 2003. Les produits d'origine: une aide au développement local. In: Delannoy, P., Hervieu, B. (Eds). *A table. Peut-on encore bien manger?*, 66–82. Paris: Editions de l'Aube.

Casabianca, F., Sylvander, B., Noël, Y., Béranger, C., Coulon, J.B., Roncin, F. 2005. Terroir et Typicité: deux concepts-clés des Appellations d'Origine Contrôlée. Essai de définitions scientifiques et opérationnelles. Symposium international Territoires et enjeux du développement régional. Lyon.

Delfosse, C. 1996. Qualité, liens au lieu et développement local. In: Casabianca, F., Valceschini, E. (Eds.). *La qualité dans l'agro-alimentaire: émergence d'un champ de recherches*, 34–40. Paris: INRA-SAD.

Dentoni, D., Menozzi, D., Capelli, M. G. 2012. Group heterogeneity and cooperation on the geographical indication regulation: The case of the "Prosciutto di Parma Consortium. *Food Policy* 37(3):207–216.

Galtier, F., Belletti, G., Marescotti, A. 2013. Factors Constraining Building Effective and Fair Geographical Indications for Coffee. Insights from a Dominican case study. *Development Policy Review* 31(5):597–615.

Giovannucci, D. Josling, T., Kerr, W., O'Connor, B., Yeung, M.T. 2009. *Guide to Geographical Indications. Linking products and their origin*. Geneve: International Trade Centre.

Hirczak, M., Moalla, M., Mollard, A., Pecqueur, B., Rambonilaza, M., Vollet, D. 2008. Le modèle du panier de biens. *Economie rurale* 6:55–70.

London Economics.2008. Evaluation of the CAP policy on protected designations of origin (PDO) and protected geographical indications (PGI). Final report, Study financed by the EU Commission. London. Downolodable at: http://londoneconomics.co.uk/blog/publication/evaluation-of-the-cap-policy-on-protected-designations-of-origin-pdo-and-protected-geographical-indications-pgi/

Pacciani, A., Belletti, G., Marescotti, A., Scaramuzzi, S. 2001. The role of typical products in fostering rural development and the effects of Regulation (eec) 2081/92. 73rd Seminar of the

European Association of Agricultural Economists (EAAE) "Policy Experiences with Rural Development in a Diversified Europe", Ancona (I), 28–30 June.
Pecqueur, B. 2001. Qualité et développement territorial: l'hypothèse du panier de biens et de services territorialisés. *Économie rurale* 261(1):37–49.
Quiñones-Ruiz, X.F., Penker, M., Belletti, G., Marescotti, A., Scaramuzzi, S. 2016. Why early collective action pays off: evidence from setting Protected Geographical Indications. *Renewable Agriculture and Food Systems*1–14. doi:10.1017/S1742170516000168.
Rangnekar, D. 2004. *The socio-economics of geographical indications.* UNCTAD-ICTSD Project on IPRs and Sustainable Development, Issue Paper, 8.
Treagar, A. 2003. From Stilton to Vimto: Using Food History to Re-think Typical Products in Rural Development. *Sociologia Ruralis* 43(2): 92.
Tregear, A., Arfini, F., Belletti, G., Marescotti, A. 2007. Regional foods and rural development: the role of product qualification. *Journal of Rural studies* 23:12–22.
Réviron, S., Chappuis, J.M. 2011. Geographical indications: collective organization and management. In Barham, E., Sylvander, B. (Eds.). *Labels of origin for food. Local development, global recognition*, 45–62. Cambridge (USA): CABI International.
Vandecandelaere, E., Arfini, F., Belletti, G., Marescotti, A. 2009. *Linking people, places and products: A guide for promoting quality linked to geographical origin and sustainable geographical indications*, ed. Food and Agriculture Organization of the United Nations (FAO) and SINER-GI, Rome.
Vermeir, I., Verbeke, W. 2006. Sustainable food consumption: Exploring the consumer "attitude–behavioral intention" gap. *Journal of Agricultural and Environmental ethics* 19(2):169–194.
Weatherell, C., Tregear, A., Allinson, J. 2003. In search of the concerned consumer: UK public perceptions of food, farming and buying local. *Journal of rural studies* 19(2):233.

Author Biographies

Professor Giovanni Belletti, PhD (M) Associate professor in Agricultural economics at the Department of Economics and Management at University of Firenze. The main research fields are economics and organisation of agriculture and agro-food system, quality issues in agro-food system with specific focus to origin products and geographical indications, rural dynamics and rural development policies, multifunctionality of agriculture and agri-environmemntal policies, agricultural diversification with specific focus to rural tourism and agritourism. He has been involved in many national and international research projects, both as a coordinator and as member of research units.

Professor Andrea Marescotti, PhD (M) Associate Professor of agri-food economics and policy at the University of Florence (italy). Main research interests range between theoretical and empirical studies on farmers' marketing strategies; coordination mechanisms between firms in agribusiness; socio-economic analysis and evaluation of geographical indications protection; short supply-chains analysis; food quality standards and globalisation.

Alessandro Brazzini, PhD (M) He received his PhD in economics and local development from the University of Tuscia and now he is collaborating with the Department of Economics and Management at University of Firenze. the main research fields are food quality and agri-food standards, with specific focus on geographical indications; coordination mechanisms between firms in agribusiness: supply-chain analysis. he has been involved in some national research activities.

Chapter 11
New World Case Study: King Island—Living the Place Brand

Jennifer Thorn

Abstract The King Island Brand is well-recognised and extends across a range of products within Australia. Internationally, the name is known in relation to cheese, seafood, beef, water, small grid renewable energy, and very recently sea salt and links golf. The King Island Brand is valuable and sustainable, by virtue of the quality of the products and well-managed production systems. The brand strength has been built over a number of years. Whilst enjoying considerable benefits from the brand value, the community has had to be constantly vigilant against possible diminution of the brand value by businesses, with no association with King Island, attempting to pass off products as produced on the island. This article discusses some of the trials the community and King Island businesses have faced, and their efforts in response to protect the brand. The introduction of Geographical Indications system for King Island is discussed as a viable alternative to the options currently available for protecting valuable brands.

11.1 Introduction

The King Island Brand is well-recognised and extends across a range of products within Australia. Internationally the name is known in relation to cheese, seafood, beef, water, small grid renewable energy, and very recently sea salt and links golf. The King Island place brand is valuable and sustainable, by virtue of the quality of the products and well-managed production systems. The brand strength has been built over a number of years. King Island is an isolated and remote community. The population is just over 1500 people.[1] The closest air and sea ports, on mainland Tasmania and mainland Australia, are approximately 230 km away. King Island is an expensive place to do business. The only passenger transport is by air, so there is

[1] Australian Bureau of Statistics Census 2011 counted 1563, on a slightly downward trend.

J. Thorn (✉)
King Island Council, King Island, Australia
e-mail: jenki@kingisland.net.au

W. van Caenegem and J. Cleary (eds.), *The Importance of Place: Geographical Indications as a Tool for Local and Regional Development*, Ius Gentium: Comparative Perspectives on Law and Justice 58, DOI 10.1007/978-3-319-53073-4_11

no free flow of labour to or from surrounding areas. Electricity is reticulated through a small grid and is relatively expensive, with no off-peak discount tariffs. Sea freight is extremely expensive, so all imports and exports incur a large freight cost.

In order to profit from their investment, it is necessary for King Island businesses to target the premium end of their markets. This represents a challenge in any market, but King Island has built a fine reputation for products that can and do attract a premium for high quality. The mild maritime climate, unpolluted seawater and air give the island a distinct advantage in growing exceptional primary produce. The combination of advantageous environmental conditions and attention to producing premium products has built a strong King Island place brand that underpins the island's economy.

The King Island community have long been aware of the value a strong place brand can bring to a regional economy and that the island's businesses have benefitted from the return on products sold into premium food markets. The community have realised tangible benefits from the intangible brand value and the range of products developed has been driven by the need to deliver quality products to the consumer (Abbing 2010). In deliberating whether the application of geographical indications might bring further advantages to the island's businesses and residents, the ownership and current use of King Island brands will need to be considered for separate industries.

11.2 Realising the Value of the King Island Brand

The King Island community well understand the concepts of provenance and place branding, but the brand is not just a place of origin label or geographical identifier of the island's products. King Island is their home. The association of the name with premium quality products across a number of industries is a result of the care they invest in their work, and underpins their livelihoods and businesses. The success of the King Island Dairy in the 1980's and 90's, in producing brie cheese and selling it as a premium product, brought the true value of product branding into the consciousness of the local community. The dairy went from strength to strength, based on the very high quality of the cheese, and with the education of the Australian palette. The island's dairy producers now had a competitive advantage, with the value-adding of their milk over-coming some of the costs of distance to the market (Sorensen 2010) and the additional sea and air-freight costs on inputs and export of product.

The quality of the King Island grass-fed beef was also becoming apparent and island beef producers started to consider the opportunities of a branded product. The influence of the sea ensured a mild climate allowing extended periods of pasture growth and a low-stress environment for livestock. With the success of the dairy brand, there was also a threat of competitors riding on the island's fortune and instances of misuse of the name King Island were reported. There was a period of constant passing off—a commonly quoted example was the sale of 'King Island Rabbit' in the Sydney market, there are no rabbits on the island and never have been. The community included Management and Protection of the King Island Brand as a key objective of the King Island Council Strategic Plan in 2004 (King Island Council 2007). 'Brand King Island' was developed through meetings and facilitated workshops, with the community and stakeholders discussing their core values and what the island meant to them. The key attributes of the King Island place brand were defined: nature and landscapes; food and produce; friendly people; pace of lifestyle; cleanliness of air; water and land. King Island Brand Management Group was formed as a Special Committee of Council in 2007, with local representatives targeting food, tourism, population growth and investment. The group aimed to protect and develop the King Island brand, champion its values and ensure continued community support.

The King Island tourism industry has benefitted greatly from brand spill-over from the successful dairy products and later the King Island beef branded product. The growing awareness of the King Island name, both domestically and internationally, drove visitation to the island. With that increasing association of 'place' with premium foods, it became apparent there was also an expectation- many visitors expected those exceptional foods would be readily available on the island, both for purchase locally, and in restaurants and cafes. Often the expectation has been that the quality of service around the hospitality would also match the 'five star' quality of the products' reputation. Meeting those expectations in the face of great off-island demand for the products, seasonality of production, and changes in business management has been a constant challenge for the business operators and King Island Brand Management Group.

King Island beef producers sought to protect the beef brand by aligning it with the King Island Coat of Arms, owned by the King Island Council. In 2005 the Coat of Arms was registered as a certification mark, with rules of use accepted by the Australian Competition and Consumer Commission (ACCC). The benefits of the certification mark have not yet been fully captured by the community. Certification marks only bind those who choose to participate (Caenegem et al. 2015). The King Island Certification Mark was used briefly but very successfully on packaging by the owner of the King Island abattoir, on 'natural young beef' sold into Japan. A connection was built between the producers and end-users, with representatives of King Island Producer Group and the processor travelling to Japan, and many visits from the customers were hosted on the island. Otherwise the corporations managing the larger volume King Island branded products, cheese and beef, have chosen not to engage with the community owned trademarks, but to manage their King Island branded products by privately owning registered trademarks and defending them as required. Subsequent owners of the King Island beef abattoir did not continue the use of the King Island Certification Mark, but registered their own trademarks. The King Island Dairy trade mark is an integral part of the King Island dairy business and has been sold a number of times, along with the processing plant and distribution network.

Southern rock lobster sourced from King Island waters is not sold under the King Island name but identified as Southern Australian. King Island waters are a distinct region and the quality of product is high and keenly sought after by the market. Despite this, there has been no attempt to develop a market for a specifically King Island branded product due to the perceived difficulty of preventing free riding on a brand using a place name. Any such investment in market development is seen to be at risk of constant passing off.

11.3 Protecting the King Island Brand

Protection of Brand King Island is crucial to further investment on the island, particularly in value-adding and market development. A consistent threat to King Island's businesses is the sale of non-genuine products with King Island named as the product place of origin. These products dilute the premium market and worse, they are often not of the same quality as the products they are imitating and thus do great damage to the genuine quality brands and to the King Island reputation. King Island Council, through the Brand Management Group, receives constant communications about the use and possible misuse of the King Island name. The population of King Island is only around 1500, but through personal and business connections an informal network of 'friends of the King Island brand' keep watch. Considerable resources for such a small local government authority have been allocated to following up these instances.

The King Island community, in their attempts to protect the King Island Brand under existing legislation, initially received little support from government agencies and the bureaucracies managing intellectual property and 'truth in labelling'. The Offices of Consumer Affairs are state-based so issues around place of origin in business names have to be followed up separately in each state. The Australian Competition and Consumer Commission (ACCC) was approached on a number of occasions regarding incidences of misleading claims but the burden of producing evidence was placed back on the King Island Brand Management Group.

In 2009 the King Island Beef Producers Group were represented at the Australian Senate Committee Hearing into Meat Marketing. The Rural and Regional Affairs and Transport Committee included in their recommendations at the conclusion of the hearing that "The ACCC should pursue clear misuse of the King Island name in accordance with section 52 of the Trade Practices Act prohibiting misleading and deceptive conduct. Indeed the committee suggests that the ACCC take a particular interest in the misuse of the King Island name to relieve the island's producers from the burden of identifying and reporting such cases" (Rural and Regional Affairs and Transport References Committee 2009). The ACCC did follow up cases referred to them by the King Island Brand Management Group and two Victorian businesses were convicted of misleading and deceptive conduct by falsely claiming that the meat it offered for sale was sourced from King Island. "Primary producers work hard to develop a regional reputation for quality produce.

Consumers are willing to pay more for that assurance of quality," ACCC Chairman, Rod Sims said. "Unfortunately, retailers that falsely associate themselves with that regional reputation for quality can quickly undermine the integrity of the reputation and the hard work of primary producers. Consumers unknowingly pay a premium for goods falsely associated with that region" (Australian Competition and Consumer Commission 2012).

Many cases of misrepresentation of King Island as the place of origin for a product have been brought to the notice of the King Island Brand Management group. In most instances the offending company has desisted after a direct approach from the group. The frequency of reported incidences has declined since successful prosecutions by the ACCC. The King Island Brand Management Group has worked with locally owned businesses around the use of trademarks for brand protection. Many businesses have attempted to register trademarks which include the King Island name, some successfully and others rejected by the Examiner of Trade Marks of IP Australia (Examiner). A number of attempts have been made to register trademarks with words only and including "King Island", that is with no image in the trademark. None of these trademarks had ever been registered until 2012 when the words "King Island dairy" were accepted by the Examiner as a trademark.

The King Island Council and business leaders are ever mindful of the risk of dis-association of the brand name from the place, which might allow production of King Island branded goods somewhere other than on King Island. This would severely undermine the profitability of local primary and secondary producers, and the whole island economy. Attempts to register a trademark consisting of just the words "King Island Dairy" had been made on a number of occasions since 1984 and always the Examiner issued an adverse report until finally accepting the registration by the Lion Corporation. King Island Brand Management Group worked with the King Island Dairy Farmers Collective Bargaining Group and registered an opposition to this registration, on the grounds that other King Island businesses should be entitled to use these words in association with their products.

In considering whether to continue with the legal challenge, the risk of costs being awarded against King Island Council and on-going legal fees was paramount. The resources available to the Council, as the local government of a population just more than 1500, were unlikely to sustain a legal battle against a company with the resources of a large multinational corporation behind it. The Council had not sought a mandate from the community for any such action and the perceptions of customers of the branded products needed to be the chief consideration. Damage to the brand in any stoush was to be avoided at all costs. Finally, King Island Council accepted a letter from the Lion corporation which committed the company to "using only King Island milk for KING ISLAND DAIRY branded products", and withdrew the opposition to the trade mark registration. The King Island Brand strength has been built over a number of years and embodies hard work and attention to detail across a number of industries. Incidences of misuse and passing-off deplete the energy behind the place brand and the enthusiasm of the brand champions.

11.4 Innovation and the King Island Brand

For the King Island Brand to remain valuable and sustainable, innovation is continually required in a range of areas: product development, supply chain management, quality assurance, brand management and communication of brand values. The established market demand for King Island goods drives product development and diversification. New products and further value-adding within industry sectors have included: development of the cheese product range by King Island Dairy; beef packaged in retail portions; and new species of seafood taken to the market, such as smoked eel, pickled octopus, and canned abalone.

Changes in market specifications, quality assurance and supply chain requirements have demanded constant innovation. Adapting to the vagaries of shipping and air-freighting product from the middle of Bass Strait is a constant challenge. Two new links golf courses have been built, taking advantage of the wild and beautiful coast. The community, Council and State Government are working together to organise a coordinated response to the increase in visitation to the island resulting from the golf developments. Island access, infrastructure, accommodation, hospitality and levels of service will all affect the visitor experience, so important to the island brand.

King Island Council, through the Brand Management Group, has taken a leading role in organising a collective response to maintaining the brand value and ensuring sustainability. This has spread the very high burden on individual producers on to a community organisation (Caenegem et al. 2015) acting as volunteers. The group aims to champion the values of King Island Brand and ensure continued community support: to develop a consistent message; to create a visual and verbal identity; and to tell King Island stories through the production, presentation, and service of the island's products.

Best practice reference lists, developed by the Brand Management Group, outline minimum standards to be applied by each industry group for products associated with the King Island brand (King Island Council 2015). The Council has developed a logo, registered as a trade mark, which King Island businesses can use if they adhere to the industry best practices defined in the reference lists. The King Island Brand Management Group organisation requires constant renewal and empowerment from the community and business, and to avoid relying too heavily on individual champions (Caenegem 2015).

A constant challenge is to update the brand values and preserve thir meaningfulness for the community. The vision of the brand must be communicated to new residents and new owners of island enterprises. Naturally, new players in the island business world come with pre-conceived ideas and with their own vision for their business. It is important to blend new ideas and developments into the existing concepts of the brand. The King Island story is constantly evolving, thus the meaning of the place brand requires constant review and discussion.

> In order to really succeed we need the support of all key stakeholders. We believe that it is important that the King Island Brand be grounded in its authenticity. King Island has stunning scenery and a pristine environment—but of course we are best known for our produce. The people and their stories enhance the authenticity of our produce. We need to nurture and maintain our position as a premium brand. When all of the key industries work together to deliver a marketing and sales plan that shares common themes, our message has greater impact in the market place.—King Island Brand Management Group presentation to stakeholders (2011).

11.5 Could Geographical Indications Be Useful?

The possibility of using Geographical Indications for King Island products presents an innovation to be considered. Such legislative backing would ensure place of origin brand protection was more manageable.

Factors in King Island's favour for establishing an effective Geographical Indications system as stated by Caenegem et al include:

- Legal certainty around the definition of King Island as a place is easy to establish. The waters of Bass Strait bound the region for land based products and for products from the sea the area of "King Island waters" are specifically defined
- King Island produces premium, high value goods
- The community generally accept that place of origin branding is crucial to the survival of King Island production
- The King Island Brand Management Group has gone some way towards overcoming the challenges in establishing a successful collective regional approach to branding a range of products.

The incentives to collectively manage the regional place brand are apparent. King Island beef producers, along with the rest of the community, have learnt the lesson of pursuing short-term individual interests over collective long-term strategy. The King Island Beef brand had the best beef brand awareness of any in Australia in 2007 (Nielson 2007).

At that time King Island Beef Producers group had strong horizontal and vertical alliances, producers participated in product marketing direct to consumers, and the King Island Certification Mark was used on packaging into Japan. Producers were at the forefront of innovation, early adopters of quality assurance systems and training programs. With subsequent changes in the ownership of the local abattoir the connection of producers through the supply chain to consumers was slowly eroded. Commitment to supply the local abattoir waned and the whole community are still suffering the consequences of losing the local export abattoir- from the impact of jobs lost from the economy, and the direct and indirect costs of shipping livestock for slaughter. Producers and the community have rallied with attempts to regain control of the supply chain in a two-pronged approach: one a multi-species abattoir, the other group pursuing development of a large beef abattoir. Both solutions require strong on-going brand management at the local level to succeed. Application of a Geographical Indications system for King Island beef would assist greatly in protecting subsequent brands developed.

Leaders in the seafood industry see real benefits that could be captured for the local economy from a Geographical Indications system. "They will not develop the King Island seafood name because there is no protection and at present they cannot leverage off the premium King Island name. If the King Island brand were protected, King Island businesses could value-add and invest effectively in marketing as King Island product" said Donna Whitehouse-Summers, Chair of the Brand Management Group in 2011. "Delivery to existing markets is at a base rate, those markets would pay more for fish marketed as King Island product. More higher end markets would open up, for example the restaurant trade would deliver much higher margins".

Island producers of other goods do gain from the spillover effect from the larger brands, but distance to market creates difficulties in establishing personal connections with the consumer which are important in establishing and holding market share. The opportunity to added value by personally telling the story behind the product, such as at farmers' markets, is not readily available to these isolated producers, disadvantaged by the cost of freight and travel. That same distance to market makes it difficult for them to check on accounts of "ripping off" the King Island name on similar product lines. The development of a Geographical Indications system could improve the opportunities for product and market development.

The brand protection afforded by a Geographical Indications system would simplify the procedures for developing brands for new products, for example King Island wallaby reportedly has a distinctly better flavour and quality than product from other areas. Investment in marketing this inherently superior product would be far more likely to occur with the protection conferred by a Geographical Indications system.

Geographical Indications cannot be traded, as a trademark can, but are embedded in the region (Caenegem et al. 2015). A system of Geographical Indications would offer King Island many benefits:

- The local community would be relieved from the need to constantly defend the King Island place name in relation to products. The definition of King Island products would be clear to all and managed by the community, with benefits returned to custodians of the place.
- Investments in quality drivers such as the development of infrastructure, care of the environment, and training could be made with more confidence of a return on investment.
- Open and frank discussions with stakeholders in existing brands should clarify and define the branding issues crucial to the regional development and maintenance of a small isolated economy, and critical to the ongoing enjoyment of the place by residents and future generations.

- Ongoing discussion amongst stakeholders on a regional level could assist in community management of other difficult issues associated with isolation- sea freight, air travel and transport, land use, environmental pressures.

11.6 Application of Geographical Indications

The practicalities of applying Geographical Indications for King Island may initially seem daunting. Considerable effort would need to be put into communication both within and between different industries. It is possible there are too many actors to effectively manage the discussions, which would be required (Caenegem et al. 2015). There may be difficulty in defining the 'terroir' of the island across a range of products. It is possible there would be opposition from existing players in the King Island Brand market.

The commitment to a Geographical Indications system needs to be part of broader planning dialogues that take place within a region (Caenegem 2015). The application of Geographical Indications, however, should and could be managed industry by industry. Successful implementation would demand the forging of common strategic objectives, with the ultimate aim of increasing the premium for products, within each industry.

In defining the 'terroir' of the island such questions as may be put include:

- Do the products possess any special characteristics linked to the nature of the place? (Caenegem et al. 2015)
- Can the benefits of a maritime climate for pasture growth and the low stress environment for livestock be discerned in the taste and quality of the end product?
- Are the attributes of the meat and dairy products consistent enough, especially in the face of variations in the management practices of different producers, to exhibit a local characteristic?
- Products are under-pinned by relevant industry quality assurance programs but is there some 'island quality'?

The dairy industry has benefited from the close proximity of the farms to the "Cheese Factory". Milk is collected daily and no older than one day when processing is started on the premium range cheeses. Reduced milk volumes enable close monitoring of the raw milk quality and batch control. There are strong social connections between many of the factory staff and the primary producers. The benefits returned to the local economy from the high quality cheese produced is well understood and a matter of pride in the community.

The consumer appetite for high quality food and growing interest in provenance has led to a proliferation of branded food products. Although the King Island Dairy brand is already established, won over many years of making consistently great product, the competitive advantage may be lost to competitors in the fine food market without innovation around the brand. The current owner of the King Island Dairy brand could use Geographical Indications to reinforce the perception in the market of the quality of King Island products. The introduction of Geographical Indications for King Island dairy products could be a useful tool, not only for protecting the brand, but as a means of creating renewed interest in the brand. The image of a sustainable community behind the product would add to the story.

The terroir for seafood would be easy to demonstrate, for example the cold water surrounding the island confers better shell colour, better meat texture and taste to Southern rock lobster. The isolation of the island is an advantage in this case, as the water and air qualities are relatively untouched by civilisation and population. The waters of King Island are already defined under regulations for taking of seafood, so there can be no question of geographical limits.

There is no doubt the place brand is strong for food-stuffs, with significant spillover effects. The quality of the cheese, beef, and wild caught seafood products has established the brand name, it has not been a result of marketing ploys. Could the perceptions around the place name itself imbue terroir to products, which could be included in the elements of a Geographical Indications system?

11.7 Conclusion

King Island producers have generally accepted the rationale that producing for commodity markets is a long-term recipe for disaster, that the cost of inputs and distance to market, and ever declining terms of trade make it imperative to produce excellent goods for high value markets. The King Island community would clearly benefit over a very long term from improved promotion and protections for their brands. The issues of intellectual property and branding are complex, and business owners and managers are often time poor, however it is absolutely sensible for them to join with community leaders and legislators to fully investigate the potential advantages a Geographical Indications system might bring to their businesses.

Acknowledgements The information contained in this article derives from the body of work by the King Island community, businesses and Brand Management Group towards developing their place brand over many years. I have gleaned the information by participating as a community member, a business operator, and as a staff member of King Island Council. I wish to acknowledge the King Island Brand Champions, forming a Special Committee of King Island Council, both for their input into this article and for their efforts as community leaders.

Bibliography

Abbing, ER 2010, Brand-Driven Innovation: Strategies For Development And Design, Lausanne: AVA Publishing, eBook Collection (EBSCOhost), EBSCOhost, viewed 15 January 2016.

Australian Bureau of Statistics 2011 Census of Population and Housing, viewed 1 January 2013 http://www.abs.gov.au/

Australian Competition and Consumer Commission 2012, ACCC acts to protect consumers and King Island producers. Media releases. Australian Competition and Consumer Commission 2012. https://www.accc.gov.au/media-release/accc-acts-to-protect-consumers-and-king-island-producers viewed 06/02/2016.
Babcock, B. A. and R. Clemens, 2004. Geographical indications and property rights: protecting value-added agricultural products. Midwest Agribusiness Trade Research and Information Center (MATRIC) Publications number 04-mbp7. Iowa State University.
Beer, AT., A. Maude and B. Pritchard, 2003. Developing Australia's Regions, Theory and Practice. UNSW Press. Sydney.
Caenegem, W., P. Drahos and J. Cleary, 2015. Australia's Regions and Agriculture: Can Geographical Indications Help? Final Report Prepared for Rural Industries Research and Development Corporation. Australia.
King Island Council 2015, King Island Management Brand Group- Logo Use. King Island Council website, viewed 11 December 2015. http://kingisland.tas.gov.au/page.aspx?u=272#e29
King Island Council 2007, Brand King Island Management Group, General Managers Report, Ordinary Council- 16 October 2007, King Island.
Nielson AC. 2007, Online Omnibus Beef Brand Awareness W2 - (OO0718).
Rural and Regional Affairs and Transport References Committee 2009, Meat Marketing Final Report. Senate Committee Report, Parliament of Australia.
Sorensen, T. 2010 Regional Development: Some Issues for Policy Makers Consultant, Economic, Commerce and Industrial Relations Group, 27 June 2000. Research Paper No. 26 1999–2000, Published by the Department of the Parliamentary Library, 2010.

Author Biography

Jennifer Thorn lives on King Island and, with her husband, has raised a family and operated a beef farming enterprise, since 1985. Involvement with the King Island place brand has been through her work as Economic Development Officer at King Island Council.